Guide to the Vascular Plants of the Blue Ridge

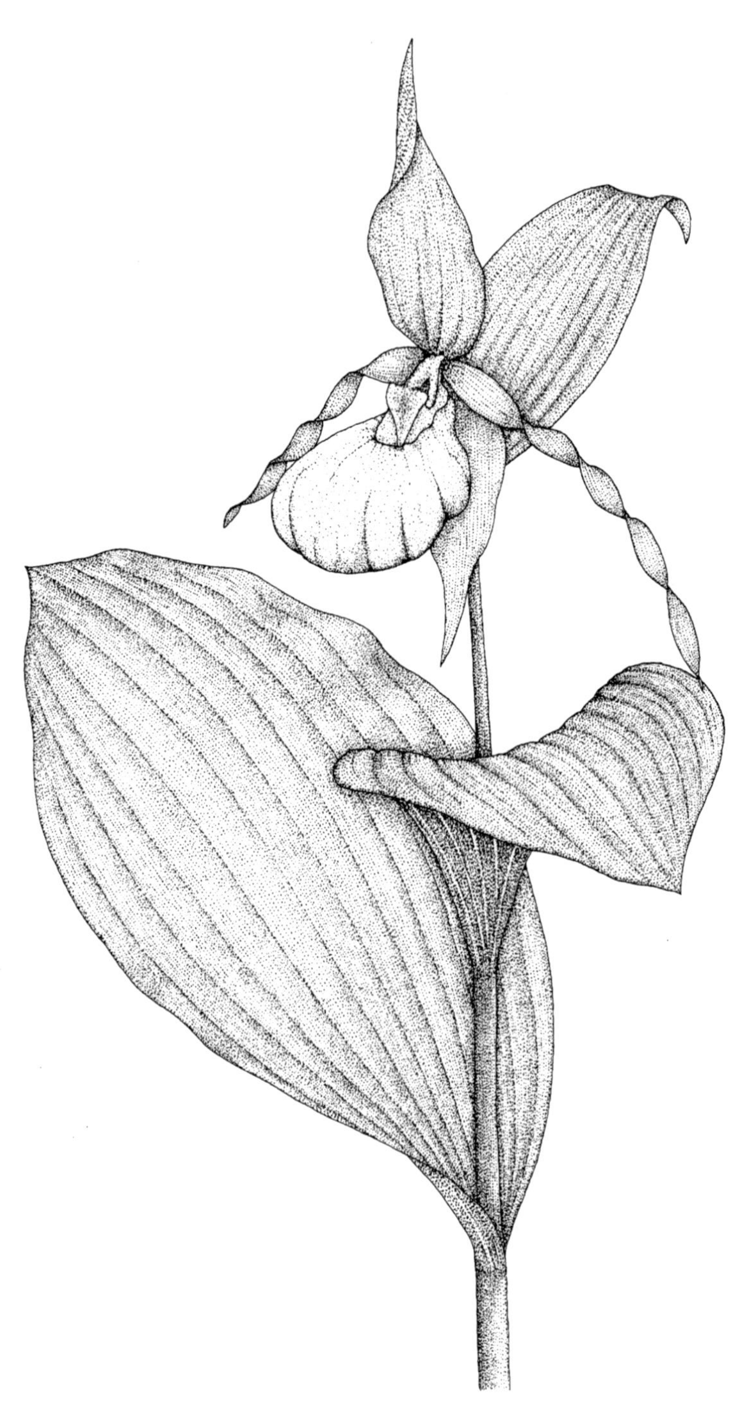

Guide to the Vascular Plants of the Blue Ridge

B. Eugene Wofford

The University of Georgia Press
Athens and London

Frontispiece: Yellow Lady's Slipper by José Panero

Athens, Georgia 30602

Set in Linotron 202 Times Roman

The paper in this book meets the guidelines for permanence and durability of the Committee on Production Guidelines for Book Longevity of the Council on Library Resources.

Printed in the United States of America

04 03 02 01 00 P 5 4 3

Library of Congress Cataloging in Publication Data

Wofford, B. Eugene.
Guide to the vascular plants of the Blue Ridge / B. Eugene Wofford.
p. cm.
Bibliography: p.
Includes indexes.
ISBN 0-8203-1049-2 (alk. paper)
ISBN 0-8203-2455-8 (pbk.: alk. paper)
1. Botany—Blue Ridge Mountains Region. 2. Plants—Identification. I. Title.
QK125.35.W63 1989 88-4879
582.0975'5—dc19 CIP

British Library Cataloging in Publication Data available

To my family and to field biologists

Contents

Illustrations

Map

Figures

Acknowledgments

In addition to my own taxonomic interpretations, I have relied upon and often adapted the proven treatments of other manuals, guides, monographs, and taxonomic and nomenclatural literatures. The most important of these are: Gleason (1952), Gleason and Cronquist (1963), Radford, Ahles, and Bell (1968), and Cronquist (1980); others include Small (1933), Fernald (1950), Strausbaugh and Core (1952–1964), Duncan (1967), Luer (1975), Mohlenbrock (1975), Batson (1977), Godfrey and Wooten (1979, 1981), Elias (1980), Kartesz and Kartesz (1980), Duncan and Kartesz (1981), and White (1982). It is my philosophy, like most before me, to use those characters that will most readily delimit taxa, and I see no value in using less-obvious character state differences solely for the sake of originality. Again, I acknowledge the use of and give full credit to the contributions of others. W. T. Batson, R. C. Clark, J. L. Collins, A. M. Evans, F. F. Fusiak, L. L. Gaddy, K. L. Hornberger, S. B. Jones, Jr., W. S. Judd, K. A. Kron, M. G. Lelong, V. E. McNeilus, T. S. Patrick, J. B. Phipps, E. E. Schilling, and D. H. Webb contributed various difficult groups to this guide and are cited throughout the text at the beginning of the keys to families and/or genera. Other taxonomic references are acknowledged in a similar fashion. I assume full responsibility and apologize for any errors or misinterpretations of the works of others.

Numerous other individuals have provided distribution data, criticisms, corrections, and suggestions. For their contributions I would especially like to thank T. S. Patrick, D. K. Smith, A. Weakley, J. D. Freeman, L. L. Gaddy, R. Kral, J. V. LaFrankie, C. Aulbach-Smith, D. Rayner, and the herbarium staffs of Clemson and Furman universities. I am also indebted to the botany staff and students at the University of Tennessee who first tested portions of the keys and provided valuable insights from their interpretations.

Special appreciation is extended to Victor Ma for carefully proofreading earlier drafts and to José Panero for providing the illustrations.

Finally, the University of Tennessee Herbarium (TENN) provided the primary reference specimens for the preparation of this guide, and I am indebted to A. J. Sharp for his long-term dedication to rebuilding and promoting TENN following the fire of 1934.

Guide to the Vascular Plants of the Blue Ridge

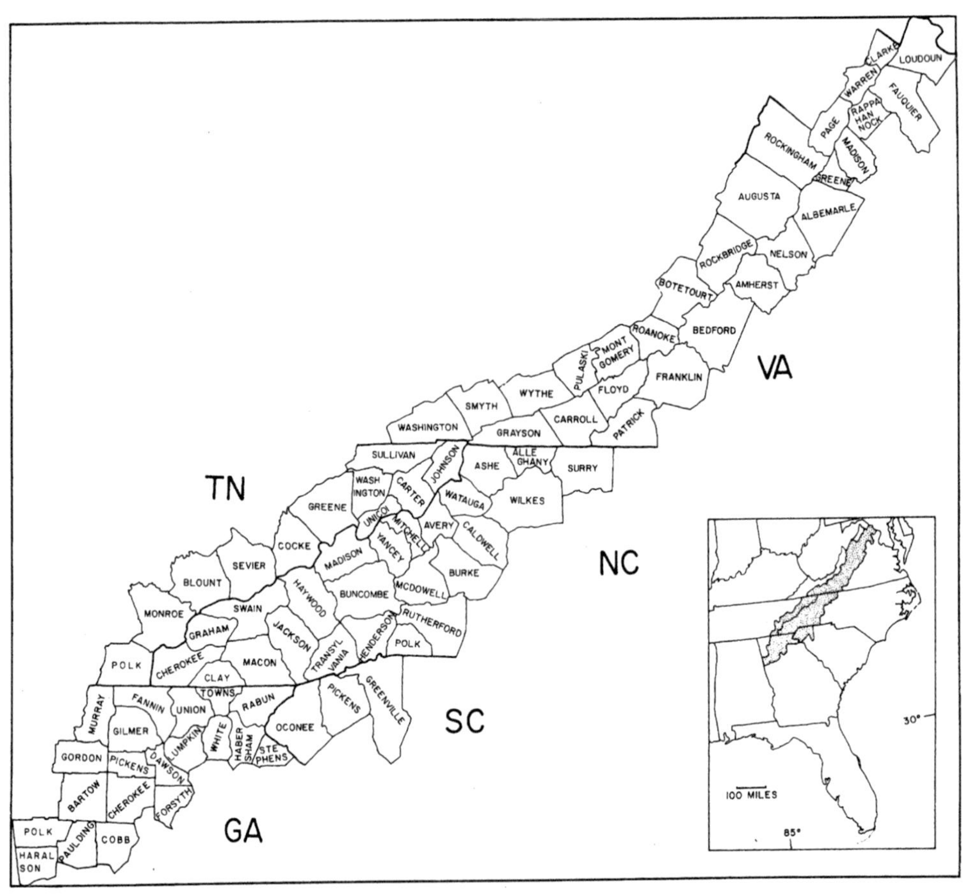
VA
TN
NC
SC
GA
LOUDOUN
CLARKE
WARREN
FAUQUIER
PAGE
RAPPA HAN NOCK
MADISON
ROCKINGHAM
GREENE
AUGUSTA
ALBEMARLE
ROCKBRIDGE
NELSON
AMHERST
BOTETOURT
ROANOKE
BEDFORD
MONT GOMERY
PULASKI
FRANKLIN
FLOYD
WYTHE
SMYTH
CARROLL
PATRICK
WASHINGTON
GRAYSON
SULLIVAN
JOHNSON
ASHE
ALLE GHANY
SURRY
WASH INGTON
CARTER
WATAUGA
WILKES
GREENE
UNICOI
MITCHELL
AVERY
CALDWELL
COCKE
YANCEY
MADISON
BURKE
SEVIER
BLOUNT
HAYWOOD
BUNCOMBE
MCDOWELL
MONROE
SWAIN
RUTHERFORD
GRAHAM
JACKSON
HENDERSON
TRANSYL VANIA
POLK
POLK
CHEROKEE
MACON
CLAY
TOWNS
FANNIN
UNION
RABUN
PICKENS
GREENVILLE
MURRAY
GILMER
OCONEE
LUMPKIN
WHITE
HABER SHAM
STE PHENS
GORDON
PICKENS
DAWSON
BARTOW
CHEROKEE
FORSYTH
POLK
PAULDING
COBB
HARAL SON
30°
100 MILES
85°

Introduction

Geomorphologists generally agree that the Blue Ridge Province extends about 550 miles from northern Georgia to southeastern Pennsylvania. The precise delimitation of its boundary is subject to debate and might also include small areas of extreme eastern West Virginia and east-central Alabama. North of Roanoke, Virginia, it is a complex ridge system varying from a single ridge less than 2 miles wide to a series of more closely spaced ridges about 10 to 14 miles wide. Elevations vary from about 1,200 to 4,100 feet. South of Roanoke, the massive mountains and high peaks broaden to nearly 80 miles, with numerous peaks exceeding 5,000 feet.

The area covered by this guide includes 85 counties from Georgia to Virginia and is perhaps the most floristically diverse region in the eastern United States (see map). Its flora has attracted the attention of botanists for nearly two centuries, yet no single treatment is available to catalog or document its floristic elements. With the exception of introduced species that have not become, or do not appear to be becoming, a part of the flora (212 excluded taxa generally known from five or fewer counties), it is represented by 161 families, 727 genera, and 2,391 species and lesser taxa. It contains a mixture of taxa from contiguous physiographic provinces, disjuncts, endemics, and numerous close relatives found primarily in or in combination with Southeast Asia, western United States, and Central America. It also contains a number of species considered to be rare or endangered at federal and/or state levels. The taking of these species for selfish reasons should be avoided to help ensure their continued existence for study and/or admiration by future generations. The smaller portions of the northern and southern extremes of the Blue Ridge have been omitted from this guide because published physiographic distribution data either are not available or would result in a discontinuity by including only those data (Wherry, Fogg, and Wahl 1979) for southeastern Pennsylvania. These omissions will not result in the exclusion of a significant number of native taxa not found elsewhere in the Blue Ridge nor render this guide unserviceable to users in contiguous physiographic provinces.

The taxonomic treatment is based primarily upon distribution data from Radford, Ahles, and Bell (1968), Harvill, Stevens, and Ware (1977), Harvill, Bradley, and Stevens (1981), Harvill, Bradley, Stevens, Wieboldt, Ware, and Ogle (1986), Duncan and Kartesz (1981), and the University of Tennessee Herbarium. Unfortunately, political boundaries do not follow physiographic ones, and a number of counties shown on the map contain floristic elements that do not strictly belong to the Blue Ridge proper. I have made no concentrated effort to obtain detailed location data within peripheral counties contiguous to other physiographic provinces but have excluded a few taxa that are not representative of the Blue Ridge flora and are not likely to be seen by beginning students or used in introductory taxonomy classes. Conversely, I have included by parenthesis within the keys a few of the more unusual elements of the shale barrens of the Ridge and Valley (e.g., *Trifolium virginicum, Eriogonum allenii, Oenothera argillicola, Clematis* spp.) and granite outcrops and associated Piedmont communities (e.g., *Arenaria alabamensis, Draba aprica, Juncus georgianus*). These peripheral taxa are known to occur within the counties defined. This approach should broaden the range of use of this guide without grossly misrepresenting the true nature of the Blue Ridge flora.

In this the first edition, I have attempted to provide a guide usable to beginning students and amateur and professional botanists. Keys are provided primarily for flowering or sporulating material except where fruits or vegetative characters are critical for the delimitation of taxa. Every effort was made to construct simple keys that can be interpreted by all users, especially beginners and amateurs. More detailed information (e.g., descriptions, distributions, illustrations) is readily available to advanced amateurs and professional botanists in monographs and regional floras, especially Radford, Ahles, and Bell (1968).

Format

The taxonomic treatments are arranged in four groups: pteridophytes, gymnosperms, monocots, and dicots. The families within each group are arranged alphabetically; the genera and species are also arranged alphabetically within their respective families and genera.

Pertinent taxonomic literature follows family and/or genus headings to indicate that this literature was, for the most part, adapted for the Blue Ridge flora. In a few cases, I have chosen a different taxonomic interpretation, and it is recommended that users of this guide consult these references for further information and conclusions.

The species and lesser taxa enumeration following each genus contain (1) scientific name based on current manuals and taxonomic and nomenclatural literature, with an asterisk following introduced taxa; (2) common name(s); (3) general habitat preferences; (4) frequency of occurrence, ranging in descending order from common (characteristic of and generally occurring in abundance throughout the region) to frequent (mostly known from more than 50 percent of the counties but not abundant throughout) to occasional (mostly known from less than 50 percent of the counties but not restricted to but a few localities) to infrequent (known only from a single or a few populations; mostly narrow endemics, disjuncts, and peripheral taxa); (5) flowering or sporulation period; (6) states of occurrence (Blue Ridge counties only), that is, Georgia (GA), North Carolina (NC), South Carolina (SC), Tennessee (TN), Virginia (VA), or ALL; and (7) pertinent taxonomic and nomenclatural synonyms, especially those differing from those given in other recent floras.

Glossary

Abaxial: Away from the axis; when referring to a leaf, the lower surface.
Acaulescent: Without an upright, leafy stem; the stem usually subterranean.
Achene: Small, dry, one-locular, one-seeded indehiscent fruit with the seed coat and ovary wall separate.
Actinomorphic: Having a symmetry such that two or more median longitudinal divisions through the flower will yield identical paired mirrored images; sometimes referred to as radial symmetry.
Acuminate: A tip whose sides are variously concave and tapering to a point.
Acute: Sharply ending in a point with margins straight or slightly convex.
Adnate: The fusion or growing together of unlike parts.
Aggregate Fruit: A compound fruit derived from the coherence of 2–many simple, superior ovaries of a single flower; ex.: blackberry (Rosaceae).
Annual: Life cycles completed in one growing season.
Annulus: A ridged row of thick-walled cells of sporangia of higher ferns that causes the sporangium to open and release its spores.
Anther: The pollen-bearing portion of a stamen.
Anthesis: The flowering period; the period when pollination takes place.
Apiculate: Abruptly ending in a small, short, usually flexible tip.
Appressed: Lying flat or pressed against the surface.
Arcuate: Slightly curved or bowed.
Areolate: An irregular pattern of depressed or raised, small, angular spaces of low relief.
Aril: An appendage or outer covering of a seed.
Aristate: Bearing a stiff awn or bristle.
Articulate: Jointed; places where separation takes place.
Attenuate: With a long, gradual taper; narrower than acute.
Auriculate: Bearing ear-shaped appendages, often at the base of leaves or petals.
Awn: A slender bristle or hair, usually at the tip of a structure.
Axil: The interior angle between any two structures.
Axile Placentation: The placentation type in which the ovules are attached along the central axis of a multiloculary ovary.
Axillary: In an axil.

Beak: A firm, prolonged, slender tip.
Berry: Pulpy or juicy, multiseeded, indehiscent fruit.

Biennial: Life cycles completed in two growing seasons.
Bifid: Two-lobed or cleft, usually at the tip of petals or leaves.
Bilabiate: Zygomorphic and distinctly divided into two "lips."
Bipinnate: Twice pinnate.
Bisbilateral: Symmetry with only two planes of division whereby the mirror images of one plane are dissimilar from the mirror images of the second plane; restricted to certain Fumariaceae.
Bisexual: A flower with both stamen(s) and pistil(s).
Bract: A reduced leaf at the base of a flower or an inflorescence.
Bracteal: Having the form or position of a bract.
Bracteate: With bracts.
Bracteoles (bractlets): A secondary bract.
Bristle: A hairlike projection; modified perianth of some Cyperaceae.
Bulblet: A small bulb, usually borne on a stem or in an inflorescence.

Calyx: The outermost whorl (sepals) of the perianth, usually green or colored as in some Ranunculaceae.
Calyx Tube: The basal or tubular portion of a fused calyx; the sepals may be fused wholly or in part.
Campanulate: Bell-shaped.
Canescent: With dense, short hairs often resulting in a gray or whitish appearance.
Capitate: A head or dense headlike cluster.
Capsule: A dry, dehiscent fruit derived from two or more carpels.
Carpel: The ovule-bearing structure of a flower; a simple pistil or one member of a compound pistil.
Cartilagenous: Tough or hard, but flexible.
Catkin (ament): A bracteate, often unisexual, apetalous, flexible spikelike or cymose inflorescence; the male inflorescence typically falling as a single unit.
Caudate: With a slender, taillike appendage.
Caudex: The woody base of a perennial plant.
Caulescent: Having an above-ground stem.
Cauline: Pertaining to or attached to the stem, as opposed to being basal.
Cell: Compartment or cavity of an ovary; a locule.
Cespitose: Growing in clumps or dense tufts.
Chaff: Thin, dry scales or bracts; often used for the bracts in the flowerhead of Asteraceae.
Ciliate: Bearing marginal hairs, especially in reference to leaves and bracts.
Clavate: Club-shaped.
Claw: The narrow, petiolelike base of petals or sepals.
Cleft: Divided into segments to near the middle.
Cleistogamous: Bearing self-pollinated flowers that do not open, as in the Violaceae, Gentianaceae, Cistaceae, etc.

Clone: Plants reproducing vegetatively by sharing a common rootstock, as in stolons, runners, rhizomes, etc.
Coma: A tuft of hairs at the tip or base of a seed.
Compound: Of two or more similar parts.
Compound Leaf: A leaf with two or more leaflets.
Compound Pistil (or ovary): A pistil derived from two or more carpels.
Connate: The fusion or joining together of similar structures.
Connivent (coherent): The touching or coming together of similar structures without actual fusion.
Cordate: Heart-shaped; with a sinus and rounded lobes, often in reference to the base of a structure.
Coriaceous: Leathery.
Corm: A solid, more or less globular, underground stem.
Corolla: Collective term for the petals or the whorl(s) of the floral envelope between the sepals and stamens; usually white or colored.
Corolla Tube: The tube or basal portion of fused petals; the petals may be fused wholly or in part.
Corona: A crown or series of petallike structures; situated between the petals and stamens in *Passiflora* or on the petals in *Hymenocallis*.
Corymb: A broad, flat-topped indeterminate inflorescence in which the pedicels are of various lengths and the outermost flowers opening first.
Corymbose: Resembling a corymb.
Costa: The midvein of a pinna or leaflet.
Cotyledon: A seed leaf.
Crenate: Margins with shallow, round, or obtuse teeth.
Crenulate: Diminutive of crenate.
Crepitant: Crackling or rattling.
Crest: An elevated ridge, usually toothed or appendaged.
Crisped: Undulate, curled, or wavy, as in the leaf margin of *Rumex crispus*.
Culm: The flowering stem of grasses and sedges.
Cuneate: Wedge-shaped, usually in reference to leaf bases.
Cupule: A cuplike structure composed of perianth or bracts at the base of flowers or fruits (ex.: Lauraceae).
Cusp: An abruptly contracted, elongate, pointed tip.
Cyathium: The reduced, cymose, cuplike inflorescence of *Euphorbia*; the entire multiflowered structure mimics a single, perfect flower.
Cyme: A determinate inflorescence, often broad and flattened, in which the central flower opens first.
Cymose: Resembling a cyme.
Cymule: A small cyme, usually few-flowered.

Decompound: More than once compound.
Decumbent: Reclining or lying on the ground, but with the tip turned upright.
Decurrent: Leaf-base tissue adnate to the petiole or stem and extending beyond.

Dehiscence: The opening at maturity of anthers or fruits by means of slits, lids, pores, or teeth.

Deltoid: Triangular.

Dentate: With coarse, sharp teeth that project outward.

Denticulate: Diminutive of dentate.

Determinate: An inflorescence of limited, definite growth; the terminal flower maturing first and thereby arresting elongation.

Dichasium: A determinate inflorescence with an older central flower and two lateral ones; the basic unit of many cymose inflorescences.

Dichotomous: Forked into two equal branches.

Diffuse: Loosely branching or spreading.

Digitate: Handlike, with individual units arising from one point.

Dimorphic: Occurring in two forms, as in fertile and vegetative leaves of different shapes.

Dioecious: Species having staminate and pistillate flowers on separate plants.

Disk: The central portion of the flowering head that bears the disk flowers of some Asteraceae.

Dissected: Divided into slender segments.

Distal: Found at or near the apex of an organ.

Distichous: Leaves or other structures arising on opposite sides of the stem; two-ranked.

Drupe: A fleshy, indehiscent fruit usually with one seed enclosed in a hard endocarp.

Eglandular: Without glands.

Elliptic: Oval; broadest near the middle and gradually tapering to both ends.

Emergent: Growing above the surface of soil or water.

Endocarp: Innermost layer of the ripened ovary wall.

Entire: A leaf margin without teeth, hairs, spines, etc.

Epipetric: Growing on rocks.

Equitant: Leaves that overlap at the base, as in *Iris*.

Erose: Minutely eroded or irregular; nearly entire.

Even Pinnate: A compound leaf with an even number of leaflets, the terminal leaflet absent.

Exfoliate: To peel off in layers or shreds.

Exserted: Extending outward and beyond, as in stamens from the corolla throat or tube.

Extrorse: Facing and opening outward, usually in reference to anther dehiscence.

Falcate: Sickle-shaped.

Farinose: Covered with a white, mealy powder.

Fascicle: A bundle or dense, close cluster.

Filament: The stalk supporting the anther of a stamen.

Filiform: Threadlike.
Fimbriate: Fringed.
Fistulose: Hollow and cylindrical.
Flabelliform: Fan- or broadly wedge-shaped.
Flexuous: Zigzagged, wavy.
Floral Tube: The fused perianth of *Dirca palustris*.
Floret: Individual, usually small, flowers as in Asteraceae and Poaceae.
Foliaceous: Resembling a leaf in texture and appearance.
Follicle: A dry, unicarpellate fruit that splits along one side at maturity.
Free Central Placentation: The placentation type in which the ovules are attached to the central column of a unilocular ovary.
Frond: The leaf of a fern; the vegetative structure in Lemnaceae.
Fruit: The mature ovary; accessory structures may be adnate to it.
Funnelform: Shaped like a funnel.
Fusiform: Thickened in the middle and tapered to both ends.

Gibbous: Swollen on one side, usually at the base.
Glabrate: Becoming glabrous or smooth with age.
Glabrous: Not hairy.
Glandular: Having secretory glands or trichomes.
Glaucous: Covered with a whitish, waxy substance.
Globose: Round, globular, or spherical.
Glom: Dense or compact cluster.
Glomerule (glomerate): A dense or compact cluster of flowers.
Glume: One of the two sterile bracts at the base of a grass spikelet.

Half-Inferior: The condition where the hypanthium or receptacle is adnate to the lower half of the ovary.
Hastate: Leaves with the general shape of an arrowhead, but with pointed basal lobes turned outward at right angles.
Head: A dense cluster of sessile flowers.
Helicoid: Branching only on one side, the main axis curving or coiling toward the unbranched side.
Heterophyllous: With leaves of two or more shapes.
Heterosporous: With spores of two different sizes; usually as small spores that can germinate into male gametophytes and much larger spores that can germinate into female gametophytes.
Hirsute: With coarse or stiff hairs.
Hirtellous: Minutely hirsute.
Hispid: With long, bristly hairs.
Homosporous: Spores all of one size.
Hood: A concave or strongly arching flower part.
Horn: Curved or straight beaklike accessory flower structure in the Asclepiadaceae.

Hyaline: Transparent.

Hypanthium: A nearly flat or cup-shaped structure produced from the fusion of sepals, petals, and stamens; it may be free from or adnate to the ovary.

Imbricate: Overlapping.

Imperfect: Unisexual; a flower with stamen(s) or pistil(s), but not both.

Included: Contained within a structure; not exserted.

Indehiscent: Not opening at maturity.

Indeterminate: An inflorescence of potentially unlimited growth; the terminal flower maturing last and thereby not arresting elongation.

Indurate: Hardened.

Indusium: A scalelike or leaflike flap of tissue arising from a leaf and covering the sorus of ferns; a *false indusium* is the rolled or folded leaf margin covering the sorus, a *true indusium* is a scalelike covering arising from the abaxial leaf surface, not the margin.

Inferior: Beneath; the ovary position in which the perianth and stamens are inserted above the ovary, thus the ovary is embedded within the hypanthium or receptacle.

Inflated: Bladdery, blown up.

Inflorescence: The flower arrangement or mode of flower bearing.

Internode: The portion of the stem between two nodes or points of leaf insertion.

Introrse: Facing or opening toward the center, usually in reference to anther dehiscence.

Involucre: One or more series of bracts immediately surrounding a flower cluster; the bell-shaped tubular indusium of the filmy ferns.

Involute: Rolled inward or toward the upper side.

Isophyllous: Leaves all more or less alike, not of two different shapes.

Keel: The two anterior united petals of certain members of the bean family; a prominent ridge.

Lacerate: Torn or irregularly cut or divided.

Lacunae: Chambered or internal air spaces.

Lamina: The blade or expanded portion of a leaf.

Lanate: Covered with woolly, intertwined hairs.

Lanceolate: Lance-shaped; narrow; broadest near the base and tapering to the tip.

Leaf Scar: Scar tissue remaining on the stem after a leaf has fallen.

Leaflet: An individual unit or secondary "leaf" of a compound leaf.

Legume: A dry fruit from a simple pistil that dehisces along two lines or sutures (ex.: bean family).

Lemma: The lower of the two bracts immediately enclosing the grass flower (*see* palea).

Lenticular: Lens-shaped; biconvex.

Lepidote: Covered with small scales.

Ligule: The appendage at the juncture of the leaf sheath and blade in Poaceae; the limb of the corolla of a ray flower in Asteraceae.

Limb: The expanded part of a corolla with fused petals.

Linear: Long and narrow with parallel sides throughout most of the length.

Lip: One of the two portions of a bilabiate or zygomorphic flower.

Locule: Compartment, cavity, or cell of an ovary, fruit, or anther.

Loment: A type of legume fruit with constrictions between the seeds.

Margin: The edge of a flat structure, usually a leaf.

Marginal Placentation: The placentation type in which the ovules or seeds are attached along a single suture; restricted to simple (unicarpellate) pistils.

Median: The middle of an organ or halfway between two points.

Membranaceous: With a thin, papery texture.

-merous: A suffix denoting the number of parts that constitute a structure.

Monoecious: With staminate and pistillate flowers on the same plant.

Monomorphic: Having the same shape; usually as in vegetative and fertile structures of the same shape.

Mucro: An abrupt point or short, spiny tip.

Nectary: A nectar-secreting gland or area.

Net Venation: A venation pattern where primary and secondary veins form a complex network or reticulum.

Neutral: Without functional stamen(s) or pistil(s).

Node: The area of a stem where one or more leaves are borne.

Nodose: Knotty; bumpy.

Nut: A hard, dry, indehiscent, one-seeded fruit derived from a 2–many carpellate ovary.

Nutlet: A small nut.

ob-: A prefix signifying inversion or the reverse of; as in oblanceolate, obovate, obovoid, etc.

Oblique: Slanted or with unequal sides, as in an elm leaf.

Oblong: Longer than broad with more or less parallel sides.

Obsolete: Rudimentary or nearly absent.

Obtuse: Blunt or rounded at the tip.

Ocrea: A tubular, nodal sheath formed by the fusion of two stipules of many Polygonaceae.

Ocreola: Secondary sheath in the inflorescence of *Polygonum*.

Odd-Pinnate: A compound leaf with an odd number of leaflets, the terminal leaflet present.

Opaque: Dull, not transparent.

Orbicular: Round.

Orifice: An opening, mouth, or outlet.

Ovary: The basal, ovule-bearing part of the pistil.

Ovate: Broadly rounded at the base and narrowed above; shaped like a hen's egg in longitudinal section.

Ovoid: A solid that is oval in outline.

Ovule: The structure(s) within the ovary that, after fertilization, become the seed(s).

Palea: The uppermost of the two bracts immediately enclosing the grass flower (*see* lemma).

Palmate: Lobed or divided in a palm- or handlike manner.

Panicle: An indeterminate branching raceme; the branches of the primary axis are racemose and the flowers are pedicellate.

Paniculate: Resembling a panicle.

Papillose: With short bumps or wartlike protuberances.

Pappus: The highly modified outer perianth series of the Asteraceae, located at the summit of the achene and possibly representing a modified calyx.

Parallel Venation: A venation pattern with veins extending in the same direction and equidistant.

Parietal Placentation: The placentation type in which the ovules are attached at two or more points along the ovary wall; restricted to compound (syncarpous) pistils.

Pectinate: Divided into narrow, comblike parts.

Pedicel: The stalk of a single flower of an inflorescence.

Pedicellate: With a pedicel.

Peduncle: The stalk supporting an inflorescence or the flower stalk of species producing solitary flowers.

Peltate: Attached away from the margin.

Pendulous: Drooping or hanging downward.

Perennial: Plants living for more than two years.

Perfect: Bisexual; a flower with both stamen(s) and pistil(s).

Perfoliate: Leaf bases that completely surround the stem, the latter appearing to pass through the former.

Perianth: The calyx and corolla (when present) collectively.

Pericarp: The mature ovary wall; may be divided into exocarp, mesocarp, and endocarp.

Perigynium: The sac or sheath surrounding the achene in *Carex*.

Petal: One segment of the corolla.

Petaloid: Resembling a petal in form and color, as in some brightly colored sepals.

Petiolate: With a petiole.

Petiole: The stalk supporting the expanded leaf blade.

Phyllary: An involucral bract subtending the receptacle in Asteraceae.

Pilose: Bearing soft, mostly erect, shaggy hairs.

Pinna: The primary division of a compound leaf; a leaflet.

Pinnate: With leaflets or pinnae on both sides of a common axis.

Pinnatifid: Deeply cut or divided in a pinnate fashion.

Pinnule: The secondary division of a compound leaf; a subleaflet, or dissected unit of a pinna.

Pistil: The ovule-bearing portion of a flower, composed of stigma, style (if present), and ovary.

Pistillate: A female flower; bearing a pistil(s), but no functional stamens.

Plano-Convex: Flat on one side and curved on the other.

Plicate: Folded; fanlike.

Plumose: Resembling a feather; a central axis bearing fine hairs or side branches.

Pome: A fleshy, indehiscent fruit derived from an inferior ovary and surrounded by an adnate hypanthium (ex.: pear, apple, and other related members of the Rosaceae).

Prismatic: Having the shape of a prism, angulate with flat sides.

Procumbent: Lying on the ground, but not rooting.

Prostrate: *See* procumbent.

Pruinose: With a waxy, powdery surface; a bloom.

Puberulent: Minutely hairy or pubescent.

Pubescence: A general term for hairs or trichomes.

Pubescent: Covered with soft hairs.

Punctate: With translucent or colored dots, depressions, or pits.

Pustulose: Blistery, pimplelike.

Pyriform: Pear-shaped.

Raceme: An elongate, unbranched, indeterminate inflorescence with pedicellate flowers.

Racemose: Resembling a raceme.

Rachis: A flower- or leaflet-bearing axis; the central axis of a compound fern blade.

Ray: Branch of an umbel, particularly in the Apiaceae; the outer or ligulate flowers of a composite (Asteraceae) head.

Receptacle: The expanded apex of a pedicel upon which the flower parts are borne.

Regular: *See* actinomorphic.

Reniform: Kidney-shaped.

Reticulate: Resembling a net.

Retrorse: Turned backward or away from the tip.

Retuse: Notched at the apex.

Revolute: Leaf margins rolled toward the lower side.

Rhizome: An underground stem with nodes, buds, and roots.

Rhombic: Diamond-shaped; with the outline of an equilateral oblique-angled figure.

Rosette: A circular cluster of leaves at or near ground level.
Rotate: Wheel-shaped; flat and circular.
Rugose: Covered with wrinkles.

Sac: A bag or pouch.
Saccate: Bag- or pouch-shaped.
Sagittate: Arrowhead-shaped; with basal lobes pointed downward.
Salverform: With a slender corolla tube and an abruptly flared, wheel-shaped limb (ex.: *Phlox*).
Samara: An indehiscent, single-seeded, dry fruit with a prominent wing (ex.: ash, elm).
Scaberulous (scabrescent): Slightly rough to the touch.
Scabrous: Rough to the touch.
Scale: Small leaves or bracts; in Cyperaceae, subtending fertile or sterile flowers; a pappus type in Asteraceae.
Scape: A nonleafy peduncle; bracts or scales may be present.
Scapose: Resembling a scape.
Scarious: Thin, dry, membranaceous, not green; in reference to bracts, leaf margins, and perianth parts.
Scimitar: Shaped like a curved sword.
Secund: An inflorescence where the flowers appear to be borne only on one side of the peduncle.
Sepal: One of the segments of the calyx.
Septate: Divided by partition(s), as in an ovary or fruit.
Septicidal: Opening along the same line as the septa (partitions) of the ovary wall.
Septum: A partition or cross wall.
Sericeous: Silky.
Serrate: A margin with sharp, forward-pointing teeth.
Sessile: Attached directly; without a petiole or pedicel.
Seta: A hairlike marginal bristle.
Setaceous: Bristlelike; stiff, straight, and hairlike.
Sheath: A tubular structure surrounding a stem or other organ.
Simple Leaf: A leaf not divided into distinct leaflets; the margin may be entire or variously divided.
Simple Pistil: A pistil derived from a single stigma, style, and placenta; a unicarpellate reproductive body.
Sinuate: Wavy along the margin.
Sinus: The space between two lobes or teeth.
Solitary: Borne singly.
Sorus: A characteristically shaped cluster of sporangia borne on either the margin or the abaxial surface of fern leaves.
Spadix: A thick, fleshy spike (ex.: Araceae).

Spathe: A leafy bract that subtends and often partially surrounds an inflorescence.

Spatulate: Spoon-shaped.

Spicate: Resembling a spike.

Spike: An elongate, unbranched, indeterminate inflorescence with sessile flowers.

Spikelet: The basic unit of the inflorescence of grasses and sedges, composed of reduced flowers and their subtending bracts.

Spinulose: With small spines.

Sporangium: The more or less spherical spore case of the ferns and fern allies.

Spore: A dormant unicellular reproductive body, spherical or bean-shaped, capable of germinating into a gametophyte plant in ferns, fern allies, and other lower plants.

Sporocarp: The beanlike structure of the heterosporous water ferns that contains the sori.

Sporophyll: A fertile leaf-bearing sporangium.

Spur: A slender, often nectar-bearing, saclike extension of some perianth part(s).

Squarrose: Recurved at the tip.

Stamen: The pollen-bearing organ of the flower.

Staminate: A male flower; bearing stamen(s), but no functional pistil(s).

Staminode: A sterile, nonfunctional, reduced or highly modified stamen.

Standard: The uppermost, usually broad petal of a bean flower.

Stellate: Starlike; trichomes with radiating branches.

Stigma: The uppermost, pollen-receiving portion of the pistil.

Stipel: Stipule of a leaflet.

Stipitate: Borne on a short stalk.

Stipule: An appendage, usually paired, at the base of a leaf petiole.

Stolon: A horizontal stem, above or below ground level, that roots at the nodes and produces new plants at the stem tip.

Stomata: Minute epidermal pores in green stems and leaves, usually bordered by two conspicuous guard cells.

Striate: With distinct, longitudinal lines or ridges.

Strigose: Covered with sharp, straight, stiff hairs, sometimes swollen at the base.

Strobilus: A conelike structure of distinctive tightly clustered sporophylls.

Style: The usually slender portion of the pistil between the stigma and the ovary.

Stylopodium: The enlarged, basal portion of the style of certain Apiaceae.

Sub-: A prefix denoting below, slightly, nearly, almost, etc.

Subulate: Awl-shaped; tapered from the base to a pointed apex.

Succulent: Plant having thick, juicy leaves and/or stems.

Suffrutescent: Somewhat shrubby; plants with woody bases and herbaceous stems.
Superior Ovary: An ovary inserted above and free from the perianth and stamens.
Suture: The point of union or separation of organs.

Tendril: A modified leaf or stem by which a plant climbs or supports itself.
Tepal: A segment of a perianth that is not clearly differentiated (except by point of insertion) into calyx and corolla.
Terete: Round in cross section.
Ternate: Occurring in threes.
Thalloid: A flattened body not clearly differentiated into stems and leaves (ex.: Lemnaceae).
Thyrse: A compact, cylindrical or ovoid panicle in which the main axis is indeterminate and the lateral branches are determinate.
Tomentose: With dense, woolly hairs.
Trichome: A bristle or hairlike epidermal projection.
Trifoliate: A plant with three leaves.
Trifoliolate: A compound leaf with three leaflets.
Trifurcate: With three prongs, spines, or forks.
Trigonous: With three angles.
Trullate: With widest axis below middle and with straight margins; trowel-shaped.
Truncate: With a nearly straight apex or base.
Tuber: A thick, underground storage stem.
Tubercle: A small, wartlike protuberance.
Tuberculate: With tubercles.
Turgid: Swollen, but filled within.

Umbel: A flat-topped or somewhat rounded indeterminate inflorescence with peduncles or pedicels originating from a common point.
Umbellate: With umbels.
Uncinate: Hooked.
Undulate: Wavy, with reference to leaf or perianth margins.
Unisexual: A flower or plant that bears either stamen(s) or pistil(s), but not both.
Urceolate: Urn-shaped; wide at the bottom and constricted at the tip.
Utricle: A small, indehiscent, one-seeded, bladdery fruit.

Valvate: Meeting at the edges but not overlapping.
Verrucose: Having a warty or knotty surface.
Verticil: A whorl; usually referring to a whorl of flowers or leaves at a node.
Verticillate: Arranged in whorls.

Villous: With long, soft, but not matted hairs.
Viscid: Sticky.

Whorl: Three or more structures, usually leaves, at a node.
Wing: A lateral petal of some members of the bean family; a thin, dry, flat appendage of an organ.

Zygomorphic: A flower that may be divided into two equal halves by only one median longitudinal division; sometimes referred to as bilaterally symmetrical.

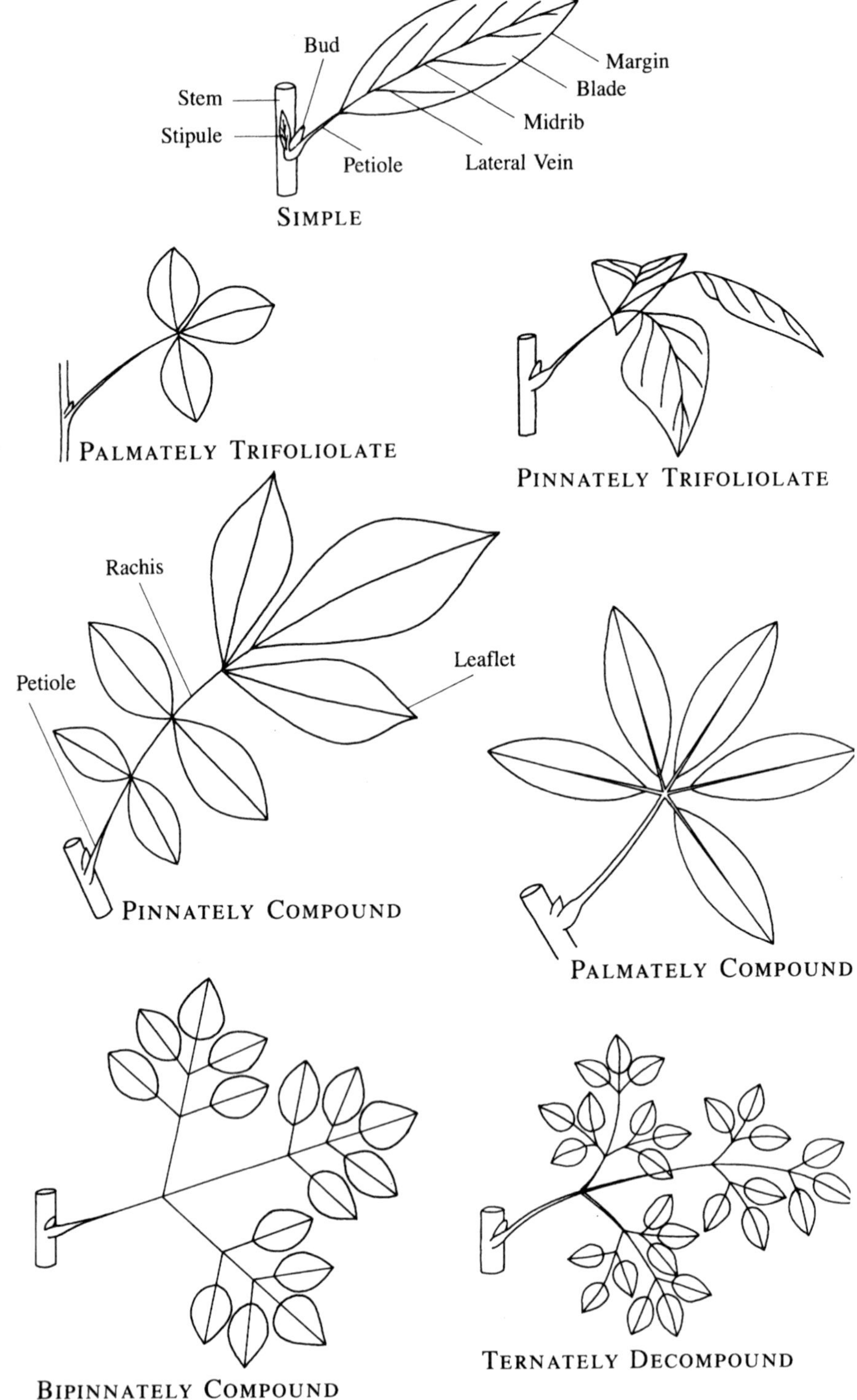
Bud
Margin
Blade
Stem
Stipule
Midrib
Petiole
Lateral Vein
SIMPLE
PALMATELY TRIFOLIOLATE
PINNATELY TRIFOLIOLATE
Rachis
Leaflet
Petiole
PINNATELY COMPOUND
PALMATELY COMPOUND
BIPINNATELY COMPOUND
TERNATELY DECOMPOUND

Leaf types and parts

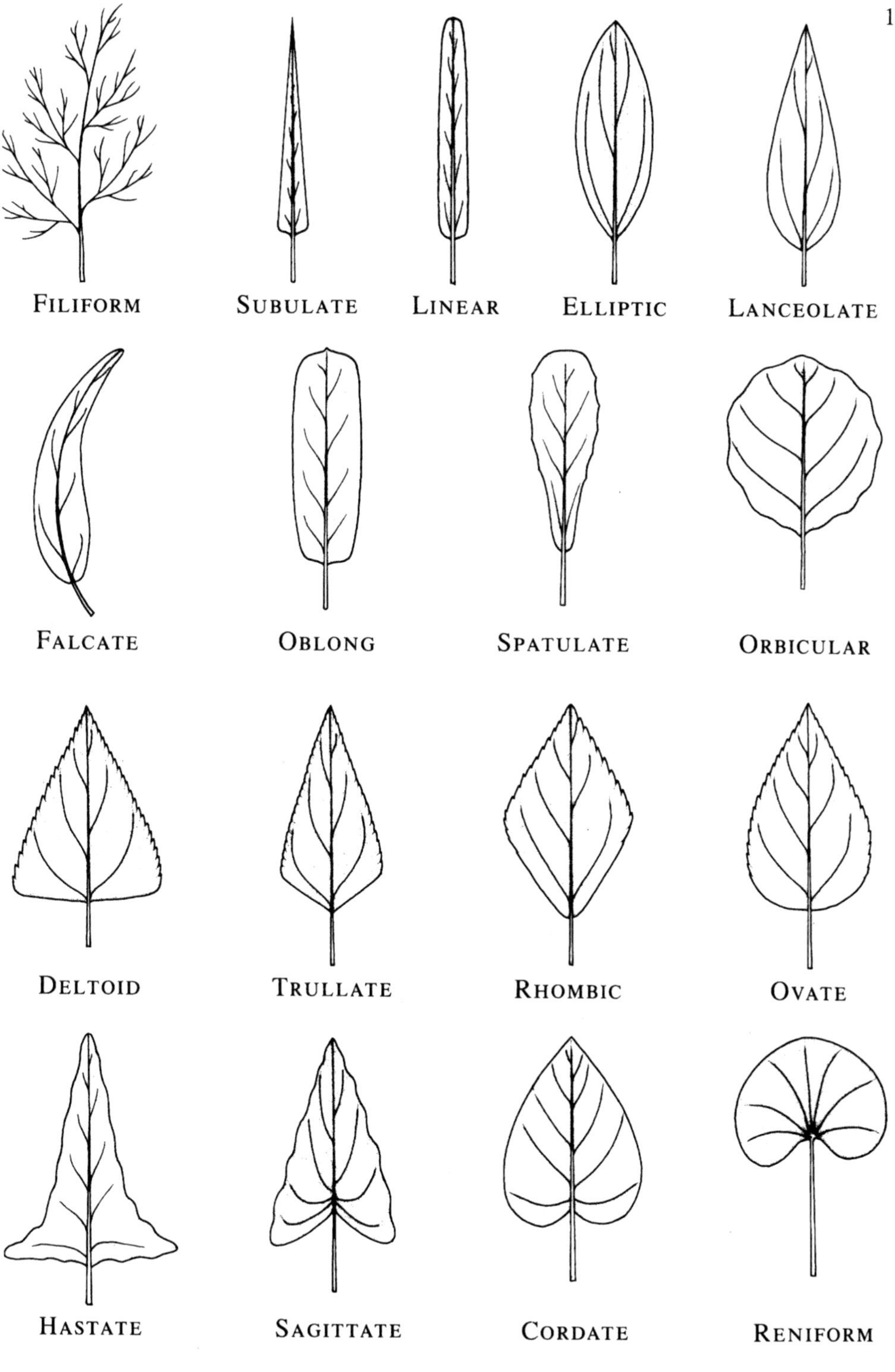

Filiform Subulate Linear Elliptic Lanceolate

Falcate Oblong Spatulate Orbicular

Deltoid Trullate Rhombic Ovate

Hastate Sagittate Cordate Reniform

Leaf shapes

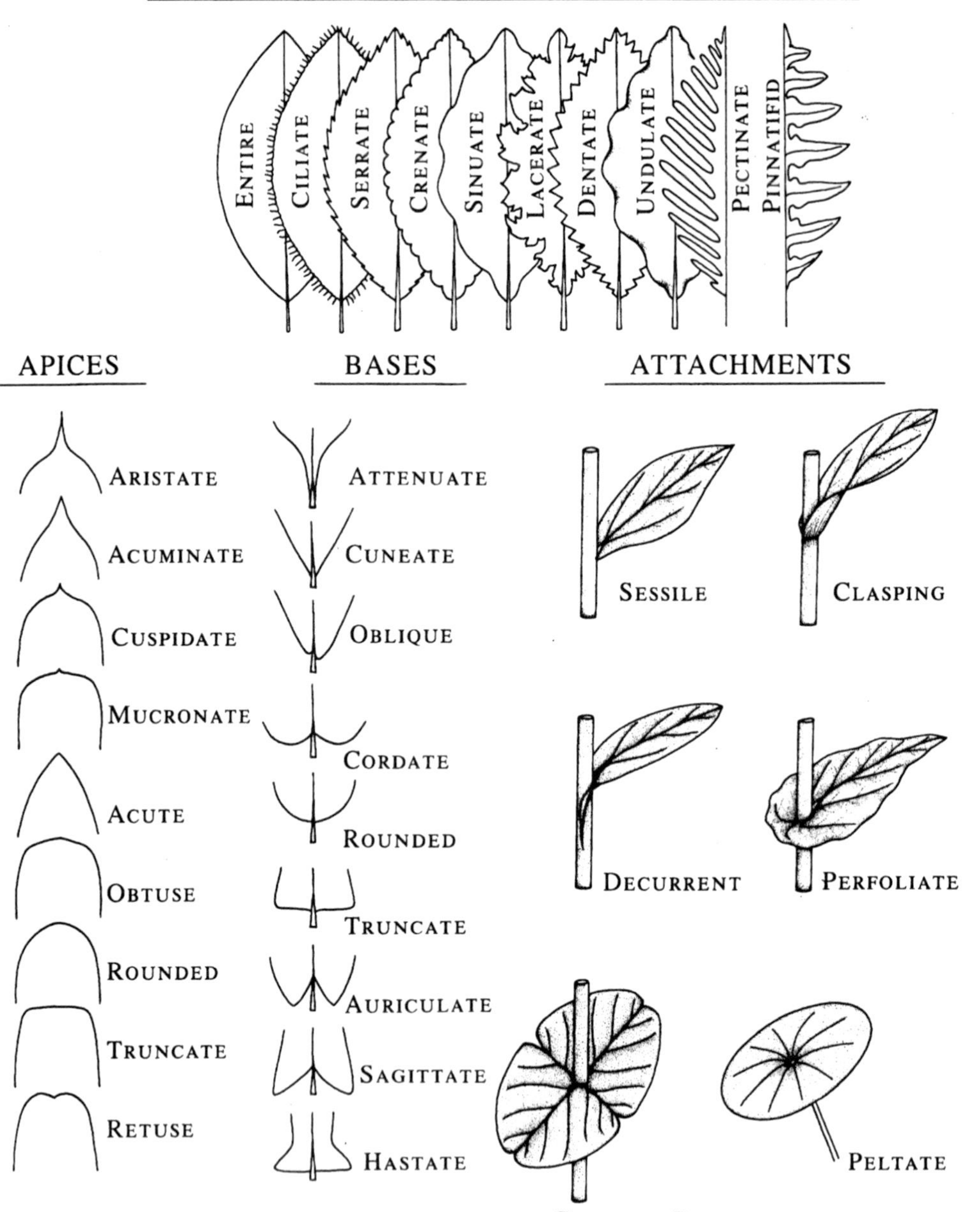

Leaf margins, apices, bases, and attachments

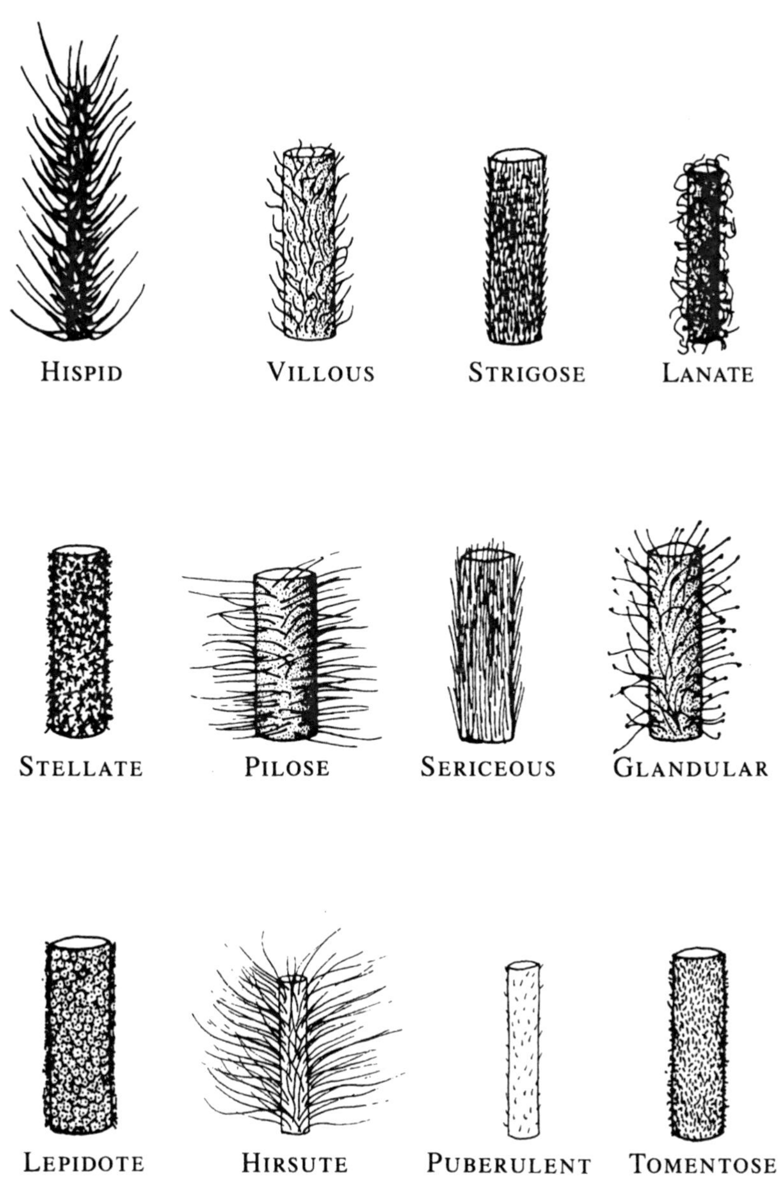

Surface features

FLOWER PARTS

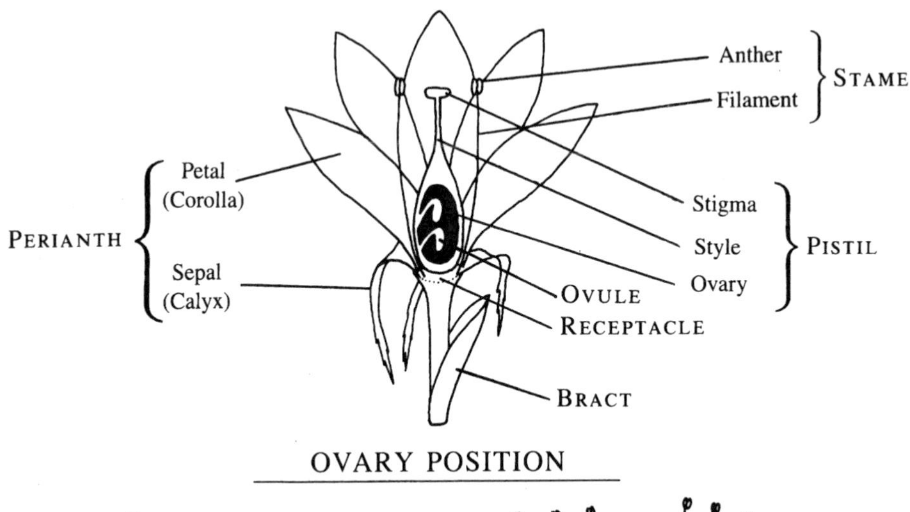

OVARY POSITION

SUPERIOR

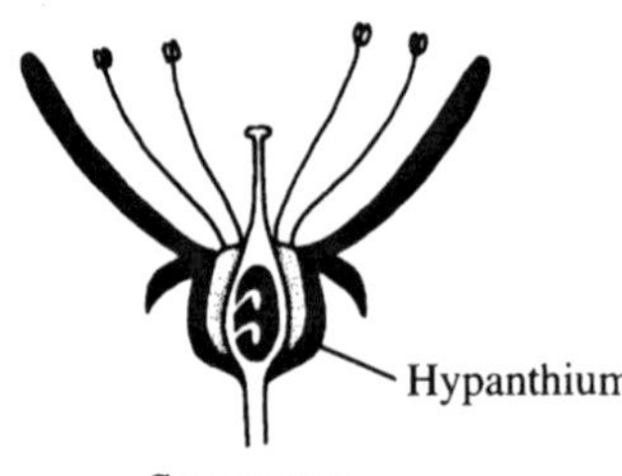

SUPERIOR

HALF-INFERIOR

INFERIOR

PLACENTATION

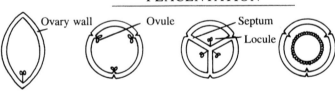

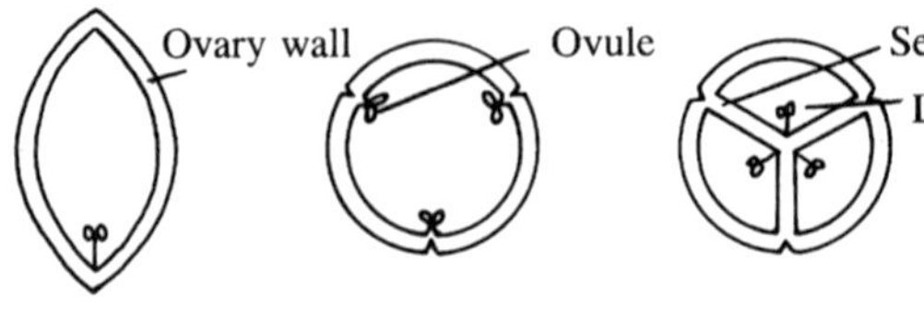

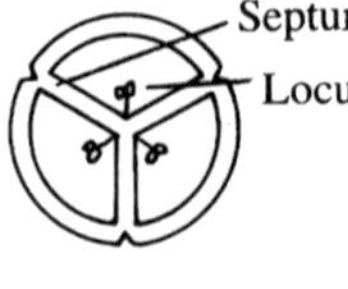

MARGINAL PARIETAL AXILE FREE CENTRAL BASAL

Floral morphology

SHAPES

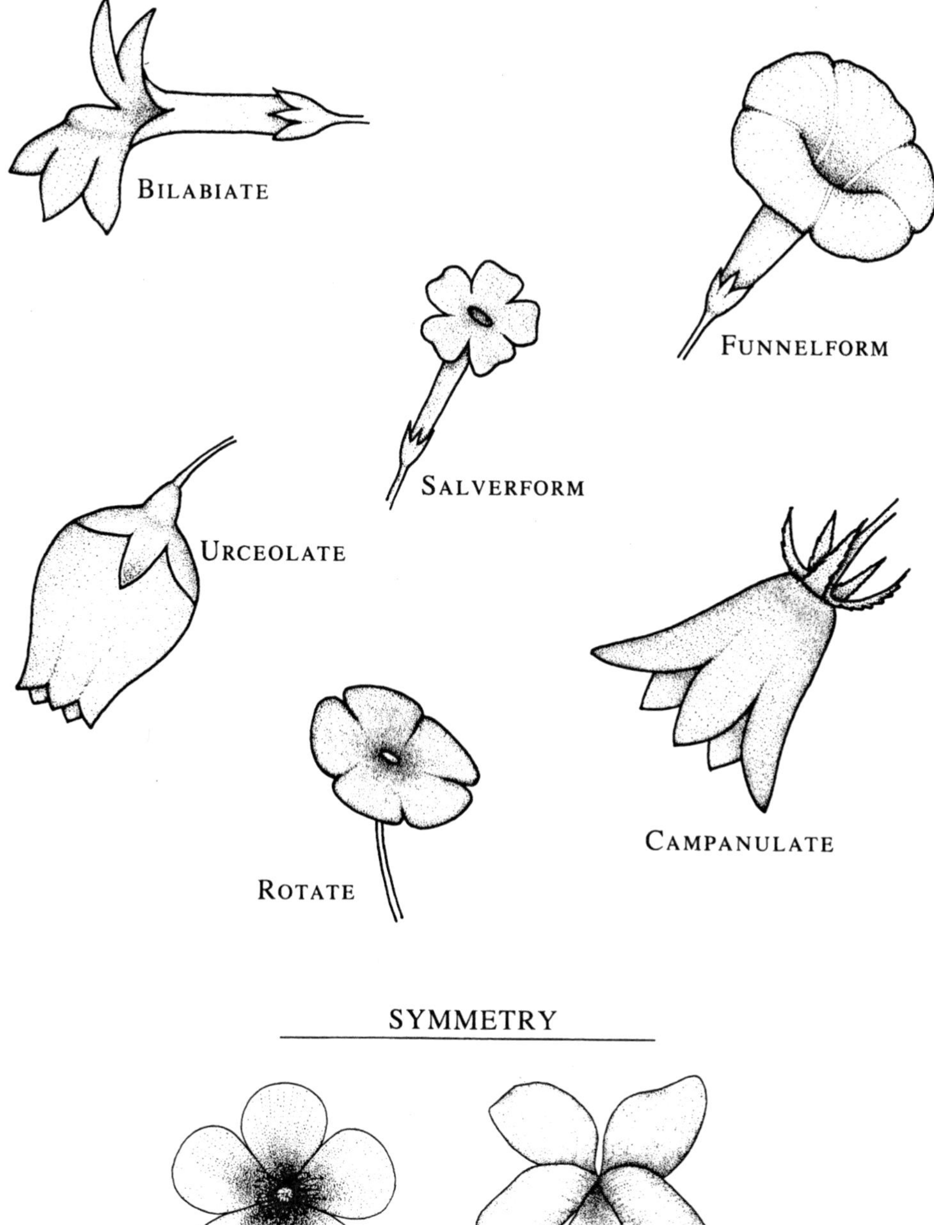

Corolla shapes and symmetry

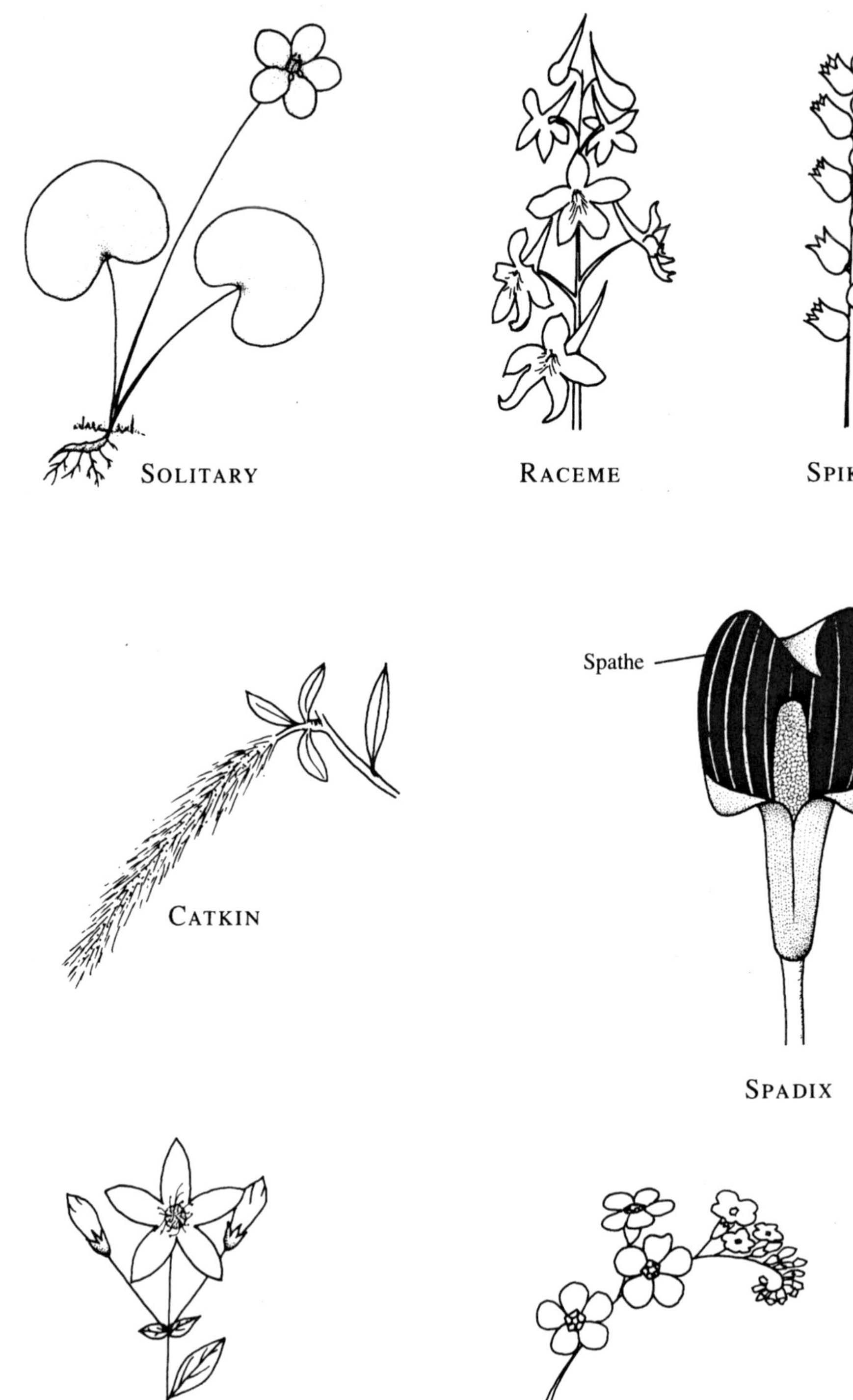

Inflorescence types

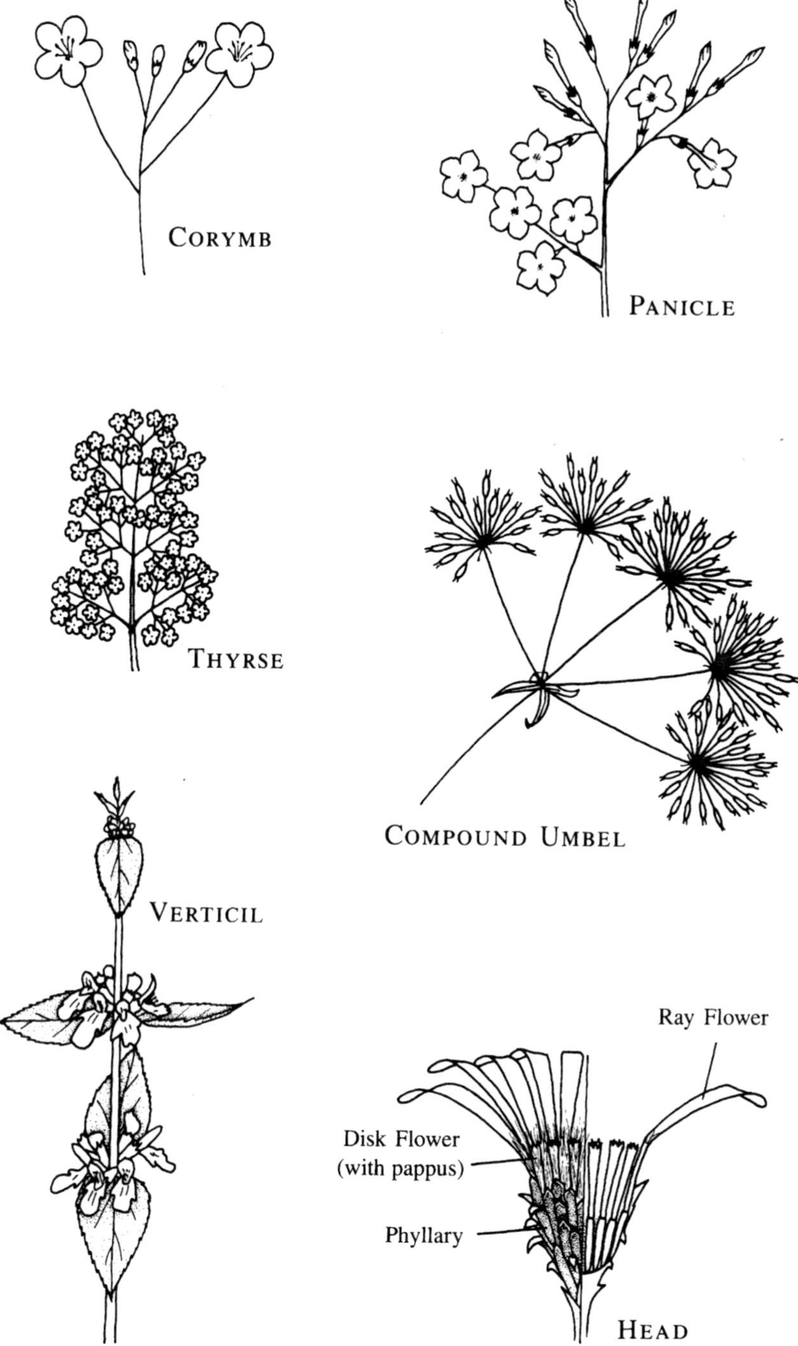

Inflorescence types *(continued)*

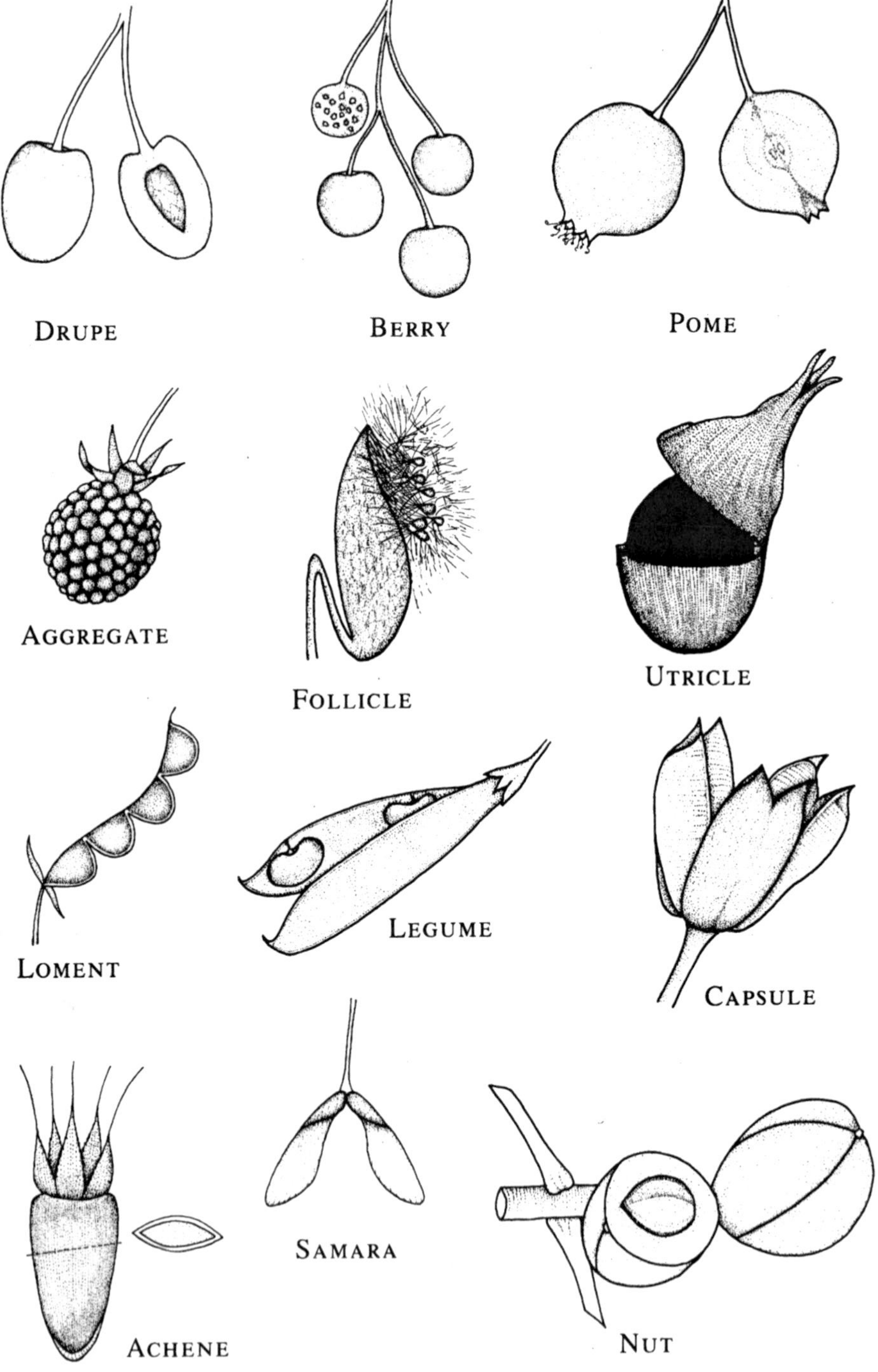
DRUPE
BERRY
POME
AGGREGATE
FOLLICLE
UTRICLE
LOMENT
LEGUME
CAPSULE
ACHENE
SAMARA
NUT

Fruit types

Keys

MAJOR PLANT GROUPS

1 Plants reproducing by spores (ferns and fern relatives) .... (p. 27) **PTERIDOPHYTES**
1 Plants reproducing by seeds (seed plants) 2
  2 Seeds borne on the surface of scales and aggregated into cones or embedded in a fleshy disk; style and stigma absent .... (p. 29) **GYMNOSPERMS**
  2 Seeds borne within a closed ovary; stigma present, style usually present **(Angiosperms)** 3
    3 Flower parts mostly in 3s or 6s; leaves usually with parallel venation; mostly herbaceous; cotyledon 1 .... (p. 29) **MONOCOTS**
    3 Flower parts mostly in 4s or 5s; leaves usually with net venation; herbaceous or woody; cotyledons 2 .... (p. 31) **DICOTS**

PTERIDOPHYTE FAMILIES

(Families arranged alphabetically beginning on page 47)

Contributed by A. M. Evans

1 Leaves needle- or scalelike, with a single midvein (microphylls), less than 1.5 cm long, or longer and grasslike with a single large sporangium imbedded in base 2
  2 Leaves tufted, grass- or rushlike, round above, flattened at the pale base and with a large sporangium imbedded in the leaf base; heterosporous; stem a soft, fleshy, 2- or 3-lobed subterranean corm .... **ISOETACEAE**
  2 Leaves needle- or scalelike, not grass- or rushlike; sporangia not imbedded in leaf base; stem stiff, elongate 3
    3 Leaves green; stems solid, alternately or dichotomously forking 4
      4 Plants homosporous, sporangia in axils of vegetative leaves or in "cones" that are round in cross section .... **LYCOPODIACEAE**
      4 Plants heterosporous, sporangia in 4-sided "cones" .. **SELAGINELLACEAE**
    3 Leaves whorled, reduced to toothed nongreen sheaths at solid nodes of hollow stems .... **EQUISETACEAE**
1 Leaves small to large, generally with branching veins; sporangia not borne in "cones" or in axils of leaves but on the leaf or leaf modification, rarely in nutlike sporocarps ... 5
  5 Plants aquatic; leaves filiform, without expanded blade; sporangia enclosed in nutlike sporocarps, heterosporous .... **MARSILEACEAE**

5 Plants terrestrial; sporangia borne on margin or on underside of leaf or on upright spike or panicle arising from the petiole, homosporous 6

6 Sporangia borne on upright spikes or panicles arising from the petiole below the vegetative blade . . . . **OPHIOGLOSSACEAE**

6 Sporangia borne on the leaf blade, on either typical vegetative leaves or strongly modified leaves 7

7 Fertile leaves or portions of leaves lacking, or appearing to lack, blade tissue; brown fertile portion at tip or middle of otherwise vegetative leaves or on separate nonlaminar leaf 8

8 Sporangia borne singly on skeletalized pinna axes, large, without well-marked annulus . . . . **OSMUNDACEAE**

8 Sporangia borne in clusters within rolled up reduced laminar segments, the dimorphic fertile leaf appearing as though without lamina; sporangia small, annulus vertical . . . . **WOODSIACEAE**

7 Sporangia borne on margin or underside of laminar parts; fertile parts green as in the vegetative parts, but may be dimorphic 9

9 Leaves climbing by vinelike midrib; rhizome terrestrial, hairy, without scales . . . . **SCHIZAEACEAE**

9 Leaves free-standing, not vinelike 10

10 Leaves delicate, only 1-cell thick between the veins, 1–10 cm long . . . . **HYMENOPHYLLACEAE**

10 Leaves more than 1-cell thick; mostly more than 10 cm long 11

11 Sori marginal or submarginal, beneath the folded, rolled over, or recurved leaf margin, or in marginal cups; indusia, if present, not kidney-shaped 12

12 Leaves large; rhizome wide-creeping, hairy and without scales . . . . **DENNSTAEDTIACEAE**

12 Leaves small to modest in size, clustered on short, ascending, scaly rhizome . . . . **SINOPTERIDACEAE**

11 Sori scattered on the underside of the leaf blade or marginal with kidney-shaped indusia 13

13 Sori without indusium, round 14

14 Leaves pinnatifid . . . . **POLYPODIACEAE**

14 Leaves decompound 15

15 Leaves tripartite . . . . **WOODSIACEAE**

15 Leaves pinnately compound . . **THELYPTERIDACEAE**

13 Sori with indusium, attached beneath or beside the sorus 16

16 Sori elongate 17

17 Sori end to end in a single row on chainlike veins on each side of the pinna axes . . . . **BLECHNACEAE**

17 Sori diagonal along veins diverging from the pinna midrib 18

18 Leaves small to midsized, evergreen; petiole dark, glossy, and wiry . . . . **ASPLENIACEAE**

18 Leaves large, deciduous; petiole dark at base only, dull and pale above . . . . **WOODSIACEAE**

16 Sori round 19

19 Indusium pocketlike, attached beside the sorus, or a ring of hair- or scalelike elements attached beneath the sporangia . . . . **WOODSIACEAE**

19 Indusium kidney-shaped (reniform), or umbrellalike (peltate), attached beneath the sorus and rising above it . . 20

20 Groove of the pinna axis with stiff needlelike hairs; leaves pale green; rhizome creeping, less than 3 mm in diameter . . . . **THELYPTERIDACEAE**

20 Groove of the pinna axis without needlelike hairs; leaves mid to dark green; rhizome ascending, more than 1 cm in diameter including heavy leaf bases . . . . **DRYOPTERIDACEAE**

GYMNOSPERM FAMILIES
(Families arranged alphabetically beginning on page 60)

1 Seed borne singly and either partially or completely enclosed in a fleshy, reddish aril . . . . **TAXACEAE**

1 Seeds (several to many) borne in woody, papery, or fleshy cones 2

2 Leaves needlelike, in fascicles of 2–5 or solitary and spirally arranged . . . . **PINACEAE**

2 Leaves scale- or awl-shaped, opposite or whorled . . . . **CUPRESSACEAE**

MONOCOT FAMILIES
(Families arranged alphabetically beginning on page 62)

1 Plant body thalloid (usually less than 1 cm wide) and not differentiated into stem and leaves; rootless or with 1–few simple roots; free-floating aquatics or occasionally on mud . . . . **LEMNACEAE**

1 Plants with stems and leaves or the leaves sometimes reduced and scalelike; roots present; terrestrial or aquatic 2

2 Stems woody 3

3 Vines; tendrils present . . . . **SMILACACEAE**

3 Plants shrubby or suffrutescent 4

4 Perianth petaloid; leaves predominantly basal, without sheathing bases . . . . **AGAVACEAE**

4 Perianth not petaloid; leaves cauline, with sheathing bases . . . . . . **POACEAE**

2 Stems herbaceous 5

5 Perianth reduced, either chaffy, scaly, bristly, or absent, at least not petallike in color or texture 6

6 Plants aquatic, the stems and flowers submerged or floating on the surface . . 7

7 Inflorescence a golden-yellow, elongate spike . . . . **ARACEAE**

7 Inflorescence various, but not golden-yellow 8

8 Leaves basal, long, and ribbonlike 9

9 Pistillate flowers solitary .... **HYDROCHARITACEAE**
9 Pistillate flowers numerous, in globose, burlike heads .... **SPARGANIACEAE**
8 Leaves cauline 10
10 Leaves alternate (the upper ones occasionally opposite) .... **POTAMOGETONACEAE**
10 Leaves opposite 11
11 Leaves serrate with broad, sheathing bases; flowers sessile .... **NAJADACEAE**
11 Leaves entire, not sheathing; flowers pedicellate .... **ZANNICHELLIACEAE**
6 Plants terrestrial or, if aquatic, at least some leaves and flowers above the surface of the water 12
12 Inflorescence a dense, brown, elongate spike (up to 40 cm) composed of thousands of flowers, the pistillate below the staminate (cattails) .... **TYPHACEAE**
12 Inflorescence otherwise 13
13 Inflorescence a single, dense, buttonlike, white or grayish head (less than 1 cm high) terminating an elongate scape .. **ERIOCAULACEAE**
13 Inflorescence a spadix, spike, cyme, or in 2–many glomerules or globose heads 14
14 Individual flowers concealed in the axils of chaffy bracts or scales with only the anthers and/or styles protruding at anthesis 15
15 Leaf sheaths open; leaves usually 2-ranked; internodes usually hollow; stems usually round .... **POACEAE**
15 Leaf sheaths closed; leaves usually 3-ranked; internodes usually solid; stems usually 3-sided .... **CYPERACEAE**
14 Individual flowers not hidden in the axils of bracts or scales ... 16
16 Inflorescence a spadix and often surrounded by a large, leafy spathe .... **ARACEAE**
16 Inflorescence cymose or in dense glomerules or globose heads 17
17 Flowers unisexual; fruit single-seeded .... **SPARGANIACEAE**
17 Flowers bisexual; fruit 3–many seeded .... **JUNCACEAE**
5 Perianth present and usually showy, either all segments similar or only the inner series petallike in color and texture 18
18 Plants submerged or floating aquatics; leaves usually less than 3 cm long .... **HYDROCHARITACEAE**
18 Plants terrestrial or, if aquatic, only their bases submerged; leaves usually more than 3 cm long 19
19 Flowers unisexual 20
20 Perianth segments dissimilar in form and color; ovaries numerous; stamens usually more than 6 .... **ALISMATACEAE**
20 Perianth segments similar in form and color; ovary 1; stamens 3 or 6 21
21 Ovary inferior; stems viny but without tendrils .... **DIOSCOREACEAE**

21 Ovary superior; stems erect or, if vines, then tendrils present 22
22 Leaves distinctly petiolate, the petioles 1 cm or more long; flowers umbellate .... **SMILACACEAE**
22 Leaves sessile or subsessile; flowers not umbellate .... **LILIACEAE**
19 Flowers bisexual 23
23 Pistils numerous (10–25) in each flower .... **ALISMATACEAE**
23 Pistil 1/flower 24
24 Ovary superior 25
25 Flowers in single, dense, conelike heads; each flower subtended by a rigid scale .... **XYRIDACEAE**
25 Flowers not borne in single, scapose heads 26
26 Flowers actinomorphic 27
27 Sepals and petals similar in form and color 28
28 Stamens 3; perianth segments 6 .... **PONTEDERIACEAE**
28 Stamen and perianth segments of the same number, usually 6, rarely 4 .... **LILIACEAE**
27 Sepals and petals dissimilar in form and color 29
29 Leaves 3, whorled .... **LILIACEAE**
29 Leaves more than 3, alternate .... **COMMELINACEAE**
26 Flowers zygomorphic 30
30 Petals separate to the base; fertile stamens 2 or 3 ....**COMMELINACEAE**
30 Petals tubular at the base; fertile stamens 6 .... **PONTEDERIACEAE**
24 Ovary inferior or half-inferior 31
31 Corolla zygomorphic; stamen(s) and pistil fused .... **ORCHIDACEAE**
31 Corolla actinomorphic; stamens free from the pistil 32
32 Stamens 3 33
33 Perianth woolly-pubescent; style 1 .... **HAEMODORACEAE**
33 Perianth glabrous; style 3-parted ...... **IRIDACEAE**
32 Stamens 6 34
34 Perianth green or purplish; leaves succulent .... **AGAVACEAE**
34 Perianth white, yellow, or pinkish; leaves not succulent .... **LILIACEAE**

DICOT FAMILIES

(Families arranged alphabetically beginning on page 135)

1 Trees, shrubs, or woody vines 2
2 Leaves or leaf scars opposite or whorled .... **KEY A**

2 Leaves or leaf scars alternate 3
3 Woody vines or cacti . . . . . KEY B
3 Trees or shrubs 4
4 Petals absent . . . . . KEY C
4 Petals present . . . . . KEY D
1 Herbs 5
5 Perianth absent . . . . . KEY E
5 Perianth present 6
6 Petals absent (the calyx sometimes petaloid) . . . . . KEY F
6 Petals present 7
7 Petals separate 8
8 Stamens numerous, more than twice the number of petals . . . . . KEY G
8 Stamens few, at least not more than twice the number of petals 9
9 Leaves opposite or whorled . . . . . KEY H
9 Leaves alternate, solitary, or mostly basal . . . . . KEY I
7 Petals fused, wholly or in part 10
10 Ovary inferior or half-inferior . . . . . KEY J
10 Ovary superior 11
11 Stamens (including staminodia) 6–many and more numerous than the lobes of the corolla . . . . . KEY K
11 Stamens (including staminodia) 2–5 and fewer than or of the same number as the lobes of the corolla 12
12 Corolla actinomorphic or nearly so . . . . . KEY L
12 Corolla distinctly zygomorphic . . . . . KEY M

KEY A

1 Leaves absent at anthesis 2
2 Corolla more than 3 cm long . . . . . **SCROPHULARIACEAE**
2 Corolla less than 3 cm long or absent 3
3 Stamens 2; ovary not lobed . . . . . **OLEACEAE**
3 Stamens 5 or more; ovary 2-lobed . . . . . **ACERACEAE**
1 Leaves present at anthesis 4
4 Leaves compound 5
5 Leaves palmately compound . . . . . **HIPPOCASTANACEAE**
5 Leaves pinnately compound 6
6 Woody vines 7
7 Stamens numerous . . . . . **RANUNCULACEAE**
7 Stamens 4, rarely 5 . . . . . **BIGNONIACEAE**
6 Trees or shrubs 8
8 Ovary inferior . . . . . **CAPRIFOLIACEAE**
8 Ovary superior 9
9 Leaflets 3; ovary 3-locular, fruit an inflated capsule . . . . . **STAPHYLEACEAE**
9 Leaflets usually 5 or more; ovary 2-locular, fruit a single or double samara 10
10 Stamens 2; style 1 . . . . . **OLEACEAE**
10 Stamens more than 2; styles 2 . . . . . **ACERACEAE**

4 Leaves simple 11
11 Plant parasitic on stems of other woody plants (mistletoe) ...... **VISCACEAE**
11 Plant not parasitic or, if so, then parasitic on roots of other plants 12
12 Perianth undifferentiated; pistils numerous .... **CALYCANTHACEAE**
12 Perianth differentiated or petals absent; pistil 1 13
13 Ovary inferior 14
14 Petals absent; flowers unisexual .... **SANTALACEAE**
14 Petals present; flowers bisexual 15
15 Petals separate 16
16 Stamens 4 .... **CORNACEAE**
16 Stamens 8–many .... **SAXIFRAGACEAE**
15 Petals fused 17
17 Perianth 4-parted; stamens 4; flowers in dense, globose heads .... **RUBIACEAE**
17 Perianth 5-parted; stamens 4 or 5; flowers variously arranged, but not in dense, globose heads ..... **CAPRIFOLIACEAE**
13 Ovary superior 18
18 Stamens numerous, more than 10 .... **CLUSIACEAE**
18 Stamens 10 or less, or rarely absent 19
19 Petals separate or absent 20
20 Leaves lobed .... **ACERACEAE**
20 Leaves not lobed 21
21 Stamens and petals of the same number, either 4 or 5 each .... **CELASTRACEAE**
21 Stamens more numerous than the petals 22
22 Leaves evergreen; stamens of equal lengths .... **ERICACEAE**
22 Leaves deciduous; stamens of unequal lengths .... **LYTHRACEAE**
19 Petals fused 23
23 Flowers actinomorphic, less than 3 cm long 24
24 Stamens 10 .... **ERICACEAE**
24 Stamens less than 10 25
25 Woody vines .... **LOGANIACEAE**
25 Trees or shrubs 26
26 Corolla 5-parted; stamens 4 .... **VERBENACEAE**
26 Corolla 4-parted; stamens 2 ...... **OLEACEAE**
23 Flowers zygomorphic, more than 3 cm long 27
27 Fertile stamens 2; capsule more than 6 cm long .... **BIGNONIACEAE**
27 Fertile stamens 4; capsule less than 6 cm long .... **SCROPHULARIACEAE**

KEY B

1 Stems thick, fleshy, padlike; leaves minute or absent (cacti) .... **CACTACEAE**
1 Stems normally developed; leaves present 2

2 Leaves compound 3
3 Leaves palmately compound . . . . **VITACEAE**
3 Leaves ternately or pinnately compound 4
4 Leaflets 3 5
5 Plants mostly climbing by aerial roots; flowers actinomorphic; fruit a drupe . . . . **ANACARDIACEAE**
5 Plants twining; flowers zygomorphic; fruit a legume . . . . **FABACEAE**
4 Leaflets more than 3 6
6 Leaves bi- or tripinnately compound . . . . **VITACEAE**
6 Leaves once-pinnately compound . . . . **FABACEAE**
2 Leaves simple 7
7 Leaves evergreen . . . . **ARALIACEAE**
7 Leaves deciduous 8
8 Plants climbing by tendrils . . . . **VITACEAE**
8 Plants climbing by twining 9
9 Flowers zygomorphic, 2 cm or more long . . . . . . **ARISTOLOCHIACEAE**
9 Flowers actinomorphic, less than 2 cm long 10
10 Pistils 12 or more; stamen filaments united; fruit an aggregate . . . . **SCHISANDRACEAE**
10 Pistils 1–6; stamen filaments not united; fruit a capsule or drupe 11
11 Leaves pinnately veined, the margin serrate; fruit a capsule . . . . **CELASTRACEAE**
11 Leaves palmately veined, entire or shallowly lobed; fruit a drupe . . . . **MENISPERMACEAE**

KEY C

1 Both sepals and petals absent 2
2 Plants dioecious 3
3 Leaves resin dotted and aromatic . . . . **MYRICACEAE**
3 Leaves not resin dotted, rarely aromatic . . . . **SALICACEAE**
2 Plants monoecious 4
4 Staminate flowers in globose heads; leaves palmately veined . . . . **HAMAMELIDACEAE**
4 Staminate flowers in elongate catkins; leaves pinnately veined . . . . **BETULACEAE**
1 Sepals present, at least in some flowers, often small and scalelike, or rarely modified into a floral tube 5
5 Perianth modified into a floral tube; bark leathery; twigs jointed . . . . **THYMELAEACEAE**
5 Perianth not modified into a floral tube; bark not leathery; twigs not jointed 6
6 Leaves compound 7
7 Trees; flowers unisexual; fruit a nut . . . . **JUGLANDACEAE**
7 Low shrub; flowers bisexual; fruit a follicle . . . . **RANUNCULACEAE**
6 Leaves simple or absent at anthesis 8
8 At least some flowers in catkins, globose heads, or dense spikes 9
9 At least some flowers in globose heads or dense spikes 10
10 Axillary bud enclosed by the leaf petiole . . . . **PLATANACEAE**
10 Axillary bud not enclosed by the leaf petiole 11

11 Sap with a milky latex .... **MORACEAE**
11 Sap watery, without a milky latex 12
12 Stamens white, conspicuous; leaves stellate-pubescent beneath; fruit a capsule .... **HAMAMELIDACEAE**
12 Stamens small, not showy; leaves glabrous or with simple trichomes; fruit a nut or drupe 13
13 Plant dioecious; fruit a drupe; leaves entire .... **SANTALACEAE**
13 Plant monoecious; fruit a nut; leaves serrate .... **FAGACEAE**
9 At least some flowers in catkins 14
14 Sap with a milky latex .... **MORACEAE**
14 Sap watery, without a milky latex 15
15 Both staminate and pistillate flowers in catkins ... **BETULACEAE**
15 Pistillate flowers not in catkins 16
16 Staminate flowers appearing in the spring; fruit subtended by a cup (acorn) or surrounded by a spiny involucre (chestnut) .... **FAGACEAE**
16 Staminate flowers appearing in the fall; fruit surrounded by a non-spiny bract and 2 bractlets .... **BETULACEAE**
8 Flowers solitary, or in umbels, panicles, fascicles, racemes, or axillary clusters 17
17 Perianth and young branches with silvery, peltate scales .... **ELAEAGNACEAE**
17 Perianth and young branches glabrous or with simple trichomes 18
18 Stamens more numerous than the calyx lobes 19
19 Calyx 6-parted; leaves and stems aromatic; ovary superior; flowers yellowish; anthers opening by uplifting lids ...... **LAURACEAE**
19 Calyx 5-parted; leaves and stems not aromatic; ovary inferior; flowers greenish; anthers opening longitudinally ...... **NYSSACEAE**
18 Stamens of the same number as the calyx lobes 20
20 Ovary inferior; leaves entire; fruit 2 cm or more long .... **SANTALACEAE**
20 Ovary superior; leaves serrate or slightly toothed at the base; fruit less than 2 cm long 21
21 Stamens opposite the sepals; leaves usually oblique at the base; medium to large trees .... **ULMACEAE**
21 Stamens alternate with the sepals; leaves not oblique at the base; shrubs less than 1 m tall .... **RHAMNACEAE**

KEY D

1 Petals fused 2
2 Corolla zygomorphic 3
3 Ovary simple; leaves compound or, rarely, simple .... **FABACEAE**
3 Ovary compound; leaves simple .... **ERICACEAE**
2 Corolla actinomorphic 4
4 Stamens more numerous than the lobes of the corolla 5

5 Leaves bipinnately compound .... **FABACEAE**
5 Leaves simple 6
6 Filaments united into a tube and surrounding the style .... **MALVACEAE**
6 Filaments separate or somewhat coherent, but not forming a tube and surrounding the style 7
7 Flowers 5 cm or more wide .... **THEACEAE**
7 Flowers less than 5 cm wide 8
8 Filaments fused at the base 9
9 Stamens commonly 8–10, rarely 16; flowers distinct pedicellate .... **STYRACACEAE**
9 Stamens numerous; flowers sessile or nearly so .... **SYMPLOCACEAE**
8 Filaments separate 10
10 Flowers bisexual; stamens 10 or fewer; anthers opening by terminal pores or slits .... **ERICACEAE**
10 Flowers mostly unisexual; stamens usually more than 10; anthers opening lengthwise .... **EBENACEAE**
4 Stamens of the same number as the corolla lobes 11
11 Inflorescence terminal .... **ERICACEAE**
11 Inflorescence axillary 12
12 Flowers dull violet; filaments hairy, fruit a berry .... **SOLANACEAE**
12 Flowers white or greenish; filaments glabrous; fruit a drupe .... **AQUIFOLIACEAE**
1 Petals separate 13
13 Stamens numerous 14
14 Pistils 2–many, simple 15
15 Sepals and petals 5; leaves simple or compound .... **ROSACEAE**
15 Sepals 3; petals 6–12; leaves simple 16
16 Flowers white, greenish, or with a yellow spot at the base of the petals; stipule scars encircling the twig .... **MAGNOLIACEAE**
16 Flowers maroon; stipules absent .... **ANNONACEAE**
14 Pistil 1, simple or compound (the carpels sometimes loosely coherent and separating at maturity) 17
17 Filaments united into a tube and surrounding the style .... **MALVACEAE**
17 Filaments separate, not united into a tube and surrounding the style 18
18 Flower clusters arising from the middle of a large, leaflike bract .... **TILIACEAE**
18 Flower clusters not arising from the middle of a large, leaflike bract 19
19 Flowers yellow; leaves less than 1 cm long .... **CISTACEAE**
19 Flowers white, pink, or reddish; leaves more than 1 cm long .... **ROSACEAE**
13 Stamens 12 or fewer 20
20 Corolla zygomorphic or petals reduced to only 1 .... **FABACEAE**
20 Corolla actinomorphic 21
21 Leaves compound 22
22 Pistils more than 1 23
23 Stems and sometimes the petioles with prickles .... **RUTACEAE**

23 Stems and petioles without prickles .... **SIMAROUBACEAE**

22 Pistil 1 24

24 Ovary inferior .... **ARALIACEAE**

24 Ovary superior 25

25 Ovary 2-celled; fruit a flat, circular samara; leaves ternately compound .... **RUTACEAE**

25 Ovary 1-celled; fruit a drupe; leaves pinnate or, rarely, ternately compound .... **ANACARDIACEAE**

21 Leaves simple 26

26 Axillary bud enclosed by the leaf petiole .... **PLATANACEAE**

26 Axillary bud not enclosed by the leaf petiole 27

27 Leaves evergreen .... **ERICACEAE**

27 Leaves deciduous 28

28 Petals 4 29

29 Petals bright yellow, more than 1 cm long, flowers opening from fall to early spring .... **HAMAMELIDACEAE**

29 Petals white, greenish- or yellowish-white, less than 1 cm long, flowers opening in late spring and summer 30

30 Ovary inferior; leaves opposite or alternate; flowers in terminal cymes .... **CORNACEAE**

30 Ovary superior; leaves alternate; flowers axillary, either solitary or in few-flowered fascicles .... **AQUIFOLIACEAE**

28 Petals 5 or more 31

31 Ovary inferior 32

32 Leaves entire or sometimes coarsely toothed above the middle .... **NYSSACEAE**

32 Leaves serrate, dentate, or variously lobed 33

33 Petals longer than the sepals; fruit a pome .... **ROSACEAE**

33 Petals shorter than the sepals; fruit a berry .... **SAXIFRAGACEAE**

31 Ovary superior 34

34 Stamens more numerous than the petals .... **CLETHRACEAE**

34 Stamens of the same number as the petals 35

35 Flowers yellow, 6-merous; stems with simple or trifurcate spines .... **BERBERIDACEAE**

35 Flowers white or greenish; stems not spiny 36

36 Flowers in elongate, terminal racemes .... **SAXIFRAGACEAE**

36 Flowers in axillary clusters, or in terminal or lateral panicles 37

37 Stamens opposite the petals .... **RHAMNACEAE**

37 Stamens alternate with the petals .... **AQUIFOLIACEAE**

KEY E

1 Sap milky; ovary stipitate; flowers greatly reduced and contained within a flowerlike, cup-shaped cyathium . . . . **EUPHORBIACEAE**
1 Sap watery; ovary not stipitate; cyathium absent 2
 2 Flowers bisexual 3
  3 Stamens 6–8, filaments separate; plants 2 dm or more tall . . . **SAURURACEAE**
  3 Stamens 2, filaments fused; plants mat forming and attached to submerged rocks . . . . **PODOSTEMACEAE**
 2 Flowers unisexual 4
  4 Stamen 1; leaves opposite and entire . . . . **CALLITRICHACEAE**
  4 Stamens 12–16; leaves whorled and deeply dissected . . . . **CERATOPHYLLACEAE**

KEY F

1 Ovary inferior or half-inferior 2
 2 Plants aquatic, either submerged or in wet soils 3
  3 Leaves alternate or whorled, the margins serrate, pectinate, or deeply divided . . . . **HALORAGACEAE**
  3 Leaves opposite and entire 4
   4 Style 1 . . . . **ONAGRACEAE**
   4 Styles 2 . . . . **SAXIFRAGACEAE**
 2 Plants terrestrial and not in permanently wet sites 5
  5 Sepals 3; stamens 6 or 12 . . . . **ARISTOLOCHIACEAE**
  5 Sepals 5; stamens 5 or less 6
   6 Leaves opposite . . . . **NYCTAGINACEAE**
   6 Leaves alternate . . . . **SANTALACEAE**
1 Ovary or ovaries superior 7
 7 Pistils more than 1/flower; completely separate or fused to near the middle 8
  8 Pistils separate to the base; stamens and/or carpels numerous . . . . **RANUNCULACEAE**
  8 Pistils united to near the middle; stamens usually 10 or less 9
   9 Leaves entire; flowers in racemes; fruit a berry . . . . . . **PHYTOLACCACEAE**
   9 Leaves serrate; flowers cymose; fruit a capsule . . . . **SAXIFRAGACEAE**
 7 Pistil 1, not divided below the style 10
  10 Leaves evergreen 11
   11 Stamens 4; flowers unisexual . . . . **BUXACEAE**
   11 Stamens 12; flowers bisexual . . . . **ARISTOLOCHIACEAE**
  10 Leaves not evergreen 12
   12 Plants climbing vines 13
    13 Leaves opposite . . . . **CANNABACEAE**
    13 Leaves alternate . . . . **POLYGONACEAE**
   12 Plants erect or decumbent, but not viny 14
    14 Stamens numerous, 12 or more . . . . **RANUNCULACEAE**
    14 Stamens few, less than 12 15
     15 Ovary 2–many celled; seeds 2–many 16
      16 Flowers unisexual . . . . **EUPHORBIACEAE**

16 Flowers bisexual 17
17 Leaves alternate 18
18 Sepals 5 . . . . **PHYTOLACCACEAE**
18 Sepals 4 . . . . **BRASSICACEAE**
17 Leaves opposite or whorled 19
19 Styles 3–5; leaves whorled, rarely opposite . . . . **MOLLUGINACEAE**
19 Style 1; leaves opposite . . . . **LYTHRACEAE**
15 Ovary 1-celled; 1-seeded 20
20 Leaves opposite or whorled 21
21 Leaves whorled . . . . **POLYGONACEAE**
21 Leaves opposite 22
22 Stems succulent; leaves reduced to scales . . . . **CHENOPODIACEAE**
22 Stems not succulent; leaves not reduced to scales 23
23 Leaves serrate . . . . **URTICACEAE**
23 Leaves entire 24
24 Calyx petaloid, pinkish, up to 1 cm long . . . . **NYCTAGINACEAE**
24 Calyx not petaloid, green or tinged with white, less than 1 cm long 25
25 Flowers in dense, woolly spikes . . . . **AMARANTHACEAE**
25 Flowers in loose or dense cymes . . . . **CARYOPHYLLACEAE**
20 Leaves alternate (at least the median and upper) 26
26 Leaves with sheaths (ocreae) surrounding the stems at the nodes . . . . **POLYGONACEAE**
26 Leaves without sheaths at the nodes 27
27 Stipules present; leaves either pinnately compound or deeply 3-lobed . . . . **ROSACEAE**
27 Stipules absent; leaves simple, entire, sinuate, toothed, or hastate 28
28 Calyx 4-parted . . . . **URTICACEAE**
28 Calyx 5- or, rarely, 3-parted 29
29 Plants with stinging trichomes . . . **URTICACEAE**
29 Plants without stinging trichomes 30
30 Sepals and bracts dry, scarious, or membranaceous . . . . **AMARANTHACEAE**
30 Sepals and bracts green or greenish and herbaceous . . . . **CHENOPODIACEAE**

KEY G

1 Plants aquatic, at least partially submerged 2
2 Pistils separate 3
3 Leaves peltate 4

4 Flowers 2–2.5 cm wide; petioles and lower leaf surfaces with a gelatinous sheath .... **CABOMBACEAE**

4 Flowers 15–25 cm wide; petioles and lower leaf surfaces without a gelatinous sheath .... **NELUMBONACEAE**

3 Leaves not peltate .... **RANUNCULACEAE**

2 Pistils partly united .... **NYMPHAEACEAE**

1 Plants terrestrial or growing in wet sites, but not submerged 5

5 Stigma umbrella-shaped, more than 2 cm wide; leaves tubular; plants carnivorous .... **SARRACENIACEAE**

5 Stigma less than 2 cm wide; leaves not tubular; plants not carnivorous 6

6 Filaments united into a tube and surrounding the style .... **MALVACEAE**

6 Filaments separate or, if partially united, then not surrounding the style 7

7 Pistils more than 1, separate 8

8 Hypanthium present, the petals and/or stamens attached to a flat, saucer- or cup-shaped receptacle; stipules usually present .... **ROSACEAE**

8 Hypanthium absent, the petals and stamens separate to the base; stipules absent .... **RANUNCULACEAE**

7 Pistil 1, either simple or compound 9

9 Pistil simple 10

10 Leaves bipinnately compound .... **FABACEAE**

10 Leaves ternately compound .... **RANUNCULACEAE**

9 Pistil compound 11

11 Plants with milky or colored sap .... **PAPAVERACEAE**

11 Plants with watery sap 12

12 Leaves alternate or basal 13

13 Flowers unisexual; leaves densely stellate beneath .... **EUPHORBIACEAE**

13 Flowers bisexual; leaves usually not stellate beneath 14

14 Sepals differing in size, the outer 2 much smaller than the inner 3 .... **CISTACEAE**

14 Sepals similar in size 15

15 Leaves simple; petals 5 .... **PORTULACACEAE**

15 Leaves trifoliolate; petals 4 .... **CAPPARACEAE**

12 Leaves opposite 16

16 Calyx oblique at the base; leaves not glandular-punctate; petals pinkish .... **LYTHRACEAE**

16 Calyx not oblique at the base; leaves glandular-punctate; petals yellow .... **CLUSIACEAE**

KEY H

1 Ovary inferior 2

2 Submerged aquatics .... **HALORAGACEAE**

2 Plants terrestrial 3

3 Leaves whorled, compound .... **ARALIACEAE**

3 Leaves opposite, simple 4

4 Petals 2 or 4 5

5 Flowers in dense heads and subtended by 4 large, white bracts .... **CORNACEAE**

5 Flowers not in dense heads and not subtended by 4 large, white bracts .... **ONAGRACEAE**

4 Petals 5 .... **SAXIFRAGACEAE**

1 Ovary superior 6

6 Sepals 2 .... **PORTULACACEAE**

6 Sepals 4 or more 7

7 Plants succulent .... **CRASSULACEAE**

7 Plants not succulent 8

8 Flowers unisexual; leaves stellate beneath .... **EUPHORBIACEAE**

8 Flowers bisexual; leaves usually not stellate beneath 9

9 Petals deeply pinnatifid or fringed .... **SAXIFRAGACEAE**

9 Petals entire, erose, cleft, or bifid, but not pinnatifid or fringed 10

10 Leaves lobed, dissected, or compound 11

11 Inflorescence paniculate; leaves appearing whorled .... **BERBERIDACEAE**

11 Inflorescence solitary, umbellate, or cymose; leaves opposite 12

12 Flowers solitary, arising between the 2 leaves; fruit many-seeded .... **BERBERIDACEAE**

12 Flowers umbellate or cymose; fruit 5-seeded .. **GERANIACEAE**

10 Leaves entire or slightly toothed 13

13 Petals yellow 14

14 Leaves glandular-punctate .... **CLUSIACEAE**

14 Leaves not glandular-punctate .... **LINACEAE**

13 Petals white, greenish, pink, or crimson 15

15 Aquatics; leaves less than 1 cm long .... **ELATINACEAE**

15 Plants terrestrial; leaves usually more than 1 cm long 16

16 Style 1; hypanthium present 17

17 Anthers opening by terminal pores .... **MELASTOMATACEAE**

17 Anthers opening lengthwise .... **LYTHRACEAE**

16 Styles 2 or more; hypanthium absent 18

18 Ovary 3-celled; leaves glandular-punctate; stamens 9, the filaments fused into 3 groups of 3 stamens each .... **CLUSIACEAE**

18 Ovary 1-celled or, if 3-celled, then the leaves not glandular-punctate; filaments not united, stamens various, usually 5 or 10 .... **CARYOPHYLLACEAE**

KEY I

1 Ovary inferior 2

2 Style 1 3

3 Aquatics 4

4 Petals yellow .... **ONAGRACEAE**

4 Petals white or greenish .... **HALORAGACEAE**

3 Plants terrestrial .... **ONAGRACEAE**
2 Styles 2 or more 5
5 Flowers in racemes, panicles, cymes, or solitary .... **SAXIFRAGACEAE**
5 Flowers in umbels or heads 6
6 Styles 5, rarely 4 or 6; fruit a berry .... **ARALIACEAE**
6 Styles 2; fruit dry .... **APIACEAE**
1 Ovary superior 7
7 Leaves evergreen or plant without chlorophyll 8
8 Fertile stamens 10 .... **ERICACEAE**
8 Fertile stamens 5, staminodia 5 .... **DIAPENSIACEAE**
7 Leaves not evergreen; chlorophyll present 9
9 Flowers zygomorphic 10
10 Carpel 1; placentation marginal .... **FABACEAE**
10 Carpels 2–5; placentation axile or parietal 11
11 Sepals and petals 3 (the lateral petals lobed); ovary 5-celled .... **BALSAMINACEAE**
11 Sepals and petals 5; ovary 1- or 2-celled 12
12 Carpels 2; stipules absent .... **SAXIFRAGACEAE**
12 Carpels 3; stipules present .... **VIOLACEAE**
9 Flowers actinomorphic 13
13 Leaves compound or deeply divided into 2 segments 14
14 Petals 3 .... **LIMNANTHACEAE**
14 Petals 4 or more 15
15 Petals 4 .... **BRASSICACEAE**
15 Petals 5 or more 16
16 Pistils 2(3) or pistil 1 with 2(3) carpels 17
17 Leaves pinnately compound; hypanthium with a row of hooked bristles; stipules present .... **ROSACEAE**
17 Leaves ternately compound; hypanthium without a row of hooked bristles; stipules absent .... **SAXIFRAGACEAE**
16 Pistil 1, with either 1 or 5 carpels 18
18 Ovary 1-celled 19
19 Leaves once-pinnate; stipules present; fruit a legume .... **FABACEAE**
19 Leaves either ternately compound or divided into 2 segments; fruit a berry or capsule ..... **BERBERIDACEAE**
18 Ovary 5-celled 20
20 Leaflets 3 .... **OXALIDACEAE**
20 Leaflets pinnately divided into numerous segments .... **GERANIACEAE**
13 Leaves simple, entire, toothed, lobed or dissected 21
21 Plant viny, climbing by tendrils .... **PASSIFLORACEAE**
21 Plant not viny, tendrils absent 22
22 Flowers unisexual 23
23 Leaves succulent, glabrous .... **CRASSULACEAE**
23 Leaves not succulent, with stellate hairs .. **EUPHORBIACEAE**
22 Flowers bisexual 24
24 Leaves succulent .... **CRASSULACEAE**

24 Leaves not succulent 25
25 Ovary 1- or 2-celled 26
26 Leaves with dense glandular trichomes; plants insectivorous .... **DROSERACEAE**
26 Leaves glabrous or with nonglandular trichomes; plants not insectivorous 27
27 Sepals unequal, the outer 2 much smaller than the inner 3 .... **CISTACEAE**
27 Sepals similar in size 28
28 Sepals 6; stamens 6 .... **BERBERIDACEAE**
28 Sepals 4 or 5; stamens 2, 4, 5, 6, or 10 29
29 Sepals 4; stamens 6, rarely 2 or 4 .... **BRASSICACEAE**
29 Sepals 5; stamens 5 or 10 .... **SAXIFRAGACEAE**
25 Ovary 5–10-celled 30
30 Leaves palmately lobed or dissected; petals pink to purple, rarely white .... **GERANIACEAE**
30 Leaves entire or serrate; petals white or yellow 31
31 Leaves serrate, 5 cm or more long; petals white, rarely present .... **SAXIFRAGACEAE**
31 Leaves entire, less than 5 cm long; petals yellow .... **LINACEAE**

KEY J

1 Anthers united into a tube and surrounding the style 2
2 Flowers in dense heads and subtended by involucral bracts; calyx absent or highly modified .... **ASTERACEAE**
2 Flowers in racemes or in the axils of uppermost leaves; calyx present and not highly modified .... **CAMPANULACEAE**
1 Anthers separate 3
3 Flowers unisexual 4
4 Plant viny, climbing by tendrils .... **CUCURBITACEAE**
4 Plant erect, not viny .... **ASTERACEAE**
3 Flowers bisexual 5
5 Leaves alternate 6
6 Stamens alternate with the petals; ovary 3–5-celled ... **CAMPANULACEAE**
6 Stamens opposite the petals; ovary 1-celled .... **PRIMULACEAE**
5 Leaves opposite or whorled 7
7 Leaves whorled .... **RUBIACEAE**
7 Leaves opposite 8
8 Stamens 3 .... **VALERIANACEAE**
8 Stamens more than 3 9
9 Ovary 1-celled; stems prickly .... **DIPSACACEAE**
9 Ovary 2–5-celled; stems not prickly 10
10 Calyx lobes 2 or 4 .... **RUBIACEAE**
10 Calyx lobes 5 or more 11

11 Flowers in dense, terminal clusters .... **RUBIACEAE**
11 Flowers axillary or in a terminal pair ..... **CAPRIFOLIACEAE**

KEY K

1 Corolla zygomorphic, rarely bisbilateral 2
2 Stamens more than 10 .... **RANUNCULACEAE**
2 Stamens 10 or fewer 3
3 Stamens 10; carpel 1 .... **FABACEAE**
3 Stamens 6 or 8; carpels 2 4
4 Sepals 5, the outer 3 small, the inner 2 larger and similar to the petals in color; petals 3; leaves simple .... **POLYGALACEAE**
4 Sepals 2, often falling early; petals 4; leaves deeply divided to variously compound .... **FUMARIACEAE**
1 Corolla actinomorphic 5
5 Leaves evergreen or plants without chlorophyll 6
6 Stamens 8–10, staminodia absent; fruit a berry .... **ERICACEAE**
6 Stamens 5, staminodia 5 and rudimentary; fruit a capsule ... **DIAPENSIACEAE**
5 Leaves deciduous and with chlorophyll 7
7 Leaves compound 8
8 Leaflets 3; carpels 5; stems without prickles .... **OXALIDACEAE**
8 Leaves bipinnately compound; carpel 1; stems prickly .... **FABACEAE**
7 Leaves simple 9
9 Sepals united for more than 1/2 their length; stamens usually more than 10 .... **MALVACEAE**
9 Sepals separate; stamens 10 .... **LINACEAE**

KEY L

1 Pistils 2; plants usually with a milky sap; fruit a follicle 2
2 Leaves evergreen .... **APOCYNACEAE**
2 Leaves deciduous 3
3 Twining vines .... **ASCLEPIADACEAE**
3 Plants erect 4
4 Flowers in axillary or terminal umbels; pollen grains united into a waxy mass (pollinium); styles separate to the stigma .... **ASCLEPIADACEAE**
4 Flowers in terminal cymes; pollen grains separate; styles united .... **APOCYNACEAE**
1 Pistil 1, or the ovary deeply divided into 4 (rarely fewer) single-seeded nutlets; plants without a milky sap; fruit not a follicle 5
5 Leaves basal or essentially so 6
6 Flowers 4–merous; leaves parallel veined; corolla scarious .... **PLANTAGINACEAE**
6 Flowers 5–merous; leaves net veined; corolla white to pink 7
7 Leaves trifoliolate; flowers racemose .... **MENYANTHACEAE**
7 Leaves simple; flowers in umbels .... **PRIMULACEAE**
5 Stem leaves present, rarely scalelike or absent 8

8 Ovary deeply 2- or 4-lobed, appearing as 2 or 4 separate ovaries; style attached basally 9
9 Leaves opposite; stem usually 4-angled . . . . . **LAMIACEAE**
9 Leaves alternate; stem round . . . . . **BORAGINACEAE**
8 Ovary not deeply 2- or 4-lobed; style attached terminally 10
10 Stem leaves opposite or whorled (bracteal leaves sometimes alternate) . . . 11
11 Stamens opposite the lobes of the corolla . . . . . **PRIMULACEAE**
11 Stamens alternate with the lobes of the corolla 12
12 Ovary 1-celled . . . . . **GENTIANACEAE**
12 Ovary 2–5-celled 13
13 Ovary 3-celled; stigmas 3; stamens arising at different levels within the corolla tube . . . . . **POLEMONIACEAE**
13 Ovary 2-, 4-, or 5-celled; stigmas more or less than 3; stamens usually arising at 1 plane 14
14 Stamens of the same number as the lobes of the corolla 15
15 Flowers 4-merous . . . . . **LOGANIACEAE**
15 Flowers 5-merous 16
16 Flowers scarlet on the outside, yellow within and arising from 1-sided racemes . . . . . **LOGANIACEAE**
16 Flowers blue, yellow, white, or greenish, but not on 1-sided racemes . . . . . **SOLANACEAE**
14 Stamens fewer than the lobes of the corolla 17
17 Ovule 1 in each of the 2–4 cells . . . . . . **VERBENACEAE**
17 Ovules 2–many in each cell 18
18 Stamens 2 . . . . . **SCROPHULARIACEAE**
18 Stamens 4 19
19 Flowers blue and the leaves entire 20
20 Corolla less than 1.5 cm long . . . . . **SCROPHULARIACEAE**
20 Corolla 2–6 cm long . . . . . **ACANTHACEAE**
19 Flowers white, yellow, or pink; if blue, then the leaves serrate . . . . . **SCROPHULARIACEAE**
10 Stem leaves alternate or absent 21
21 Leaves compound 22
22 Leaves pinnate; ovary 1–3-celled 23
23 Ovary 1- or 2-celled . . . . . **HYDROPHYLLACEAE**
23 Ovary 3-celled . . . . . **POLEMONIACEAE**
22 Leaves bipinnate; ovary 1-celled . . . . . **FABACEAE**
21 Leaves entire, toothed, deeply lobed, scalelike, or absent 24
24 Leaves scalelike, 5 mm or less long, or absent 25
25 Plants parasitic, attached to host plants . . . **CONVOLVULACEAE**
25 Plants not attached to host plants . . . . . **GENTIANACEAE**
24 Leaves more than 5 mm long, not scalelike 26
26 Stamens opposite the lobes of the corolla . . . . . **PRIMULACEAE**
26 Stamens alternate with the lobes of the corolla 27
27 Ovary 1-celled 28
28 Plant trailing; anthers concealed within the corolla tube; calyx concealed by 2 leafy bracts . . . . . . **CONVOLVULACEAE**

28 Plants erect; anthers exserted; calyx not concealed within bracts . . . .**HYDROPHYLLACEAE**

27 Ovary 2- or 6-celled 29

29 Ovules 2 in each cell; plants usually trailing or twining . . . . **CONVOLVULACEAE**

29 Ovules more than 2 (usually numerous) in each cell; plants erect, decumbent, or rarely trailing 30

30 Leaves petiolate; fruit a berry or spiny capsule . . . . **SOLANACEAE**

30 Leaves sessile; capsule not spiny . . . . **SCROPHULARIACEAE**

KEY M

1 Ovary deeply 2- or 4-lobed, appearing as 2 or 4 separate ovaries; style attached basally 2

2 Stamens 2 or 4; stem usually 4-angled . . . . **LAMIACEAE**

2 Stamens 5; stem round . . . . **BORAGINACEAE**

1 Ovary not deeply 2- or 4-lobed (occasionally shallowly 4-lobed); style terminal 3

3 Calyx or corolla spurred or saccate at the base 4

4 Plants aquatic; roots absent . . . . **LENTIBULARIACEAE**

4 Plants terrestrial; roots present 5

5 Sepals 5; stamens 4 . . . . **SCROPHULARIACEAE**

5 Sepals 3; stamens 5 . . . . **BALSAMINACEAE**

3 Neither calyx nor corolla spurred or saccate at the base 6

6 Plants lacking chlorophyll, either white, yellow, brown, or purplish; leaves reduced to scales . . . . **OROBANCHACEAE**

6 Plants green; leaves not scaly 7

7 Ovary 1-celled and with a single ovule; fruit reflexed at maturity . . . . **PHRYMACEAE**

7 Ovary 2–4-celled and with 2–many ovules; fruit not reflexed at maturity . . . 8

8 Leaves alternate . . . . **SCROPHULARIACEAE**

8 Leaves opposite 9

9 Ovule 1 in each of the 2–4 cells of the ovary 10

10 Anther-bearing stamens 2 . . . . **SCROPHULARIACEAE**

10 Anther-bearing stamens 4 11

11 Flowers in dense spikes (the flowers sometimes overlapping); stamens included within the corolla tube . . . . **VERBENACEAE**

11 Flowers in panicles, racemes, or solitary; stamens equal to or exserted beyond the corolla tube . . . . **LAMIACEAE**

9 Ovules 2–many in each of the 2 cells of the ovary 12

12 Fertile stamens 2 13

13 Flowers in dense, long-peduncled, axillary spikes; anther sacs separate; plants 4–8 dm tall . . . . **ACANTHACEAE**

13 Flowers solitary or in terminal or axillary racemes; anther sacs not separate; plants usually less than 4 dm tall . . . . **SCROPHULARIACEAE**

12 Fertile stamens 4 or 5 . . . . **SCROPHULARIACEAE**

Pteridophyte Families

Contributed by A. M. Evans

ASPLENIACEAE (Spleenwort Family)

Asplenium L. (Spleenwort)

1 Leaves simple, veins areolate . . . . **A. rhizophyllum**
1 Leaves lobed or pinnately divided, veins free 2
2 Blade deeply lobed, or pinnate at base only, apex more or less caudate . . . . **A. pinnatifidum**
2 Blade pinnately divided throughout, apex not caudate 3
3 Blade 1-pinnate 4
4 Pinnae alternate, sessile, the bases usually overlapping the midrib; leaves subdimorphic, the fertile erect and taller than the spreading sterile leaves . . . . **A. platyneuron**
4 Pinnae opposite, not sessile, not overlapping the midrib; leaves monomorphic 5
5 Sori several, in 2 rows along both sides of the pinna midrib 6
6 Pinnae widest above the base, lacking auricle on upper side of base . . . 7
7 Plants mostly of acidic rocks; petioles reddish-brown; leaves arching away from the rocks; pinnae suborbicular; plants diploid . . . . **A. trichomanes** subsp. **trichomanes**
7 Plants of calcareous rocks; petioles blackish-brown; leaves appressed to rocks; pinnae oblong; plants tetraploid . . . . **A. trichomanes** subsp. **quadrivalens**
6 Pinnae widest at base, with more or less definite anterior basal auricle . . . . **A. resiliens**
5 Sori only 1 or 2, confined to the lower edge of the pinna . . . **A. monanthes**
3 Blade 2-pinnate, at least at base, the lower pinnae divided into pinnules or deep lobes 8
8 Blade ovate-triangular in outline; petiole green; indusium ciliate; on limestone . . . . **A. ruta-muraria**
8 Blade lance-ovate to lance-oblong; at least lower half of petiole blackish; indusium entire; on acidic rocks 9
9 Blades 2-pinnate throughout; petiole dark only in lower half . . . . **A. montanum**
9 Blades 2-pinnate only at base; petiole and lower blade midrib blackish . . . . **A. bradleyi**

A. bradleyi D. C. Eaton, Bradley's S.; on acidic rock outcrops, especially sandstone; infrequent; Jun–Oct; GA, TN, NC, VA.

A. monanthes L., Single-sorus S.; on acidic rock outcrops; infrequent; Jun–Oct; NC, SC. Disjunct from tropics.

A. montanum Willd., Mountain S.; on sandstone, igneous, and metamorphic outcrops; common; Jun–Sep; ALL.

A. pinnatifidum Muhl., Lobed S.; on sandstone, igneous, and metamorphic outcrops; infrequent; May–Oct; ALL.

A. platyneuron (L.) BSP., Ebony S.; rocky woods, old fields, thickets, and calcareous rock outcrops; common; Apr–Oct; ALL.

A. resiliens Kunze, Black-stemmed S.; calcareous rock outcrops and sinkholes; frequent; May–Oct; ALL.

A. rhizophyllum L., Walking Fern; mossy boulders and ledges, primarily on limestone, occasionally on tree bases; frequent; May–Oct; ALL. *Camptosorus rhizophyllus* (L.) Link.

A. ruta-muraria L., Wall-rue; on calcareous rock outcrops; infrequent; May–Sep; NC, TN, VA.

A. trichomanes L. subsp. **quadrivalens** Meyer amend. Lovis; on calcareous rock outcrops; rare; Jun–Oct; VA.

A. trichomanes L. subsp. **trichomanes**, Maidenhair S.; mostly on acidic rock outcrops; frequent; Jun–Oct; ALL.

Several sterile *Asplenium* hybrids are found in the region. They can be recognized by their aborted spores and intermediate morphology between their parents and include: *A. montanum* × *pinnatifidum* (*A.* × *trudellii* Wherry), *A. bradleyi* × *montanum* (*A.* × *gravesii* Maxon), *A. pinnatifidum* × *platyneuron* (*A.* × *kentuckiense* McCoy), *A. platyneuron* × *rhizophyllum* (*A.* × *ebenoides* R. R. Scott), *A. platyneuron* × *trichomanes* (*A.* × *virginicum* Maxon).

BLECHNACEAE (Chain Fern Family)

Woodwardia J. E. Smith (Chain Fern)

1 Leaves pinnate at base, pinnatifid above, dimorphic, the fertile more erect and contracted . . . . **W. areolata**

1 Leaves pinnate-pinnatifid throughout, monomorphic . . . . **W. virginica**

W. areolata (L.) Moore, Netted C. F.; wet woods, swamps, and acid bogs; common; May–Sep; ALL. *Lorinseria areolata* (L.) Presl.

W. virginica (L.) J. E. Smith, Virginia C. F.; swamp woods and pond margins; infrequent; GA, NC, TN, VA. *Anchistea virginica* (L.) Presl.

DENNSTAEDTIACEAE (Hay-scented Fern Family)

1 Sori elongate and continuous beneath folded under leaf margin . . . . **Pteridium**

1 Sori round, small, marginal, the indusium a cup surrounding the sporangia . . . . **Dennstaedtia**

Dennstaedtia Bernh. (Hay-scented Fern)

D. punctilobula (Michx.) Moore; dry, rocky, open slopes and woods, acid soils; common; Jun–Sep; ALL.

Pteridium Gled. ex Scop. (Bracken Fern)

P. aquilinum (L.) Kuhn; old fields, clearings, second-growth woods; common; Jul–Sep; ALL. Two poorly defined varieties occur in our area: the more northern var. *latiusculum* (Desv.) Underw. has the longest entire apical segments about 4× as long as broad and the margins moderately pubescent; the more southern var. *pseudocaudatum* (Clute) Heller has the longest entire apical segments 6–15× as long as broad and the margins glabrous.

DRYOPTERIDACEAE (Wood Fern Family)

1 Indusium kidney-shaped (reniform), attached along the narrow sinus (groove); pinna margins occasionally spinulose . . . . **Dryopteris**

1 Indusium umbrella-shaped (peltate), attached in the center of the sorus; pinna margins spinulose . . . . **Polystichum**

Dryopteris Adans. (Wood Fern)

1 Sori marginal; leaf leathery, gray-green . . . . **D. marginalis**

1 Sori medial; leaf lamina herbaceous to chartaceous, bright to dark green 2

2 Leaf blade pinnate-pinnatifid 3

3 Leaves dimorphic; fertile leaves erect, linear to linear-lanceolate, 2–3× as long as the spreading sterile leaves; scales at petiole base tan . . . . **D. cristata**

3 Leaves not, or only slightly, dimorphic; blade lanceolate to ovate; scales at petiole base dark brown 4

4 Blade lanceolate, tapering gradually to apex, the lowest pinnae $\frac{1}{2}$–$\frac{2}{3}$× as long as longest pinnae; basal pinnae triangular, with basal segments longer than adjacent ones; petiole base scales with narrow, dark median band . . . . **D. celsa**

4 Blade ovate to ovate-oblong, narrowing abruptly near apex, the lowest pinnae nearly as long as longest median pinnae; basal pinnae oblong with basal segments shorter than adjacent ones; petiole base scales dark with narrow pale margin . . . . **D. goldiana**

2 Leaf blade 2- to 4-pinnate 5

5 Leaf evergreen, with minute stalked glands particularly on indusium, rachis, and pinna midrib; basal pinnae with innermost lower pinnule shorter than adjacent outer pinnule; longest lower pinnule rarely more than 2× as long as longest upper pinnule; plants of rich mesic woods . . . . **D. intermedia**

5 Leaf deciduous, without stalked glands (except occasionally on indusium); basal pinnae with innermost inferior pinnule usually longer than adjacent pinnule; longest inferior pinnule more than 2× as long as superior pinnules 6

6 Indusium glabrous; pinnae 2-pinnate and slightly ascending; blade ovate, about $\frac{1}{2}$ as wide as rachis length; innermost inferior pinnule of basal pinna always the longest, usually only 2× as long and 1.5× as broad as opposite superior pinnule; plants of marshes, acid woods, and bogs . . . . **D. carthusiana**

6 Indusium occasionally glandular; pinnae 3-pinnatifid to 3-pinnate, nearly perpendicular to rachis; blade broadly ovate, about $\frac{2}{3}$ as broad as rachis length; innermost lower pinnule of basal pinnae usually longer than adjacent pinnule, and 2–4× as long and about 2× as broad as opposite upper pinnule; plants of high elevations . . . . **D. campyloptera**

D. campyloptera Clarkson, Mountain W. F.; high-elevation rocky valleys and wooded summits; frequent; Jul–Sep; NC, TN, VA.

D. carthusiana (Villars) H. P. Fuchs, Spinulose W. F.; bogs and acid woods; occasional; Jun–Sep; NC, TN, VA. *D. spinulosa* (O. F. Muell.) Watt.

D. celsa (Palmer) Small, Log Fern; wooded-mountain seepage slopes, stream banks, river bottoms; infrequent; Jun–Sep; NC, TN, VA.

D. cristata (L.) A. Gray, Crested W. F.; bogs, marshes, and swamp woods; infrequent; Jul–Sep; NC, TN, VA.

D. goldiana (Hooker) A. Gray, Goldie's Fern; rich, wooded, rocky seepage slopes and sinks; infrequent; Jun–Sep; ALL.

D. intermedia (Muhl.) Gray, Intermediate W. F.; rich woods and shaded rocky slopes; common; Jun–Sep; ALL.

D. marginalis (L.) A. Gray, Marginal W. F.; rocky woods and slopes; common; Jun–Sep; ALL.

Sterile hybrids with aborted spores and morphology intermediate between the parents can occasionally be found with the parents. The most likely are: *D. carthusiana* × *cristata* (*D.* × *uliginosa* Druce), *D. carthusiana* × *intermedia* (*D.* × *triploidea* Wherry), *D. carthusiana* × *marginalis* (*D.* × *pittsfordensis* Slosson), *D. celsa* × *intermedia* (*D.* × *separabilis* Small), *D. celsa* × *marginalis* (*D.* × *leedsii* Wherry), *D. cristata* × *intermedia* (*D.* × *boottii* Underw.), *D. goldiana* × *marginalis* (*D.* × *neowherryi* Wagner).

Polystichum Roth (Christmas Fern)

P. acrostichoides (Michx.) Schott; dry to mesic wooded sites; common; Jun–Sep; ALL.

EQUISETACEAE (Horsetail Family)

Equisetum L. (Horsetail, Scouring Rush)

1 Stems dimorphic; the fertile stems unbranched and nonchlorophyllous, the vegetative stems with regular whorls of branches . . . . **E. arvense**

1 Stems monomorphic, unbranched, occasionally with few uneven branches 2

 2 Stems deciduous; leaf sheaths about 1.5× as long as wide; cones rounded at apex . . . . **E. fluviatile**

 2 Stems firm, perennial, at least at base; cones apiculate 3

  3 Stems perennial; leaf sheaths about as long as wide, mostly with basal and apical dark bands; spores gray-green, normal . . . . **E. hyemale**

  3 Stems perennial near base only; leaf sheaths about 1.5× as long as wide, with dark band only at apex, or both apical and basal bands near base of stem; spores yellow, granular, aborted . . . . **E. × ferrissii**

E. arvense L., Field H.; moist roadsides, stream banks, old fields; common; Mar–Apr; GA, NC, TN, VA.

E. × ferrissii Clute, Ferriss' S. R.; wet places, gravelly or sandy lake or river margins; infrequent; Apr–Jun; NC, TN, VA. A sterile hybrid spreading vegetatively.

E. fluviatile L., Water H.; wet places, shallow water; infrequent; Jun–Aug; VA.

E. hyemale L., S. R.; wet roadsides, waste places, river and lake shores; common; Jun–Sep; GA, NC, TN, VA. Represented in our area by var. **affine** (Engelm.) A. A. Eaton.

HYMENOPHYLLACEAE (Filmy Fern Family)

1 Sporangia borne on bristle not extending beyond apex of pouchlike deeply 2-parted involucre . . . . **Hymenophyllum**
1 Bristle usually extending beyond apex of funnelform involucre . . . . **Trichomanes**

Hymenophyllum Smith (Filmy Fern)

H. tunbrigense (L.) J. E. Smith, Tunbridge F. F.; continuously moist metamorphic rock outcrops in deep gorge; very rare; Jun–Sep; Pickens Co., SC.

Trichomanes L. (Filmy Fern)

1 Leaves decompound, more than 5 cm long, without black marginal hairs . . . . **T. boschianum**
1 Leaves simple, less than 2 cm long, with stiff black marginal hairs . . . . . . **T. petersii**

T. boschianum Sturm, Appalachian F. F.; in deep shade in moist crevices, shallow caves, or grottoes in sandstone or other acidic rock outcrops; infrequent; Jun–Sep; GA, NC, SC, TN.

T. petersii A. Gray, Dwarf F. F.; moist sandstone, igneous or metamorphic boulders and outcrops; infrequent; Jun–Aug; GA, NC, SC, TN.

ISOETACEAE (Quillwort Family)

Isoetes L. (Quillwort)

1 Leaf bases white; leaves erect or recurved and twisted; plants aquatic or amphibious 2
  2 Leaves straight, erect; sporangial wall without pigmented cells; megaspores 0.4–0.6 mm in diameter . . . . **I. engelmannii**
  2 Leaves recurved, twisted; sporangial wall with scattered brown pigmented cells; megaspores 0.6–1.0 mm in diameter . . . . **I. macrospora**
1 Leaf bases brown; leaves recurved; plants amphibious in temporary pools . . . . **I. virginica**

I. engelmannii A. Braun, Engelmann's Q.; amphibious or submerged in ditches, ponds, or slow-water streams; infrequent; Jul–Sep; ALL.

I. macrospora Dur., Large-spored Q.; submerged, rooted in gravels in shoals of cold streams; infrequent; Jul–Sep; TN.

I. virginica Pfeiffer, Virginia Q.; amphibious in temporary pools; infrequent; Jul–Sep; VA.

LYCOPODIACEAE (Club-moss Family)

Lycopodium L. (Club-moss)

1 Sporophylls like the sterile leaves or only slightly reduced, in annual bands along the stem; vegetative reproduction by leafy gemmae near stem apex 2
  2 Leaves oblanceolate, the apical portion toothed; mainly in rocky forest soils; stomata on abaxial leaf surface . . . . **L. lucidulum**
  2 Leaves awl-shaped, margins not, or hardly, toothed; mainly on acidic rock outcrops; stomata on both leaf surfaces 3

3 Leaves dark shiny green, bulbous at base, entire, mostly appressed; high elevations . . . . **L. selago**

3 Leaves paler, duller green, flat at base, indistinctly toothed at apex, mostly spreading; mid-elevations . . . . **L. porophilum**

1 Sporophylls differing from sterile leaves, either broader and shorter, or more spreading, aggregated into terminal cones; without gemmae 4

4 Plants of open marshy or boggy habitats; herbaceous; pale or yellow-green, dull; rhizome dying back annually to an underground vegetative tuber at apex 5

5 Leaves of creeping rhizomes entire; fertile erect stems 3–10 cm tall, mostly solitary; strobili 0.8–4.0 cm long, 6–10 mm broad; sporophylls with ciliate-denticulate margins; vegetative leaves of erect stem entire . . . . . . **L. inundatum**

5 Leaves of creeping rhizomes with ciliate margin; erect stems 5–40 cm tall, mostly more than 1/creeping rhizome branch; strobili 2–8 cm long 6

6 Erect stems (including leaves) 4 mm or less in diameter; leaves of the peduncle and strobilus appressed and entire or, rarely, with a few cilia at the leaf base; horizontal rhizomes rooting to the ground throughout . . . . **L. appressum**

6 Erect stems (including leaves) 6 mm or more in diameter; leaves of the peduncle incurved and those of the strobilus spreading, both with ciliate margins; horizontal rhizomes mostly arching and only rooting intermittently . . . . **L. alopecuroides**

4 Plants of drier forest or rocky soils; rigid, bright to dark green, shiny, evergreen; rhizome trailing, perennial 7

7 Erect leafy stems 10 mm or more in diameter, not treelike or fanlike; leaves with apical setae 8

8 Strobili 1/erect stem, sessile; vegetative leaves with subulate-aristate apex . . . . **L. annotinum**

8 Strobili 1–6/erect leafy stem, on a bracteate peduncle; vegetative leaves with a long apical cilium . . . . **L. clavatum**

7 Erect leafy stems 2–8 mm in diameter, in a treelike or fanlike pattern; leaves without apical setae 9

9 Strobili sessile at stem apices; leaves more or less awl-shaped and distinct; leafy branches usually 5–8 mm broad 10

10 Leaves of the main vertical axis appressed, branchlets with 6-ranked leaves on dorsiventral branchlets, 1 adaxial and 1 abaxial rank, and 2 pairs of lateral ranks; abaxial leaves mostly shorter and more appressed than spreading lateral ones 11

11 Abaxial leaves of horizontal branches about ½ as long as lateral leaves . . . . **L. obscurum** var. **obscurum**

11 Abaxial leaves about the same length as the lateral leaves . . . . **L. obscurum** var. **isophyllum**

10 Leaves of the main vertical axis widely spreading, particularly near the lower lateral branches; branchlets essentially symmetrical; leaves 6-ranked, 2 lateral ranks and pairs of adaxial and abaxial ranks, equal in length and spreading . . . . **L. dendroideum**

9 Strobili on long, narrow peduncles; leaves scalelike and appressed; leafy branches usually 3 mm or less broad 12

12 Rhizomes buried 1–5 cm; branches blue-green in color, with annual constrictions . . . . **L. tristachyum**

12 Rhizomes superficial; branches dark green in color, without constrictions . . . . **L. digitatum**

L. alopecuroides L., Foxtail C.; open acidic seeps and mountain bogs; infrequent; Jul–Sep; GA, SC, VA.

L. annotinum L., Stiff C.; rich rocky woods and bogs in acid soils at high elevations; infrequent; Aug–Oct; VA, and an old collection from Great Smoky Mts., TN.

L. appressum (Chapm.) Lloyd & Underw., Southern Bog C.; bogs and wet meadows in acid soil; infrequent; Jul–Sep; GA, NC, TN, VA.

L. clavatum L., Running C.; open woods, rocky slopes and bog margins in acid soil; infrequent; Aug–Oct; ALL.

L. dendroideum Michx., Round-branch Ground Pine; rocky woods and clearings at high elevations; infrequent; Aug–Oct; NC, TN, VA.

L. digitatum Dillen. ex A. Braun, Running-pine; dry woods, slopes, and pinelands in acid soil; common; Aug–Oct; ALL. Hybridizes with *L. tristachyum*. *L. flabelliforme* (Fern.) Blanch.

L. inundatum L., Bog C.; acid bogs and gravelly seeps; infrequent; Jul–Sep; NC, VA.

L. lucidulum Michx.; Shining C.; moist cool woods; common; Jul–Nov; ALL.

L. obscurum L. var. **isophyllum** Hickey; heath and grassy balds at high elevations; infrequent; Aug–Oct; NC, TN.

L. obscurum L. var. **obscurum**, Ground Pine; heath balds, laurel thickets, dry woods, and bog margins in acid soils; common; Aug–Oct; ALL.

L. porophilum Lloyd & Underw., Rock C.; shaded moist sandstone and granitic cliffs; infrequent; Jul–Sep; NC, SC, TN, VA.

L. selago L., Fir C.; exposed, high-elevation rocky summits, in crevices of igneous or metamorphic rocks, also rarely on bog margins and seepage slopes; infrequent; Jul–Sep; GA, NC, TN, VA. Hybridizes with *L. lucidulum* (*L.* × *buttersii* Abbe).

L. tristachyum Pursh, Ground Cedar; dry, exposed sandy woods and rocky slopes; frequent; Jul–Oct; ALL.

MARSILEACEAE (Water-clover Family)

Pilularia L. (Pillwort)

P. americana A. Braun, American P.; amphibious on mud and pond margins; infrequent; Jul–Sep; TN.

OPHIOGLOSSACEAE (Adder's-tongue Family)

1 Sterile and fertile segments of leaf dissected . . . . **Botrychium**
1 Sterile and fertile segments of leaf simple . . . . **Ophioglossum**

Botrychium Swartz (Grape Fern)

1 Common petiole of sterile and fertile leaf segments rising well above ground; fertile segment joining petiole 0.0–0.5 cm below basal sterile pinnae; leaves herbaceous or fleshy, appearing in spring to summer 2

2 Sterile blade broadly deltoid, more or less ternate, 2–3-pinnate, herbaceous . . . . **B. virginianum**

2 Sterile blade narrowly deltoid or elliptic, pinnatifid to pinnate (occasionally 2-pinnate), fleshy 3

3 Sterile blade oblong to ovate, with oblong, obovate, or fan-shaped blunt or rounded segments; the expanding basal bud with both sterile and fertile portions erect, ascending or diverging 4

4 Sterile blade petioled, borne from near base to near summit of common petiole, simple or pinnate, when pinnate then with fan-shaped to narrowly obovate segments . . . . **B. simplex**

4 Sterile blade sessile or short-petioled, borne from above middle to summit of common petiole, pinnate or bipinnate, with oblong, oblong-ovate or narrowly obovate segments . . . . **B. matricariifolium**

3 Sterile blade deltoid, sessile at apex of common petiole; segments lanceolate; expanding basal bud with both sterile and fertile blades abruptly reflexed . . . . **B. lanceolatum**

1 Common petiole of sterile and fertile leaf portions terminating at or near ground level; fertile portion joining petiole more than 0.5 cm below sterile pinnae; leaves coriaceous, overwintering 5

5 Ultimate pinnules with strong central vein; leaves normally 1/year; plants of rich, cool loamy woods or old field sites 6

6 Ultimate segments lanceolate to lance-ovate, acute at apex, or highly dissected, the prominent terminal segments prolonged; sterile blades becoming bronze in autumn 7

7 Blades mostly 2-pinnate, the outline of the small lateral lobes of pinnae, or pinnules of more dissected pinnae, oblong and somewhat rounded . . . . **B. biternatum**

7 Blades mostly 3-pinnate, outline of the small lateral lobes of pinnae, or pinnules of more dissected pinnae, rhomboidal and angular, or whole sterile blade highly dissected into delicate linear segments . . . . **B. dissectum**

6 Ultimate segments of sterile blade ovate, obovate, rhombic, or oval with blunt or rounded apex, not highly dissected, terminal pinnule segments not greatly prolonged 8

8 Sterile blade fleshy or coriaceous; segments less than 2× as long as broad, often imbricate; spores shed in Aug and Sep . . . . **B. multifidum**

8 Sterile blade thinner, submembraneous; segments more than 2× as long as broad, not imbricate; spores shed in Sep and Oct . . . . **B. oneidense**

5 Ultimate pinnules with central veins poorly developed or absent; leaves often 2 or more/season; plants of dry, sandy pinewoods . . . . **B. jenmanii**

B. biternatum (Sav.) Underw., Southern G. F.; rich or alluvial woods, brushy fields, pinelands, swamps; common; Aug–Oct; ALL.

B. dissectum Spreng.; rich or alluvial woods, brushy old fields, thickets; common; Aug–Oct; ALL.

B. jenmanii Underw., Jenman's G. F.; dry brushy woods, old fields, pinelands; infrequent; Aug–Oct; GA, NC, SC, TN. *B. alabamense* Maxon.

B. lanceolatum (Gmel.) Angstr., Lance-leaved Moonwort; rich woods, old fields and swamp margins in subacid soils; infrequent; Jul–Sep; NC, VA. Represented in our area by subsp. **angustisegmentum** (Pease & Moore) Clausen.

B. matricariifolium A. Braun, Daisy-leaved Moonwort; old fields, rich woods and thickets, in subacid soils; infrequent; Jun–Jul; NC, TN, VA.

B. multifidum (Gmel.) Rupr., Leather G. F.; rich and alluvial woods, meadows and thickets; infrequent; Jul–Sep; NC, TN, VA.

B. oneidense (Gilb.) House, Blunt-lobed G. F.; wet woods and shady thickets in acid soils; infrequent; Jul–Oct; NC, TN, VA.

B. simplex E. Hitchc., Least Moonwort; old fields, meadows and open woods in subacid soils; infrequent; May–Jun; NC, VA.

B. virginianum (L.) Swartz, Rattlesnake Fern; thickets and rich woods; common; Apr–Jun; ALL.

Ophioglossum L. (Adder's-tongue)

1 Veins of leaf blade in 2 series, with prominent areoles enclosing fainter areoles with included veinlets .... **O. engelmannii**

1 Veins of blade not of markedly different size or prominence, without included veinlets .... **O. pycnostichum**

O. engelmannii Prantl, Engelmann's A.; cedar barrens, pastures, open woods in calcareous soils; occasional; Mar–Jun; TN, VA.

O. pycnostichum (Fern.) Löve & Löve, Southeastern A.; wet flood-plain woods in neutral soils; infrequent; Apr–Jul; ALL. *O. vulgatum* L. var. *pycnostichum* Fern.

OSMUNDACEAE (Royal Fern Family)

Osmunda L.

1 Leaves dimorphic, the vegetative without sporangia, the fertile without lamina; matted trichomes covering young leaves remaining in a small patch on underside of pinna base .... **O. cinnamomea**

1 Fertile pinnae without lamina, borne on leaves of mostly vegetative green pinnae; collar at pinna base with few or no trichomes 2

  2 Leaf blade pinnate-pinnatifid; fertile pinnae in the middle of blade .... **O. claytoniana**

  2 Leaf blade bipinnate; fertile pinnae terminal .... **O. regalis**

O. cinnamomea L., Cinnamon Fern; swamps, marshes, stream banks, bogs, and rich woods; common; Apr–May; ALL.

O. claytoniana L., Interrupted Fern; loamy woods and swamps; frequent; Apr–Jun; ALL.

O. regalis L., Royal Fern; swamps, bogs, stream banks, and moist woods; common; Apr–Jun; ALL. Represented in our area by var. **spectabilis** (Willd.) A. Gray.

POLYPODIACEAE (Polypody Family)

Polypodium L.

1 Leaves densely scaly .... **P. polypodioides**

1 Leaves sparsely scaly on the midribs only .... **P. virginianum**

P. polypodioides (L.) Watt, Resurrection Fern; epiphytic and epipetric, usually in dry calcareous sites; frequent; Jun–Oct; ALL. Represented in our area by var. **michauxianum** Weath.

P. virginianum L., Rock Polypody; epipetric in moist sites, occasionally epiphytic; common; Jun–Oct; ALL.

SCHIZAEACEAE (Curly-grass Family)

Lygodium Swartz (Climbing Fern)

L. palmatum (Bernh.) Swartz, American C. F.; woods and thickets in acid soils; infrequent; Jul–Sep; ALL.

SELAGINELLACEAE (Spike-moss Family)

Selaginella Beauv. (Spike-moss)

1 Plants herbaceous, pale green; leaves heterophyllous, 2 rows of larger spreading lateral leaves and 2 of appressed-ascending leaves on top of stem; plants of wet soils . . . . **S. apoda**
1 Plants leathery, gray-green; leaves isophyllous; plants matted, on acidic rock outcrops 2
  2 Setae at leaf tips long and twisted . . . . **S. tortipila**
  2 Setae at leaf tips short and straight . . . . **S. rupestris**

S. apoda (L.) Spring, Meadow S.; wet loamy meadows and margins of ponds and streams; frequent; Jun–Oct; ALL.
S. rupestris (L.) Spring, Rock S.; dry, more or less exposed acidic rock ledges and cliffs; infrequent; Jun–Sep; ALL.
S. tortipila A. Braun, Twisted-hair S.; more or less exposed acidic rock ledges and cliffs; frequent; Jul–Sep; GA, NC, SC.

SINOPTERIDACEAE (Maiden-hair Fern Family)

1 Sori round or oblong, distinct and separate along leaflet margins . . . . **Adiantum**
1 Sori continuous along leaf margins 2
  2 Ultimate segments 1–4 mm long; margins weakly to strongly recurved; leaves densely hairy . . . . **Cheilanthes**
  2 Ultimate segments 8 mm or more long; margins strongly reflexed; leaves sparsely hairy or glabrous . . . . **Pellaea**

Adiantum L. (Maiden-hair Fern)

1 Leaves dichotomously divided, fanlike . . . . **A. pedatum**
1 Leaves pinnate, with a strong central axis . . . . **A. capillus-veneris**

A. capillus-veneris L., Southern M. F.; shady calcareous slopes and wet ledges; infrequent; Jun–Jul; GA, TN, VA.
A. pedatum L., Northern M. F.; rich, rocky, wooded slopes; common; Jun–Aug; ALL.

Cheilanthes Swartz (Lip Fern)

1 Leaf segments glabrous or sparsely puberulous . . . . **C. alabamensis**
1 Leaf segments villose to tomentose 2
  2 Leaf segments hairy but without scales, blades 2–3-pinnate; spores 32 or 64/sporangium 3
    3 Leaves bipinnate-pinnatifid; segments villose on abaxial surface; spores 64/sporangium . . . . **C. lanosa**
    3 Leaves 3-pinnate; segments tomentose on abaxial surface; spores 32/sporangium . . . . **C. feei**

2 Leaf segments tomentose and scaly; blades 3-pinnate; spores 32/sporangium 4
4 Leaves 20–45 cm long, abaxial tomentum buff to silver-gray colored . . . . **C. tomentosa**
4 Leaves 15–22 cm long, abaxial tomentum chestnut-colored . . . . **C. castanea**

C. alabamensis (Buckl.) Kunze, Alabama L. F.; dry, calcareous rock ledges; infrequent; Jun–Sep; GA, NC, TN.

C. castanea Maxon, Chestnut L. F.; dry rock ledges and shale barrens; infrequent; Jun–Sep; VA.

C. feei Moore, Slender L. F.; dry calcareous ledges; very local; Jun–Sep; Pulaski Co., VA.

C. lanosa (Michx.) D. C. Eaton, Hairy L. F.; on ledges and cliffs of acidic rock formations; infrequent; Jun–Sep; ALL.

C. tomentosa Link, Woolly L. F.; on ledges and cliffs of acidic rock formations; infrequent; Jun–Sep; ALL.

Pellaea Link (Cliff-brake)

1 Petiole and midrib villose; blade lime-green; basal pinnae of the most-dissected leaves usually with more than 3 segments . . . . **P. atropurpurea**
1 Petiole and midrib glabrous or sparsely hairy; blade blue-green; basal pinnae of the most-dissected leaves mostly with 1–3 segments . . . . **P. glabella**

P. atropurpurea (L.) Link, Purple C.; crevices in calcareous cliffs and ledges; common; May–Sep; ALL.

P. glabella Mett. ex Kuhn, Smooth C.; crevices in limestone cliffs and ledges; infrequent; May–Sep; TN, VA.

THELYPTERIDACEAE (Marsh Fern Family)

Thelypteris Schmidel

1 Leaf blades broadly triangular, not more than 2× as long as broad; sori without indusia 2
2 Leaf blade about 2× as long as broad; rachis wing interrupted between the 2 basal pairs of pinnae; rachis with many dark scales beneath . . . . **T. phegopteris**
2 Leaf blade about as broad as long; rachis wing continuous between the 2 basal pairs of pinnae; rachis with few pale scales beneath . . . . **T. hexagonoptera**
1 Leaf blades lanceolate to oblong-lanceolate, if triangular, then more than 3× as long as broad; sori with or without reniform indusia 3
3 Leaf outline conspicuously narrowed to the base, the lower pinnae widely spaced and gradually reduced to mere auricles . . . . **T. noveboracensis**
3 Leaf base not tapered; basal pinnae shorter but not reduced to mere auricles 4
4 Lateral veinlets of pinna lobes mostly forked; abaxial surface of the blade without glands . . . . **T. palustris**
4 Lateral veinlets of pinna lobes unbranched, or rarely forked; abaxial surface of blade with scattered silvery to golden resinous glands . . . . **T. simulata**

T. hexagonoptera (Michx.) Weath., Broad Beech Fern; rich woods and river bottoms; common; Apr–Aug; ALL. *Phegopteris hexagonoptera* (Michx.) Fee.

T. noveboracensis (L.) Nieuwl., New York Fern; woods, swamps, field margins; common; May–Aug; ALL.

T. palustris Schott., Marsh Fern; marshes, swamps, and bogs; infrequent; Jun–Sep; NC, TN, VA. Represented in our area by var. **pubescens** (Laws.) Fern. *T. thelypteroides* (Michx.) Holub.

T. phegopteris (L.) Slosson, Northern Beech Fern; high-elevation forests and on wet acidic rock cliffs; infrequent; Jun–Aug; NC, TN, VA. *Phegopteris connectilis* (Michx.) Watt.

T. simulata (Davenp.) Nieuwl., Bog Fern; acid bogs; very local; Jul–Sep; NC.

WOODSIACEAE (Cliff Fern Family)

1 Leaves strongly dimorphic, fertile leaves with beadlike segments with little apparent blade tissue . . . . **Onoclea**

1 Leaves monomorphic, or fertile leaves similar to, but more constricted than, the sterile 2

2 Sori elongate, some J-shaped; with indusium 3

3 Leaves 2–3-pinnate; sori crescent- or J-shaped, some arching across veinlet and nearly round . . . . **Athyrium**

3 Leaves pinnate to pinnate-pinnatifid; sori nearly straight 4

4 Leaves pinnate; midrib glabrous . . . . **Homalosorus**

4 Leaves pinnate-pinnatifid; midrib silvery pubescent . . . . **Deparia**

2 Sori round; with or without indusium 5

5 Indusium present, either hoodlike or stellate; leaf blade lanceolate in outline . . . 6

6 Indusium attached around one side of the sorus, hoodlike, the apex truncate or pointed . . . . **Cystopteris**

6 Indusium attached beneath the sporangia, circling the sorus as a ring of scale- or hairlike lobes . . . . **Woodsia**

5 Indusium absent; leaf blade ternate, deltoid in outline . . . . **Gymnocarpium**

Athyrium Roth (Lady Fern)

1 Blade widest near base; indusium margins mixed ciliate and glandular-ciliate; petiole scales early deciduous, mostly less than 5 mm long and 1 mm wide; spores reticulate . . . . **A. filix-femina** subsp. **asplenioides**

1 Blade widest near middle; indusium margins toothed or ciliate, but not glandular; petiole scales persistent, less than 1 cm long and 1.5 mm wide; spores smooth or papillate . . . . **A. filix-femina** subsp. **angustum**

A. filix-femina (L.) Roth subsp. **angustum** (Willd.) Clausen, Northern L. F.; mountain woods or clearings at high elevations; infrequent; Jul–Sep; TN, VA.

A. filix-femina (L.) Roth subsp. **asplenioides** (Michx.) Hultén, Southern L. F.; moist to swampy woods, clearings, stream banks, and seepage slopes; common; Jun–Sep; ALL. *A. asplenioides* (Michx.) A. A. Eat.

Cystopteris Bernh. (Brittle Fern)

1 Leaf blades linear-lanceolate; bulblets present on some leaf axes; indusium with stalked glands 2

2 Longest leaves attenuate, more than 40 cm long; bulblets present, smooth, green, 2–3 mm in diameter, usually on both rachis and costa . . . . **C. bulbifera**

2 Longest leaves acute, less than 40 cm long; bulblets, if present, deformed, scaly and dark, less than 1.5 mm long and on rachis only . . . . **C. tennesseensis**

1 Leaf blades ovate-lanceolate; bulblets absent; indusium glabrous 3
3 Blade 3–4× longer than broad; basal pinnules sessile, truncate to obtuse at base; indusium lanceolate . . . . **C. fragilis**
3 Blade 2–2.5× longer than broad; basal pinnules short-stalked to sessile, obtuse to cuneate at base; indusium round to ovate 4
4 Rhizome long-creeping, the apex extending 1–3 cm beyond the wide-spaced petiole bases . . . . **C. protrusa**
4 Rhizome short-creeping, extending little beyond the close-spaced petiole bases . . . . **C. tenuis**

C. bulbifera (L.) Bernh., Bulblet Fern; shaded, wet, calcareous ledges; infrequent; May–Aug; NC, TN, VA.

C. fragilis (L.) Bernh., B. F.; on acidic outcrops and cliffs; infrequent; Jun–Sep; VA.

C. protrusa (Weath.) Blasdell, Lowland B. F.; terrestrial in rich woods; common; May–Aug; ALL. *C. fragilis* (L.) Bernh. var. *protrusa* Weath.

C. tennesseensis Shaver, Tennessee B. F.; wet calcareous outcrops; infrequent; Jun–Aug; NC, VA.

C. tenuis (Michx.) Desv., Mackay's B. F.; on acidic, wet, rock ledges and outcrops; infrequent; May–Aug; NC, TN, VA. *C. fragilis* (L.) Bernh. var. *mackayii* Laws.

Deparia Hook & Grev. (Silvery Glade Fern)

D. acrostichoides (Sw.) Kato; rich woods and seepage slopes; frequent; Jun–Sep; ALL. *Athyrium thelypterioides* (Michx.) Desv.

Gymnocarpium Newman (Oak Fern)

G. dryopteris (L.) Newm.; cool rocky woods; infrequent; Jun–Sep; NC, VA.

Homalosorus Small ex Pichi Sermolli (Glade Fern)

H. pycnocarpos (Spreng.) Pic. Ser.; bottoms, wet woods, and seepage slopes in calcareous to circumneutral soils; infrequent; Jul–Sep; ALL. *Athyrium pycnocarpon* (Spreng.) Tidstr.

Onoclea L. (Sensitive Fern)

O. sensibilis L.; marshes, swamps, and river bottoms; frequent; Jul–Sep; ALL.

Woodsia R. Brown (Cliff Fern)

1 Petiole jointed, breaking off to produce an even stubble 1–3 cm long; indusium of 10–20 hairlike segments . . . . **W. ilvensis**
1 Petiole not jointed, not forming an even stubble; indusium of 3–6 ribbonlike segments 2
2 Pinnae without scales, a few broad scales on the midrib . . . . **W. obtusa**
2 Pinnae with numerous long whitish scales . . . . **W. scopulina**

W. ilvensis (L.) R. Brown, Rusty C. F.; cliffs; infrequent; Jun–Sep; NC, VA.

W. obtusa (Spreng.) Torrey, Blunt-lobed C. F.; calcareous rocky woods, banks, old walls, epipetric or terrestrial; common; Jun–Sep; ALL.

W. scopulina D. C. Eaton, Mountain C. F.; shale and sandstone cliffs and ledges; infrequent; Jun–Sep; NC, TN, VA. *W. scopulina* D. C. Eaton var. *appalachiana* Taylor.

Gymnosperm Families

Hardin, J. W. 1971. Studies of the southeastern United States flora. II. The gymnosperms. *J. Elisha Mitchell Sci. Soc.* 87:43–50.

CUPRESSACEAE (Cypress Family)

1 Seed cone woody; seeds winged; twigs flattened . . . . **Thuja**
1 Seed cone berrylike; seeds wingless; twigs 4-angled . . . . **Juniperus**

Juniperus L.

1 Leaves all awl-shaped, in whorls of 3; decumbent shrubs . . . . **J. communis**
1 Both awl-shaped and scalelike leaves present, mostly opposite; trees . . . **J. virginiana**

J. communis L., Mountain/Ground Juniper; dry rocky slopes; infrequent; Mar–Apr; NC, SC, VA. Represented in our area by var. **depressa** Pursh.

J. virginiana L., Eastern Red Cedar; fields and dry woodlands; common; Jan–Mar; ALL.

Thuja L. (Arbor Vitae, White Cedar)

T. occidentalis L.; calcareous seeps and outcrops; occasional; Mar–Apr; NC?, TN, VA.

PINACEAE (Pine Family)

1 Leaves needlelike, borne in fascicles of 2–5 . . . . **Pinus**
1 Leaves linear, solitary 2
  2 Leaves 4-angled . . . . **Picea**
  2 Leaves flat 3
    3 Seed cones erect, 4 cm or more long; leaves sessile . . . . **Abies**
    3 Seed cone pendant, less than 4 cm long; leaves attached to a persistent, decurrent peg . . . . **Tsuga**

Abies Mill. (Fir)

1 Bracts of mature cones shorter than the scales or slightly exserted, but not reflexed . . . . **A. balsamea**
1 Bracts of mature cones longer than the scales and reflexed . . . . **A. fraseri**

A. balsamea (L.) Mill., Balsam F.; moist boreal forests; infrequent; Apr–May; VA.

A. fraseri (Pursh) Poir., Fraser F./She Balsam; high-elevation boreal forests; occasional; May–Jun; NC, TN, VA. The balsam woolly adelgid has had a serious, negative impact on our populations in recent years.

Picea Dietr. (Spruce)

P. rubens Sarg., Red S./He Balsam; high-elevation boreal forests; occasional; May–Jun; NC, TN, VA.

Pinus L. (Pine)

1 Needles 5(6)/fascicle, sheaths deciduous; cone scales thin apically ...... **P. strobus**
1 Needles 2–4/fascicle, sheaths persistent; cone scales thick apically (often prickly) .. 2
  2 Leaves mostly 3/fascicle 3
    3 Seed cone less than 7 cm long, pedunculate; needles stiff .... **P. rigida**
    3 Seed cone more than 7 cm long, sessile; needles flexible .... **P. taeda**
  2 Leaves mostly 2/fascicle 4
    4 Seed cones light brown and lustrous, scales with a stout, upwardly curved prickle .... **P. pungens**
    4 Seed cones dull brown or gray, scales with a slender prickle 5
      5 Small branchlets with loose, blackish bark, the inner bark pale orange; a few 3-needled fascicles present .... **P. echinata**
      5 Small branchlets with smooth, gray bark; only 2-needled fascicles present .... **P. virginiana**

P. echinata Mill., Shortleaf P.; fields and dry woods; frequent; Mar–Apr; ALL.

P. pungens Lamb., Table Mountain P.; dry, rocky, upland woods; frequent; Apr–May; ALL.

P. rigida Mill., Pitch P.; dry upland woods; frequent; Apr–May; ALL.

P. strobus L., White P.; dry to moist woods; frequent; Apr–May; ALL.

P. taeda L., Loblolly P.; roadsides, fields, and dry woods; occasional; Mar–Apr; GA, SC, TN, VA.

P. virginiana Mill., Virginia/Scrub P.; fields and dry woods; frequent; Mar–May; ALL.

Tsuga (Endl.) Carr. (Hemlock)

1 Leaves less than 1.5 cm long, distichous, the margins minutely serrulate; seed cones less than 2 cm long .... **T. canadensis**
1 Leaves more than 1.5 cm long, spirally arranged, the margins entire; seed cones more than 2 cm long .... **T. caroliniana**

T. canadensis (L.) Carr., Eastern/Canada H.; moist woods and stream banks; frequent; Mar–Apr; ALL.

T. caroliniana Engelm., Carolina H.; rocky woods and bluffs; occasional; Mar–Apr; ALL.

TAXACEAE (Yew Family)

Taxus L. (Yew)

T. canadensis Marsh., American Y.; rich woods, rarely bog margins; infrequent; Apr–May; NC, VA.

Monocot Families

AGAVACEAE (Century Plant Family)

1 Ovary superior; perianth white; leaves evergreen . . . . **Yucca**
1 Ovary inferior; perianth green or purplish; leaves succulent, not evergreen . . . . **Manfreda**

Manfreda Salisb. (Agave)

M. virginica (L.) Rose; dry rocky woods; infrequent; May–Jul; NC, SC, TN. *Agave virginica* L.

Yucca L. (Yucca)

Y. filamentosa L.; dry woods and bluffs, often along stream banks; occasional; Jun–Jul; NC, SC, TN, VA. *Yucca flaccida* Haw., recognized by some authors, has narrower leaves, pubescent branches, and petals 2–3 cm broad.

ALISMATACEAE (Water Plantain Family)

1 Pistils in a single cycle on a flat receptacle; stamens 6; flowers bisexual . . . . . **Alisma**
1 Pistils in a dense globose head; stamens usually numerous; flowers mostly unisexual with staminate above and pistillate below . . . . **Sagittaria**

Alisma L. (Water Plantain)

A. subcordatum Raf.; shallow water of marshes, ponds, wet ditches; frequent; Apr–Nov; NC, SC, TN, VA.

Sagittaria L. (Arrowhead)

Beal, E. O., and J. W. Wooten. 1982. Review of the *Sagittaria engelmanniana* complex (Alismataceae) with environmental correlations. *Systematic Botany* 7:417–432.

1 Leaf blades sagittate or hastate at the base 2
2 Sepals of mature flowers appressed; petals usually with a purple or green spot at base; filaments pubescent . . . . **S. calycina**
2 Sepals of mature flowers reflexed; petals without purple at base; filaments glabrous 3
3 Beak of mature achene projecting at right angles from the body . . . . . **S. latifolia**
3 Beak of mature achene ascending to erect 4
4 Bracts thick and herbaceous, rounded apically; achenes with facial resin ducts . . . . **S. engelmanniana**

4 Bracts papery, sharply pointed apically; achenes without facial resin ducts .. 5
5 Petiole sharply 5-winged; heads up to 1.5 cm in diameter ..... **S. australis**
5 Petiole often ribbed, but not sharply 5-winged; heads 1.7–2.2 cm in diameter .... **S. brevirostra**
1 Leaf blades without basal lobes (rarely so in *S. rigida*) 6
6 Fruiting heads sessile .... **S. rigida**
6 Fruiting heads pedicellate 7
7 Leaf blades 1 cm or less wide .... **S. graminea**
7 Leaf blades 1.2–2 cm wide .... **S. fasciculata**

S. australis (J. G. Sm.) Small; margin of shallow waters; frequent; Jun–Oct; ALL. *S. engelmanniana* J. G. Sm. subsp. *longirostra* sensu Bogin.

S. brevirostra Mackenzie & Bush; margins of shallow waters; infrequent; Jun–Oct; VA. *S.engelmanniana* J. G. Sm. subsp. *brevirostra* (Mackenzie & Bush) Bogin.

S. calycina Engelm.; pond margins; infrequent; Jun–Aug; TN. *S. montevidensis* Cham. & Schlecht. subsp. *calycina* (Engelm.) Bogin.

S. engelmanniana J. G. Sm.; lakeshores; infrequent; Jun–Aug; VA.

S. fasciculata E. O. Beal; swamps, bogs, and sluggish streams; infrequent; May–Jul; NC, SC.

S. graminea Michx.; shallow waters; infrequent; May–Oct; VA.

S. latifolia Willd.; marshes and streams; frequent; Jul–Oct; ALL. *S. latifolia* Willd. var. *obtusa* (Muhl. ex Willd.) Wiegand. Pubescent plants may be segregated as *S. latifolia* Willd. var. *pubescens* (Muhl.) J. G. Sm.

S. rigida Pursh; marshes and ponds; infrequent; Jul–Oct; VA.

ARACEAE (Arum Family)

1 Spathe absent or obscure 2
2 Spadix lateral on a leaflike axis; leaves narrow, irislike, up to 12 dm long .... **Acorus**
2 Spadix terminal; leaves broadly oblong to elliptic .... **Orontium**
1 Spathe present and surrounding spadix 3
3 Leaves compound .... **Arisaema**
3 Leaves simple 4
4 Leaves hastate; spathe green .... **Peltandra**
4 Leaves ovate or cordate; spathe purple or mottled .... **Symplocarpus**

Acorus L. (Calamus, Sweetflag)

A. americanus (Raf.) Raf.; marshes, swamps, and wet meadows; frequent; May–Jun; NC, TN, VA. *A. calamus* sensu auctt. non L.

Arisaema Martius

1 Leaf divided into 3–5 leaflets; distal portion of spadix not extending beyond the spathe .... **A. triphyllum**
1 Leaf divided into 7–15 leaflets; distal portion of spadix extending beyond the spathe .... **A. dracontium**

A. dracontium (L.) Schott, Green Dragon; mesic and damp woods; occasional; Apr–Jun; NC, SC, TN, VA.

A. triphyllum (L.) Schott, Jack-in-the-Pulpit, Indian Turnip; mesic woods; common; Mar–May; ALL. *A. quinatum* (Nutt.) Schott. Highly variable in spathe color and size and shape of the leaflets.

Orontium L. (Golden Club)

O. aquaticum L.; shallow streams, bogs, and pools; occasional; Mar–May; NC, SC, TN, VA.

Peltandra Raf. (Green Arum)

P. virginica (L.) Schott & Endl.; marshes and stream banks; occasional; May–Jun; NC, SC, TN, VA.

Symplocarpus Salisb. ex Nutt. (Skunk Cabbage)

S. foetidus (L.) Nutt.; bogs and wet meadows; occasional; Feb–Mar; NC, TN, VA.

COMMELINACEAE (Spiderwort Family)

1 Corolla actinomorphic; flowers subtended by leaflike bracts 2
  2 Fertile stamens 2 or 3; flowers usually solitary, roseate . . . . **Murdannia**
  2 Fertile stamens 6; flowers cymose, usually blue . . . . **Tradescantia**
1 Corolla zygomorphic; flowers subtended by a conspicuous folded spathe . . . . **Commelina**

Commelina L. (Dayflower)

1 Margin of spathe united basally; plants not rooting at the nodes 2
  2 Leaf sheaths ciliate with dark brown hairs; lowermost petal blue . . . . . **C. virginica**
  2 Leaf sheaths ciliate with whitish hairs; lowermost petal white . . . . **C. erecta**
1 Margin of spathe free at the base; plants rooting at the nodes 3
  3 Lowermost petal white; anthers 6; annuals . . . . **C. communis**
  3 Lowermost petal blue; anthers 5; perennials . . . . **C. diffusa**

C. communis L.*; moist waste areas; common; May–Oct; NC, SC, TN, VA.
C. diffusa Burman f.; low woods and wet areas; infrequent; May–Oct; NC, SC, TN, VA.
C. erecta L.; dry woodlands; frequent; Jun–Oct; NC, SC, TN, VA.
C. virginica L.; low woods and wet areas; occasional; Jun–Oct; NC, SC, TN, VA.

Murdannia Royle

M. keisak (Hassk.) Hand.-Maz.*; shorelines and marshes; infrequent; Aug–Oct; NC, SC, TN, VA. *Aneilema keisak* Hassk.

Tradescantia L. (Spiderwort)

1 Uppermost leaf blades usually more than 2 cm wide and wider than opened leaf sheaths . . . . **T. subaspera**
1 Uppermost leaf blades usually less than 2 cm wide and as narrow as or narrower than the opened leaf sheaths 2
  2 Stems densely pubescent with some trichomes more than 2 mm long . . . . **T. hirsuticaulis**

2 Stems glabrous or with trichomes less than 2 mm long 3
3 Pedicels and sepals pubescent . . . . **T. virginiana**
3 Pedicels glabrous; sepals glabrous or with a tuft of hairs at the tip . . . **T. ohiensis**

T. hirsuticaulis Small; dry woods and outcrops; infrequent; Apr–Jul; NC, SC.
T. ohiensis Raf.; woodlands and waste areas; frequent; Apr–Jul; NC, SC, TN, VA.
T. subaspera Ker-Gawl.; woodlands; common; May–Jul; ALL. *T. subaspera* Ker-Gawl. var. *montana* (Shuttlew. ex Small & Vail) Anderson & Woodson.
T. virginiana L.; woods and open areas; occasional; May–Sep; VA.

Hybrids are known to occur between *T. ohiensis* and *T. subaspera*.

CYPERACEAE (Sedge Family)

1 Achene enclosed within a saclike structure (perigynium) with only a terminal opening through which the style protrudes at anthesis 2
2 Leaf blades flat, without a midrib; flowers white . . . . **Cymophyllus**
2 Leaf blades keeled, with a midrib; flowers predominantly greenish . . . . **Carex**
1 Achene not enclosed within a saclike structure 3
3 Achene white or light gray, hard, and bony (and without a tubercle at the summit); flowers unisexual . . . . **Scleria**
3 Achene not white, hard, and bony; at least some flowers bisexual 4
4 Spikelet scales distichous (2-ranked) 5
5 Perianth bristles present; inflorescence axillary; tubercle present . . . **Dulichium**
5 Perianth bristles absent; inflorescence terminal; tubercle absent . . . . . **Cyperus**
4 Spikelet scales spirally imbricate or with only 1 or 2 scales 6
6 Style base enlarged, persistent on the summit of the achene as a tubercle . . 7
7 Spikelets solitary; leaves consisting of bladeless sheaths only . . . **Eleocharis**
7 Spikelets few to many; at least some leaves with expanded blades 8
8 Spikelets 1- or 2-flowered, the lowermost scales empty; cauline leaves present . . . . **Rhynchospora**
8 Spikelets few- to many-flowered, with 1 empty basal scale; leaves basal . . . . **Bulbostylis**
6 Achene without a distinct tubercle, either rounded or apiculate at the summit 9
9 Perianth bristles present 10
10 Perianth bristles more than 8, up to 2 cm long . . . . **Eriophorum**
10 Perianth bristles 1–8, less than 1 cm long 11
11 Achene stipitate, subtended by 3 sepallike, apiculate scales alternating with 3 bristles . . . . **Fuirena**
11 Achene not stipitate, subtended by bristles only . . . . **Scirpus**
9 Perianth bristles absent 12
12 Styles broadened at the base, often fimbriate . . . . **Fimbristylis**
12 Styles not broadened at the base, not fimbriate 13
13 Achenes round; spikelets 1–4-flowered, subtended by 2–many empty scales . . . . **Cladium**
13 Achenes flattened or 3-angled; spikelets usually more than 2-flowered, subtended by 1 empty scale . . . . **Scirpus**

Bulbostylis Kunth

1 Spikelets in dense, headlike, terminal fascicles, rarely solitary or in 2s or 3s .... **B. barbata**
1 Spikelets in open, umbellate cymes, rarely subsessile or subcapitate 2
  2 Spikelet scale notched or truncate at the apex, the keel tip rarely longer than the base of the notch; achene transversely latticed .... **B. capillaris**
  2 Spikelet scale obtuse at the apex, the keel tip equal to and usually exceeding the scale; achene minutely papillose .... **B. ciliatifolia**

B. barbata (Rottb.) Clarke; sandy clearings; infrequent; Jul–Oct; SC.
B. capillaris (L.) Clarke; sandy, disturbed sites and outcrops; frequent; Jul–Oct; ALL.
B. ciliatifolia (Ell.) Fern.; sandy clearings; infrequent; Jul–Oct; SC. Represented in our area by var. **coarctata** (Ell.) Kral.

Carex L.

Contributed by V. E. McNeilus

Bryson, C. T. 1980. A revision of the North American *Carex* section *Laxiflorae* (Cyperaceae). Ph.D. diss., Mississippi State University.
Radford, A. E. 1968. *Carex* L., pp. 218–255. In A. E. Radford, H. E. Ahles, and C. R. Bell, *Manual of the vascular flora of the Carolinas*. Chapel Hill: Univ. of North Carolina Press.

1 Achenes lenticular; stigmas 2 2
  2 Spikes all alike 3
    3 Some or all spikes with staminate flowers or their remnants at the tips, pistillate flowers below (androgynous) .... **Key A**
    3 Some or all spikes with staminate flowers or their remnants at the base, pistillate flowers above (gynecandrous) 4
    4 Perigynia not thickened at base, margins winged .... **Key B**
    4 Perigynia thickened at base, margins not winged .... **Key C**
  2 Spikes not all alike .... **Key D**
1 Achenes trigonous; stigmas 3 5
  5 Spikes usually 1/culm .... **Key E**
  5 Spikes more than 1/culm 6
    6 Style not jointed at base, indurated, persistent .... **Key F**
    6 Style jointed at base, not persistent 7
      7 Lowest pistillate bracts with definite tubular sheaths (more than 2 mm long) 8
        8 Pistillate spikes usually short, erect 9
          9 Body of perigynia with impressed nerves, or with less than 3 nerves .... **Key G**
          9 Body of perigynia with many raised nerves .... **Key H**
        8 Pistillate spikes elongate, usually drooping .... **Key I**
      7 Lowest pistillate bracts without distinct tubular sheaths (2 mm or less long) 10
        10 Perigynia pubescent or puberulent .... **Key J**
        10 Perigynia glabrous .... **Key K**

KEY A

1 Central rachis branched near base (inflorescence compound); usually more than 12 perigynia/spike 2
  2 Culms crepitant when squeezed, drying flat; perigynia with spongy base 3
    3 Perigynia lanceolate, broadest near the truncate or retuse base, the beak equaling or longer than the body 4
      4 Leaf sheaths cross-rugulose ventrally, orifice without thickened band . . . . **C. stipata**
      4 Leaf sheaths not cross-rugulose ventrally, orifice with thickened band . . . . **C. laevivaginata**
    3 Perigynia distinctly ovate, rounded at the base, the beak shorter than the body . . . . **C. conjuncta**
  2 Culms not crepitant when squeezed, drying round 5
    5 Beak of perigynia usually as long as the body; leaf blades longer than culm . . . . **C. vulpinoidea**
    5 Beak of perigynia shorter than body; leaf blades shorter than culm . . . . **C. annectens**
1 Central rachis unbranched (inflorescence simple), usually 2–12 perigynia/spike 6
  6 Leaf sheaths loose, septate; leaves 4–10 mm wide . . . . **C. sparganioides**
  6 Leaf sheaths tight, nonseptate; leaves 1–5 mm wide 7
    7 Inflorescence capitate-ovoid; spikes densely aggregated 8
      8 Perigynia about 1/2 as wide as long, rounded or cuneate at base . . . . **C. cephalophora**
      8 Perigynia 3/5–3/4 as wide as long, widest at the truncated base . . . . **C. leavenworthii**
    7 Inflorescence spicate or moniliform; spikes distinct and often separated by an exposed internode 9
      9 Perigynia spreading or reflexed, bases spongy-thickened 10
        10 Beak of perigynia serrulate . . . . **C. rosea**
        10 Beak of perigynia smooth . . . . **C. retroflexa**
      9 Perigynia ascending, bases not spongy-thickened . . . . **C. muhlenbergii**

KEY B

1 Perigynia lanceolate in outline, 1/6–1/3 as wide as long 2
  2 Leaf blades 1–3 mm wide; perigynia 4–6.5 mm long . . . . **C. scoparia**
  2 Leaf blades 3–7 mm wide; perigynia 3–5 mm long 3
    3 Spikes separate on an elongate inflorescence; perigynia distinctly spreading at the tips . . . . **C. projecta**
    3 Spikes overlapping and crowded, perigynia appressed or ascending . . . . **C. tribuloides**
1 Perigynia of broader shape, normally at least 2/5 as wide as long 4
  4 Pistillate scales as long as or longer than the perigynia 5
    5 Body of perigynia obovate, broadest at or beyond the summit of achene . . . . **C. albolutescens**
    5 Body of perigynia ovate or suborbicular, broadest at about the middle of the achene 6

6 Perigynia widest at $^1/_4$–$^1/_3$ its length from the base .... **C. normalis**
6 Perigynia widest at $^1/_3$–$^1/_2$ its length from the base 7
7 Perigynia brown at maturity, at least in the basal half, nerveless or obscurely 1–3 nerved on the upper (adaxial) face .... **C. aenea**
7 Perigynia permanently silvery-green or silvery-brown, distinctly 5–7-nerved on the upper (adaxial) face .... **C. argyrantha**
4 Pistillate scales distinctly shorter than the perigynia 8
8 Perigynia less than 4 mm long **and** less than 2 mm wide 9
9 Body of perigynia orbicular to obovate, rounded above and abruptly prolonged into a beak .... **C. festucacea**
9 Perigynia ovate or oblong in general shape, gradually tapering into a beak 10
10 Perigynia with erect or ascending beaks, winged to the base 11
11 Perigynia widest at $^1/_4$–$^1/_3$ of their length (beak included) from the base .... **C. normalis**
11 Perigynia widest at $^1/_3$–$^2/_5$ of their length (beak included) from the base 12
12 Spikes crowded with erect or closely ascending perigynia .... **C. tribuloides**
12 Spikes well separated (moniliform) with spreading or ascending perigynia .... **C. tenera**
10 Perigynia with spreading to recurved beaks, wings of the perigynia very narrow or obsolete below the middle .... **C. cristatella**
8 Perigynia more than 4 mm long **or** more than 2 mm wide 13
13 Perigynia $^1/_2$–$^3/_5$ as wide as long, faintly nerved on the lower (abaxial) face .. .. **C. suberecta**
13 Perigynia $^3/_5$–$^3/_4$ as wide as long, distinctly nerved on the lower (abaxial) face 14
14 Perigynia several-nerved on both faces, body obovate .... **C. alata**
14 Perigynia nerveless or faintly nerved on the upper face, body broadly ovate or orbicular .... **C. brevior**

KEY C

1 Perigynia horizontal or reflexed 2
2 Perigynia ellipsoid, beak smooth .... **C. seorsa**
2 Perigynia broadly ovate to lanceolate, beak serrulate 3
3 Leaf blades 1.3 mm or less broad .... **C. howei**
3 Leaf blades more than 1.3 mm broad 4
4 Beak of perigynia more than $^1/_2$ the length of the ovoid to lanceolate body .... **C. muricata**
4 Beak of perigynia less than $^1/_2$ the length of the broadly ovoid body 5
5 Perigynia sharply nerved over both faces 6
6 Pistillate scales keeled dorsally, acute .... **C. incomperta**
6 Pistillate scales rounded dorsally, obtuse .... **C. atlantica**
5 Perigynia nerveless over upper face of the achene .... **C. interior**
1 Perigynia ascending 7
7 Perigynia lanceolate with long, bidentate beaks .... **C. bromoides**

7 Perigynia ovate, without long, bidentate beaks 8
8 Spikes 3–several; perigynia 5–10 . . . . **C. brunnescens**
8 Spikes 1–3; perigynia 1–5 . . . . **C. trisperma**

KEY D

1 Pistillate scales acuminate to aristate; all pistillate spikes drooping . . . . **C. crinita**
1 Pistillate scales obtuse to acute; the upper pistillate spikes erect 2
2 Beak of perigynium twisted or bent; lowest pistillate spike drooping . . . . . **C. torta**
2 Beak of perigynium straight; lowest pistillate spike erect . . . . **C. stricta**

KEY E

1 Style indurated, persistent, not jointed at base 2
2 Style straight in lower 1/3; achene 2–2.5 mm long . . . . **C. typhina**
2 Style curved in lower 1/3; achene 3 mm long . . . . **C. squarrosa**
1 Style jointed at base, not persistent, not indurated 3
3 Lowest pistillate scale undifferentiated; leaves 1 mm wide . . . . **C. leptalea**
3 Lowest pistillate scale foliaceous, dilated at base; leaves 2–4 mm wide 4
4 Body of perigynium subglobose above the stipelike base, abruptly prolonged into the slender beak . . . . **C. jamesii**
4 Body of perigynium obovoid-oblong, gradually tapering into the beak . . . . **C. willdenowii**

KEY F

1 Perigynia lanceolate, gradually tapering into a beak 2
2 Teeth of perigynium strongly reflexed . . . . **C. collinsii**
2 Teeth of perigynium straight . . . . **C. folliculata**
1 Perigynia ovoid, obovoid, or subglobose 3
3 Perigynia at least finely puberulent 4
4 Perigynia densely short pubescent; leaf sheaths purple at orifice . . . . **C. trichocarpa**
4 Perigynia sparingly pubescent-hispidulous; leaf sheaths not purple at orifice . . . . **C. grayi**
3 Perigynia glabrous 5
5 Perigynia coarsely ribbed, 7–9-veined 6
6 Beak of perigynium scabrous on the margins; pistillate scales acute, essentially awnless . . . . **C. bullata**
6 Beak of perigynium smooth; pistillate scales with a slender rough awn, much longer than the body 7
7 Staminate scales acuminate . . . . **C. schweinitzii**
7 Staminate scales with long, rough awns 8
8 Pistillate spikes 14–20 mm wide; beak of perigynium shorter than body, perigynia 6–10 mm long . . . . **C. lurida**
8 Pistillate spikes 8–13 mm wide; beak of perigynium longer than body, perigynia 5–7 mm long . . . . **C. baileyi**
5 Perigynia closely ribbed, 11–19-veined 9
9 Perigynia obovoid 10

10 Spikes unisexual . . . . **C. frankii**
10 Spikes bisexual 11
11 Achenes less than 1/2 as wide as long; style much curved near its base . . . . **C. squarrosa**
11 Achenes at least 1/2 as wide as long; style straight at the base . . . . **C. typhina**
9 Perigynia ovoid 12
12 Pistillate spikes subglobose 13
13 Plants with long rhizomes . . . . **C. louisianica**
13 Plants without long rhizomes . . . . **C. intumescens**
12 Pistillate spikes cylindric or subcylindric 14
14 Pistillate spikes 10–17 mm wide, more than 3× as long as wide; perigynia little inflated, densely crowded 15
15 Perigynia coriaceous, teeth 1.5–2 mm long, widely divergent . . . . **C. comosa**
15 Perigynia membranous, teeth 0.4–1 mm long, nearly parallel . . . . **C. hystricina**
14 Pistillate spikes 15–35 mm wide, less than 3× as long as wide; perigynia much inflated 16
16 Plants with long rhizomes; peduncle of staminate spike much surpassing the uppermost pistillate spike . . . . **C. louisianica**
16 Plants with short rhizomes; peduncle of staminate spike about equal to the uppermost pistillate spike . . . . **C. lupulina**

KEY G

1 Perigynia with 2 ribs, otherwise nerveless 2
2 Perigynia conspicuously obovoid, with a long stipelike base; leaves 2–3 mm wide . . . . **C. pedunculata**
2 Perigynia fusiform, rarely obovoid; leaves 3–6 mm wide . . . . **C. leptonervia**
1 Perigynia with numerous impressed nerves 3
3 Body of perigynium tapered at base 4
4 Perigynia 3.5–4.3 mm long, broadest at about the middle; sheaths of the bracts usually glabrous; achene with minute, straight beak . . . . **C. oligocarpa**
4 Perigynia 4.3–5.9 mm long, broadest well above the middle; sheaths of the bracts hispidulous; achene with strongly bent beak . . . . **C. hitchcockiana**
3 Body of perigynium rounded at base 5
5 Midvein of pistillate scales conspicuously long and rough; perigynia 2.5–3.8 mm long . . . . **C. conoidea**
5 Midvein of pistillate scales smooth or only slightly roughened; perigynia 3.5–5.4 mm long 6
6 Lowest pistillate scale awnless or with an inconspicuous awn to 1 mm long, the whole scale much shorter than the perigynium; perigynia usually dull . . . . **C. flaccosperma**
6 Lowest pistillate scale with awns about as long as or longer than the body, the whole scale distinctly longer than the perigynium; perigynia lustrous 7
7 Leaf blades 2–4 mm wide; basal sheaths conspicuously tinged with purple; perigynia 1/3–2/5 as wide as long . . . . **C. amphibola**

7 Leaf blades 4–8 mm wide; basal sheaths green or brown; perigynia $^2/_5$–$^1/_2$ as wide as long . . . . **C. grisea**

KEY H

1 Leaf blades capillary, involute . . . . **C. eburnea**
1 Leaf blades not capillary, not involute 2
2 Perigynia 5–6.6 mm long . . . . **C. careyana**
2 Perigynia shorter 3
3 Beaks of perigynia 1.3–2.2 mm long, orifice long-oblique . . . . **C. polymorpha**
3 Beaks of perigynia shorter or the orifice not oblique 4
4 Body of perigynium rounded, nearly globular . . . . **C. granularis**
4 Body of perigynium tapered at base, distinctly longer than wide 5
5 Elongate rhizome present 6
6 Leaves of basal tufts with blades . . . . **C. meadii**
6 Leaves of basal tufts with sheaths only 7
7 Leaf blades 1–2 mm wide; culms 1–2.5 mm thick at base; rootstocks 1 mm or less thick . . . . **C. woodii**
7 Leaf blades 3.5–5.5 mm wide; culms 3–6 mm thick at base; rootstocks thicker than 1 mm . . . . **C. biltmoreana**
5 Elongate rhizome absent 8
8 Pistillate spikes with 1–3 staminate flowers at the base . . . **C. laxiculmis**
8 Pistillate spikes without staminate flowers at the base 9
9 Staminate spikes purplish 10
10 Leaf blades 10–25 mm wide . . . . **C. plantaginea**
10 Leaf blades 3–6 mm wide . . . . **C. austrocaroliniana**
9 Staminate spikes greenish or brownish 11
11 Leaf blades of sterile culms 10–25 mm wide; blades of fertile culms less than 5 mm wide . . . . **C. platyphylla**
11 Leaf blades of sterile and fertile culms similar in width 12
12 Peduncle of staminate spike overtopped by upper pistillate spike 13
13 Pistillate scales truncate; pistillate bracts often concealing pistillate spikes . . . . **C. albursina**
13 Pistillate scales not truncate; pistillate bracts not usually concealing pistillate spikes 14
14 Culms short, 1–3 dm tall; pistillate spikes often hidden in the leaves . . . . **C. abscondita**
14 Culms taller; pistillate spikes not hidden in the leaves 15
15 Body of perigynia fusiform; peduncle of staminate spike overtopped by upper 2 pistillate spikes . . . . **C. crebriflora**
15 Body of perigynia obovoid; peduncle of staminate spike overtopped by only 1 pistillate spike 16
16 Beak of perigynia elongate, straight, or only slightly bent . . . . **C. laxiflora**
16 Beak of perigynia short, abruptly bent . . . . **C. blanda**

12 Peduncle of staminate spike not overtopped by upper pistillate spike 17
17 Leaf blades whitish, coriaceous; perigynia whitish, giving the plant a greenish-white appearance . . . . . **C. striatula**
17 Leaf blades not whitish or coriaceous; perigynia not whitish; plant not having a greenish-white appearance 18
18 Perigynia fusiform . . . . . **C. styloflexa**
18 Perigynia obovoid 19
19 Perigynia 2.5–3.5 mm long; pistillate peduncles pendulous . . . . . **C. digitalis**
19 Perigynia 3.5–4.5 mm long; pistillate peduncles not pendulous 20
20 Base of plant brownish; pistillate spikes tightly flowered . . . . . **C. gracilescens**
20 Base of plant conspicuously purplish; pistillate spikes loosely flowered 21
21 Basal bracts lustrous wine-colored; perigynia 2–7/spike; plants of noncalcareous habitats . . . . . **C. manhartii**
21 Basal bracts dark purple; perigynia 5–10/spike; plants of calcareous habitats . . . . . **C. purpurifera**

KEY I

1 Perigynia 5–7 mm long 2
2 Plants with conspicuous shaggy rhizomes to 1 cm in diameter . . . **C. cherokeensis**
2 Plants without conspicuous shaggy rhizomes 3
3 Internodes between perigynia 2–6 mm long . . . . . **C. debilis**
3 Internodes between perigynia 1–1.5 mm long . . . . . **C. venusta**
1 Perigynia less than 5 mm long 4
4 Perigynia densely white-hirsute; achenes brown with dark red spots . . . . . **C. roanensis**
4 Perigynia glabrous; achenes without dark red spots 5
5 Leaf blades and sheaths glabrous 6
6 Perigynia strongly angled, tapered into a triangular beak . . . . . **C. prasina**
6 Perigynia obscurely trigonous, beakless . . . . . **C. gracillima**
5 Leaf blades and sheaths pubescent 7
7 Perigynia 1 mm broad; leaf blades 1.5–2.5 mm wide . . . . . **C. aestivalis**
7 Perigynia more than 1 mm broad; leaf blades 3 mm or more wide . . . . . **C. oxylepis**

KEY J

1 Body of perigynium tightly filled by the achene 2
2 Many culms short and hidden among the leaves 3
3 Lowest pistillate spikes on elongate, capillary peduncles . . . . . **C. umbellata**

3 Lowest pistillate spikes not on elongate, capillary peduncles .... **C. nigromarginata**
2 None of the culms short and hidden among the leaves 4
4 Plants rhizomatous, not cespitose 5
5 Body of perigynium suborbicular or short ovoid .... **C. pensylvanica**
5 Body of perigynium ellipsoid .... **C. physorhyncha**
4 Plants cespitose, not rhizomatous 6
6 Pistillate scales wider than and about 1.5–2× longer than the perigynia .... **C. amplisquama**
6 Pistillate scales not significantly wider or longer than the perigynia 7
7 Body of perigynium suborbicular or short ovoid .... **C. communis**
7 Body of perigynium slenderly ellipsoid to fusiform 8
8 Perigynia much exceeding the scales, copiously hirsute ...... **C. peckii**
8 Perigynia about as long as the scales, faintly puberulent 9
9 Culms weak, loosely spreading, arching, recurved, or reclining; head usually dense ovoid, 1–2 cm long .... **C. emmonsii**
9 Culms firmer, erect, or strongly ascending; head linear-cylindric, 1–4.5 cm long .... **C. artitecta**
1 Body of perigynium not tightly filled by the achene 10
10 Staminate spikes more than 1; perigynia very pubescent, tending to obscure the veins .... **C. lasiocarpa**
10 Staminate spikes 1, less pubescent; veins not obscured or perigynium nerveless 11
11 Perigynium sparsely puberulent; leaves and sheaths glabrous .... **C. scabrata**
11 Perigynium pubescent; leaves and sheaths pubescent 12
12 Perigynia nerveless; leaf blades 3–7 mm wide .... **C. hirtifolia**
12 Perigynia nerved; leaf blades 1–4 mm wide 13
13 Leaf blades 1–2 mm wide; plants confined to open rocky crevices above 4,500 ft. .... **C. misera**
13 Leaf blades 2–4 mm wide; plants of mesic woodlands below 4,500 ft. 14
14 Terminal spike 2–4 cm long; anthers 2 mm long when dry .... **C. virescens**
14 Terminal spike 0.9–2.2 cm long; anthers about 1 mm long when dry .... **C. swanii**

KEY K

1 Terminal spike staminate 2
2 Lowest pistillate spike elongate, drooping .... **C. barrattii**
2 Lowest pistillate spike short, erect .... **C. pallescens**
1 Terminal spike bisexual 3
3 Perigynia flattened, compressed 4
4 Perigynia broadly obovate, essentially as wide as long .... **C. shortiana**
4 Perigynia longer than wide 5
5 Perigynia rounded at summit, without a beak .... **C. complanata**
5 Perigynia acute at summit, with definite beak .... **C. buxbaumii**

3 Perigynia round in cross section, turgid 6
6 Perigynia 3–4 mm long; pistillate scales long-acuminate, 3–6 mm long . . . . **C. bushii**
6 Perigynia 2–3 mm long; pistillate scales acute, 2–3 mm long . . . **C. caroliniana**

C. **abscondita** Mackenzie; rich low woods; occasional; Apr–Jun; GA, NC, SC, VA.
C. **aenea** Fern.; grassy balds; Jun–Aug; known only from Roan Mt., NC, TN?.
C. **aestivalis** M. A. Curtis; rich woods and seepage slopes; occasional; May–Jun; ALL.
C. **alata** Torr.; marshes and low woods; occasional; May–Jun; GA, SC.
C. **albolutescens** Schweinitz; low fields, meadows, marshes; frequent; May–Jun; NC, SC, TN, VA. *C. longii* Mackenzie.
C. **albursina** Sheldon; rich woods over calcareous rock; infrequent; Apr–Jun; NC, TN, VA.
C. **amphibola** Steud.; rich low woods; frequent; May–Jun; TN, VA.
C. **amplisquama** F. J. Hermann; exposed open areas; infrequent; Jul–Aug; GA.
C. **annectens** (Bickn.) Bickn.; marshes, ditches, low woods; May–Jun; NC, SC, TN, VA.
C. **argyrantha** Tuckerman; dry sandy soil; infrequent; Jun–Aug; NC, VA.
C. **artitecta** Mackenzie; dry woods; infrequent; Apr–May; ALL.
C. **atlantica** Bailey; bogs and swamps; frequent; May–Jun; GA, NC, VA.
C. **austrocaroliniana** Bailey; rich mesic mountain woods; frequent; Apr–May; GA, NC, SC, TN.
C. **baileyi** Britt.; bogs, swamps, streamsides; infrequent; Jun–Jul; NC, TN, VA.
C. **barrattii** Schweinitz & Torr.; bogs, marshes; infrequent; Apr–May; GA, NC, VA.
C. **biltmoreana** Mackenzie; in wet shallow soil on granite-gneiss outcrops; infrequent; May–Jun; GA, NC, SC.
C. **blanda** Dewey; low, rich mesic woods; common; Apr–Jun; NC, SC, TN, VA.
C. **brevior** (Dewey) Mackenzie ex Lunell; dry woods and edges of fields; infrequent; May–Jun; VA.
C. **bromoides** Willd.; low wet woods and marshes; occasional; May–Jul; GA, NC, TN, VA.
C. **brunnescens** (Pers.) Poir.; grass and heath balds at high elevations; occasional; Jun–Jul; GA, NC, TN, VA.
C. **bullata** Schkuhr; bogs, marshes; infrequent; May–Jun; NC, TN, VA.
C. **bushii** Mackenzie; meadows and edges of low woods; infrequent; May–Jun; NC, VA.
C. **buxbaumii** Wahl.; bogs and swamps; infrequent; Jun–Jul; NC, VA.
C. **careyana** Dewey; rich mesic woods over limestone; infrequent; May–Jun; VA.
C. **caroliniana** Schweinitz; rich low woods and meadows; frequent; May–Jun; NC, SC, TN, VA.
C. **cephalophora** Willd.; mesic woods and waste places; frequent; May–Jul; ALL.
C. **cherokeensis** Schweinitz; low fields and woods; frequent; May–Jun; GA, NC, TN.
C. **collinsii** Nutt.; bogs and marshes; infrequent; Jun–Jul; NC.
C. **communis** Bailey; dry woods; frequent; May–Jun; ALL.
C. **comosa** Boott; low wet woods and marshes; infrequent; Apr–Jun; VA.
C. **complanata** Torr. & Hook.; low mesic woods and roadsides; frequent; May–Jun; NC, TN, VA. *C. hirsutella* Mackenzie.
C. **conjuncta** Boott; mesic woods; infrequent; May–Jul; TN, VA.
C. **conoidea** Willd.; bogs; infrequent; May–Jun; NC, VA.

C. **crebriflora** Wieg.; rich low woods; infrequent; Apr–Jun; GA, SC, VA.
C. **crinita** Lam.; low wet woods, ponds, ditches; frequent; May–Jun; ALL.
C. **cristatella** Britt.; low meadows; infrequent; May–Jun; NC, VA.
C. **debilis** Michx.; low meadows and woods; common; May–Jun; ALL.
C. **digitalis** Willd.; rich mesic woods; common; Apr–Jun; ALL.
C. **eburnea** Boott; rocky limestone ledges and crevices; occasional; May–Jun; NC, SC, TN, VA.
C. **emmonsii** Dewey; dry to mesic woods; infrequent; Apr–May; TN, SC, VA.
C. **festucacea** Willd.; low woods and meadows; frequent; May–Jun; NC, VA.
C. **flaccosperma** Dewey; meadows and low woods; occasional; May–Jun; NC, TN, VA.
C. **folliculata** L.; bogs, marshes; occasional; May–Jul; NC, TN, VA.
C. **frankii** Kunth; wet woods and meadows; common; May–Jul; NC, SC, TN, VA.
C. **gracilescens** Steud.; rich woods; infrequent; May–Jun; TN, VA.
C. **gracillima** Schweinitz; rich low woods; frequent; Apr–Jun; NC, SC, TN, VA.
C. **granularis** Muhl. ex Willd.; low moist woods; infrequent; May–Jun; TN, VA.
C. **grayi** Carey; wet woods; infrequent; May–Jun; SC, TN, VA.
C. **grisea** Wahl.; mesic woodlands; occasional; May–Jun; NC, SC, TN, VA. *C. corrugata* Fern.
C. **hirtifolia** Mackenzie; dry woods; infrequent; May–Jun; VA.
C. **hitchcockiana** Dewey; rich moist woods; infrequent; Jun–Jul; VA.
C. **howei** Mackenzie; bogs and marshes; occasional; May–Jun; NC, VA.
C. **hystricina** Muhl. ex Willd.; swamps, wet meadows, shorelines; infrequent; Jun–Jul; VA.
C. **incomperta** Bickn.; bogs, marshes, wet meadows; occasional; May–Jun; NC, SC, TN.
C. **interior** Bailey; swamps, bogs, wet meadows; occasional; May–Jun; TN, VA.
C. **intumescens** Rudge; mesic woods, marshes; frequent; May–Jul; NC, SC, TN, VA.
C. **jamesii** Schweinitz; rich mesic slopes; infrequent; May–Jun; TN, VA.
C. **laevivaginata** (Kukenthal) Mackenzie; marshes and low woods; frequent; May–Jun; NC, SC, TN, VA.
C. **lasiocarpa** Ehrh.; open swamps, marshes, and shorelines; infrequent; Jun–Jul; VA.
C. **laxiculmis** Schweinitz; rich mesic woods; frequent; Apr–Jun; ALL.
C. **laxiflora** Lam.; rich low woods; frequent; May–Jun; NC, SC, TN, VA.
C. **leavenworthii** Dewey; dry to mesic woods; infrequent; May–Jun; NC, TN, VA.
C. **leptalea** Wahl.; bogs, swamps, marshes; occasional; May–Jun; NC, SC, TN, VA.
C. **leptonervia** Fern.; rich woods and seepage slopes; occasional; May–Jun; NC, SC, TN, VA.
C. **louisianica** Bailey; bogs and wet woodlands; infrequent; May–Jul; TN.
C. **lupulina** Willd.; low woods and marshes; occasional; Jun–Sep; NC, TN, VA.
C. **lurida** Wahl.; marshes, ditches, meadows; common; Jun–Sep; ALL.
C. **manhartii** Bryson; mesic wooded slopes; infrequent; Apr–May; GA, NC, SC.
C. **meadii** Dewey; marshes and wet meadows; infrequent; May–Jun; NC.
C. **misera** Buckl.; exposed rocky crevices at high elevations; infrequent; Jun–Jul; GA, NC, TN.
C. **muhlenbergii** Willd.; mesic woods and fields; common; May–Jun; NC, SC, TN, VA.
C. **muricata** L.; bogs, seepage slopes, marshes; infrequent; Jun–Jul; NC, SC, TN, VA.
C. **nigromarginata** Schweinitz; dry to mesic woods; occasional; Mar–May; NC, SC, TN, VA.

C. **normalis** Mackenzie; mesic woods and meadows; occasional; May–Jun; GA, NC, TN, VA.
C. **oligocarpa** Willd.; rich woods over calcareous rock; occasional; May–Jun; NC, SC, TN, VA.
C. **oxylepis** Torr. & Hook.; low mesic woods; infrequent; May–Jun; TN.
C. **pallescens** L.; mesic woods and meadows; infrequent; Jun–Jul; VA.
C. **peckii** Howe; calcareous rocky slopes and open woods; infrequent; May–Jul; GA.
C. **pedunculata** Willd.; rich mesic woods usually over limestone; infrequent; May–Jun; NC, SC, TN, VA.
C. **pensylvanica** Lam.; mesic to dry woods; common; Apr–Jun; ALL. *C. lucorum* Willd. ex Link.
C. **physorhyncha** Lieb.; dry to mesic woods; infrequent; May–Jun; GA, NC, TN, VA.
C. **plantaginea** Lam.; rich mesic woods; frequent; Apr–May; ALL.
C. **platyphylla** Carey; rich mesic woods; infrequent; Apr–Jun; NC, TN, VA.
C. **polymorpha** Muhl.; dry open woods, mostly in acid soils; infrequent; May–Jun; VA.
C. **prasina** Wahl.; seepage slopes, swamps, streamsides; frequent; May–Jun; NC, SC, TN, VA.
C. **projecta** Mackenzie; low mesic woods; infrequent; May–Jun; NC, TN.
C. **purpurifera** Mackenzie; rich, usually, calcareous woods; infrequent; May–Jun; GA, NC, TN.
C. **retroflexa** Willd.; dry to mesic low woods; frequent; May–Jun; NC, SC, TN, VA.
C. **roanensis** F. J. Hermann; mesic to wet mountain woods; Jun–Jul; known only from Roan Mt., TN, and vicinity.
C. **rosea** Willd.; mesic woods; common; May–Jun; NC, SC, TN, VA.
C. **scabrata** Schweinitz; shaded saturated soils; occasional; May–Jul; ALL.
C. **schweinitzii** Dewey; bogs, marshes; infrequent; Jun–Jul; NC, VA.
C. **scoparia** Schkuhr ex Willd.; bogs, marshes, wetlands; occasional; May–Jun; ALL.
C. **seorsa** Howe; mesic woods; occasional; May–Jun; SC, TN.
C. **shortiana** Dewey; moist woods and meadows; infrequent; May–Jun; VA.
C. **sparganioides** Willd.; rich mountain woods; occasional; May–Jun; GA, NC, TN, VA.
C. **squarrosa** L.; low wet woods and streamsides; occasional; Jun–Jul; TN, VA.
C. **stipata** Muhl. ex Willd.; marshes, ditches, bottomlands; frequent; May–Jun; NC, SC, TN, VA.
C. **striatula** Michx.; rich low woods; occasional; May–Jun; ALL.
C. **stricta** Lam.; swales and marshes; frequent; Jun–Jul; NC, SC, TN, VA.
C. **styloflexa** Buckl.; bogs, marshes, low woods; frequent; May–Jun; ALL.
C. **suberecta** (Olney) Britt.; swamps and wet meadows; infrequent; May–Jun; VA.
C. **swanii** (Fern.) Mackenzie; mesic woods; common; May–Jun; ALL.
C. **tenera** Dewey; low moist woods; occasional; May–Jun; GA, VA.
C. **torta** Boott; rocky streambeds; common; Apr–May; NC, SC, TN, VA.
C. **tribuloides** Wahl.; low mesic woods and roadsides; common; May–Jun; GA, TN, VA.
C. **trichocarpa** Schkuhr; marshes and wet meadows; infrequent; May–Jul; NC, VA.
C. **trisperma** Dewey; bogs and marshes; infrequent; Jun–Jul; NC, TN, VA.
C. **typhina** Michx.; low wet woods; infrequent; Jun–Jul; NC, VA.
C. **umbellata** Schkuhr ex Willd.; dry to mesic woods; occasional; May–Jun; ALL.
C. **venusta** Dewey; bogs and marshes; occasional; May–Jun; NC, SC, TN.

C. virescens Willd.; rich mesic woods; common; May–Jun; ALL.
C. vulpinoidea Michx.; marshes, ditches, low woods; common; May–Jun; ALL.
C. willdenowii Schkuhr; rich mesic woods; infrequent; May–Jun; GA, NC, SC, VA.
C. woodii Dewey; dry rich woods; infrequent; May–Jun; NC, SC, VA.

The largest and perhaps most taxonomically difficult genus in our area. Whole plants with mature perigynia (along with patience, luck, and experience) are required for positive identification. Several varieties recognized by some authors have been excluded from this treatment until additional biosystematic data are available to clarify their taxonomic positions. The most important of these are: *C. crinita* Lam. var. *gynandra* (Schweinitz) Schweinitz & Torr., *C. crinita* Lam. var. *mitchelliana* (M. A. Curtis) Gleason, *C. debilis* Michx. var. *pubera* Gray, *C. debilis* Michx. var. *rudgei* Bailey, *C. muricata* L. var. *angustata* Carey, and *C. muricata* L. var. *ruthii* (Mackenzie) Gleason.

Cladium P. Browne

C. mariscoides (Muhl.) Torr., Twig Rush; bogs; infrequent; Jul–Sep; NC, VA.

Cymophyllus Mackenzie (Fraser's Sedge)

C. fraseri (Andr.) Mackenzie; rich woods; infrequent; May–Jun; NC, TN, VA. *Carex fraseri* Andr.

Cyperus L. (Flat Sedge, Galingale, Umbrella Sedge)

1 Achene lenticular; styles 2-cleft 2
  2 Spikelets with 1 or 2 achenes 3
    3 Rhizomatous, mat-forming perennials; scale keel smooth . . . . **C. brevifolioides**
    3 Densely tufted annuals; scale keel denticulate . . . . **C. tenuifolius**
  2 Spikelets with 3–many achenes 4
    4 Scale margins white . . . . **C. flavicomus**
    4 Scale margins not white 5
      5 Achenes black, transversely wrinkled . . . . **C. flavescens**
      5 Achenes grayish, brown, or reddish-brown, without transverse wrinkles 6
        6 Scales yellowish at maturity; achene grayish-iridescent at maturity . . . . **C. polystachyos**
        6 Scales reddish-purple at maturity; achene greenish, brown or reddish-brown at maturity 7
          7 Style divided nearly to the base, those of the basal flower as long as or longer than the next 2 scales above . . . . **C. diandrus**
          7 Style divided to about the middle, those of the basal flower shorter than the next 2 scales above . . . . **C. bipartitus**
1 Achene trigonous; styles usually 3-cleft 8
  8 Scale tips extending into a recurved awn giving the inflorescence a bristly appearance . . . . **C. aristatus**
  8 Scale tips not extended into recurved awns 9
    9 Spikelet at maturity breaking into joints, each composed of the achene, its scale, and the next lower internode, which clasps the base of the achene . . . **C. odoratus**
    9 Spikelet shedding scales and achenes individually, the next lower internode not clasping the base of the achene **or** the entire spikelet falling at maturity 10
    10 Spikelets pinnately arranged on an elongate axis mostly radiating horizontally such that the axis of the inflorescence is visible 11

11 Scales 1–1.5 mm long; base of the plant strongly suffused with reddish-purple pigmentation .... **C. erythrorhizos**

11 Scales more than 1.5 mm long; base of plant weakly pigmented or without reddish-purple pigmentation 12

12 Spikelets round or nearly so in cross section; scales 4–6 mm long ... . **C. refractus**

12 Spikelets flattened **or** the scales less than 4 mm long 13

13 Scales 3.5–4.5(6) mm long, conspicuously keeled ... **C. strigosus**

13 Scales 3 mm or less long, rounded on the back .... **C. esculentus**

10 Spikelet-bearing axis short, the spikelets closely clustered in globose, subcapitate, cylindric, or obconic heads; the spikelet-bearing axis not readily visible 14

14 All, or most, of the spikelets (except the uppermost) turned downward 15

15 Culms smooth 16

16 Spikelets usually more than 3-flowered; uppermost spikelets held upright .... **C. lancastriensis**

16 Spikelets usually less than 3-flowered; most, if not all, of the uppermost spikelets either horizontal or turned downward .... **C. retrofractus**

15 Culms scabrous or puberulent 17

17 Peduncle smooth; longest bract longer than the longest peduncle .... **C. dipsaciformis**

17 Peduncle scabrous or puberulent; longest bract equal to or shorter than the longest peduncle .... **C. plukenetii**

14 Spikelets, except for the lowermost, spreading or erect, but not obviously turned downward 18

18 Spikelets not more than 2.5× longer than wide; lowermost scale not overlapping the base of the opposing scale 19

19 Achene linear, usually curved or falcate ..... **C. pseudovegetus**

19 Achene elliptic or ovate, occasionally asymmetric, but not falcate .... **C. acuminatus**

18 Spikelets 3× or more longer than wide; lowermost scale usually overlapping the base of the opposing scale 20

20 Spikelets 1- or 2-(3)flowered 21

21 Longest scale 3.5 mm or more long; spikes globose .... **C. ovularis**

21 Longest scale 2–2.5 mm long; spikes ellipsoid .. **C. retrorsus**

20 Most, if not all, of the spikelets 4–many-flowered 22

22 Spikes mostly with fewer than 10 spikelets, the spikelets spreading or fingerlike 23

23 Scales 9–13-nerved, usually suffused with reddish-purple pigmentation .... **C. compressus**

23 Scales 3–5-nerved, without reddish-purple pigmentation .... **C. dentatus**

22 Spikes mostly with more than 10 spikelets, the spikelets closely appressed into subcapitate or cylindric spikes 24

24 At least the lower 2/3 of the spikelets turned downward .... **C. lancastriensis**

24 Only the lowermost or none of the spikelets turned downward, at least the majority of the spikelets erect or horizontally spreading 25

25 Scales nearly round, about as wide as or wider than long (in side view) .... **C. houghtonii**

25 Scales ovate or elliptic, longer than wide (in side view) 26

26 Scale tips closely appressed to the scale above ... . **C. globulosus**

26 Scale tips spreading, not appressed to the scale above 27

27 Leaves flat; margins of bracts scabrous .... **C. filiculmis**

27 Leaves folded; margins of bracts smooth .... **C. grayi**

C. **acuminatus** Torr. & Hook.; marshes; infrequent; Jun–Aug; VA.

C. **aristatus** Rottb.; sandy exposed sites; occasional; Jul–Sep; NC, SC, VA.

C. **bipartitus** Torr.; wet openings; frequent; Jul–Sep; NC, TN, VA. *C. rivularis* Kunth.

C. **brevifolioides** Thieret & Delahoussaye; wet clearings; infrequent; Jun–Sep; NC, TN, VA.

C. **compressus** L.; damp sandy clearings; infrequent; Jul–Sep; SC.

C. **dentatus** Torr.; wet clearings; infrequent; Jul–Oct; TN, VA.

C. **diandrus** Torr.; wet clearings; infrequent; Jun–Sep; VA.

C. **dipsaciformis** Fern.; sandy woods and fields; occasional; Jun–Sep; NC, SC, TN.

C. **erythrorhizos** Muhl.; wet ditches and marshes; infrequent; Jul–Sep; NC, SC, VA.

C. **esculentus** L., Nut Grass; fields and waste places; frequent; Jul–Oct; NC, SC, TN, VA.

C. **filiculmis** Vahl; dry woods and fields; frequent; Jul–Oct; NC, TN, VA.

C. **flavescens** L.; damp openings; common; Jul–Sep; ALL.

C. **flavicomus** Michx.; damp openings; infrequent; Jul–Oct; SC. *C. albomarginatus* Mart. & Schrad. ex Nees.

C. **globulosus** Aubl.; dry or wet clearings; occasional; Jul–Oct; NC, SC, TN, VA.

C. **grayi** Torr.; dry sandy clearings; infrequent; Jul–Sep; VA.

C. **houghtonii** Torr.; dry sandy soils; infrequent; Jul–Sep; VA.

C. **lancastriensis** Porter; sandy clearings; occasional; Jul–Sep; NC, SC, VA.

C. **odoratus** L.; wet clearings; infrequent; Jul–Sep; VA.

C. **ovularis** (Michx.) Torr.; sandy woods and clearings; occasional; Jul–Sep; SC, TN, VA.

C. **plukenetii** Fern.; sandy clearings; infrequent; Jul–Oct; SC.

C. **polystachyos** Rottb.; marshes and wet ditches; infrequent; Jul–Oct; SC, TN, VA. Represented in our area by var. **texensis** (Torr.) Fern.

C. **pseudovegetus** Steudel; marshes and wet ditches; infrequent; Jul–Oct; TN, VA.

C. **refractus** Engelm. ex Steud.; dry woods and fields; occasional; Jul–Sep; NC, SC, TN, VA.

C. **retrofractus** (L.) Torr.; dry woods; occasional; Jul–Sep; NC, SC, VA.

C. **retrorsus** Chapm.; dry woods; occasional; Jul–Oct; NC, TN.
C. **strigosus** L.; damp openings; common; Jul–Oct; ALL.
C. **tenuifolius** (Steud.) Dandy; damp clearings; frequent; Jul–Oct; ALL.

Dulichium L. C. Rich. ex Pers. (Three-way Sedge)

D. **arundinaceum** (L.) Britt.; marshes, bogs, and pond margins; occasional; Jul–Sep; ALL.

Eleocharis R. Br. (Spike Rush)

1 Spike about as broad as the supporting stem 2
2 Stem 4-sided; spike up to 5 cm long, many-flowered . . . . . **E. quadrangulata**
2 Stem triangular; spike 1–2.5 cm long, few-flowered . . . . . **E. robbinsii**
1 Spike distinctly broader than the supporting stem 3
3 Achene lenticular; stigmas 2 4
4 Tubercle flat, broadly triangular, its base as broad as the summit of the achene; tufted annuals 5
5 Bristles longer than the achene . . . . . **E. obtusa**
5 Bristles shorter than or as long as the achene . . . . . **E. engelmannii**
4 Tubercle bulbous thickened, its base constricted; cespitose or rhizomatous perennials 6
6 Spike subtended by a single, sterile basal scale that completely encircles the stem . . . . . **E. erythropoda**
6 Spike subtended by 2 or 3 sterile basal scales, none encircling the stem . . . . . **E. palustris**
3 Achene trigonous or nearly round; stigmas usually 3 7
7 Achene with longitudinal ridges and transverse lines . . . . . **E. acicularis**
7 Achene smooth or variously sculptured, but without longitudinal ridges and transverse lines 8
8 Tubercle not constricted at the base, confluent with and not differentiated from the achene tip 9
9 Stems less than 8 cm tall; achenes light brown . . . . . **E. parvula**
9 Stems more than 8 cm tall; achenes dark brown to black . . . **E. melanocarpa**
8 Tubercle constricted at the base, clearly differentiated from the achene body 10
10 Achene surface smooth . . . . . **E. intermedia**
10 Achene surface variously sculptured 11
11 Achene, including tubercle, less than 1.5 mm long 12
12 Stems 4–6-angled, slender, usually less than 0.8 mm wide . . . . . **E. tenuis**
12 Stems flattened, about 1 mm wide . . . . . **E. elliptica**
11 Achene, including tubercle, more than 1.5 mm long 13
13 Tubercle as broad and as long as the achene . . . . . **E. tuberculosa**
13 Tubercle about 1/2 the width and length of the achene . . . **E. tortilis**

E. **acicularis** (L.) R. & S.; muddy shorelines and wet ditches; occasional; Jun–Sep; NC, VA.

E. **elliptica** Kunth; open wet areas; infrequent; May–Sep; VA. Represented in our area by var. **compressa** (Sull.) Drap. & Mohlenbr.

E. engelmannii Steud.; open wet areas; occasional; Jul–Sep; NC, VA.
E. erythropoda Steud.; open wet areas; occasional; Jul–Sep; NC, TN, VA. *E. calva* Torr.
E. intermedia Schultes; open wetlands; infrequent; Jun–Sep; VA.
E. melanocarpa Torr.; wet sandy areas; infrequent and more common on the Coastal Plain; Jun–Sep; VA.
E. obtusa (Willd.) Schultes; shorelines and marshes; common; Jun–Oct; NC, SC, TN, VA. Our most commonly encountered species.
E. palustris (L.) R. & S.; marshes and bogs; infrequent; Jul–Oct; NC, TN, VA. *E. smallii* Britt., *E. macrostachya* Britt.
E. parvula (R. & S.) Link ex Buff. & Fingerh.; marshes and pond margins; infrequent; Jul–Sep; VA (Saltville). A Coastal Plain disjunct.
E. quadrangulata (Michx.) R. & S.; shorelines and marshes, often partially submerged; infrequent; Jun–Oct; GA, NC, SC, VA.
E. robbinsii Oakes; pond margins; infrequent; Jul–Sep; VA.
E. tenuis (Willd.) Schultes; bogs and marshes; occasional; Jun–Sep; NC, SC, TN, VA.
E. tortilis (Link) R. & S.; swamps and bogs; infrequent; Jul–Sep; SC.
E. tuberculosa (Michx.) R. & S.; ponds and marshes; infrequent; Jul–Sep; NC, SC.

Eriophorum L. (Cotton Grass)

E. virginicum L.; bogs and wet meadows; occasional; Jul–Sep; ALL.

Fimbristylis Vahl

1 Stigmas 3; achene trigonous 2
2 Spikelets oblong to lanceolate . . . . **F. autumnalis**
2 Spikelets subglobose . . . . **F. miliacea**
1 Stigmas 2; achene lenticular 3
3 Spikelets 1–1.5 cm long; perennials up to 2 m tall . . . . **F. spadicea**
3 Spikelets 3–8 mm long; annuals up to 0.5 m tall . . . . **F. annua**

F. annua (All.) R. & S.*; disturbed sites; infrequent; Jul–Sep; VA. *F. baldwiniana* (Schultes) Torr.
F. autumnalis (L.) R. & S.; wet fields and waste areas; common; Jul–Oct; ALL.
F. miliacea (L.) Vahl*; disturbed sites; infrequent; Jul–Sep; NC, SC.
F. spadicea (L.) Vahl; wet meadows; infrequent; Jul–Sep; SC.

Fuirena Rottb. (Umbrella Grass)

F. squarrosa Michx.; flood plains and wet ditches; infrequent; Jul–Oct; GA, NC, SC, TN.

Rhynchospora Vahl (Beak Rush)

1 Perianth bristles barbed downwardly 2
2 Bristles 8–12; spikelet scales white or pinkish . . . . **R. alba**
2 Bristles 6; spikelet scales brownish 3
3 Spikelets 1-fruited . . . . **R. chalarocephala**
3 Spikelets 2- or 3-fruited (if 1-fruited, a sterile bract also present) 4
4 Leaves capillary, the blades less than 1 mm wide; glomerules lanceolate with 1–10 spikelets; achene elliptic, about 2× as long as wide, without a pale margin . . . . **R. capillacea**

4 Leaf blade broader, about 1–3(–7) mm wide; glomerules hemispheric with numerous spikelets; achene obovate or obpyriform, less than 2× as long as wide and with a pale margin 5
5 Achene (excluding tubercle) 1.8–2 mm long, its center with a pale, prominent bulge . . . . **R. glomerata**
5 Achene (excluding tubercle) about 1.5 mm long, its center rounded but without a pale, prominent bulge . . . . **R. capitellata**
1 Perianth bristles barbed upwardly 6
6 Achene surface smooth or nearly so . . . . **R. gracilenta**
6 Achene surface pitted or appearing honeycombed 7
7 Bristles longer than the achene body . . . . **R. caduca**
7 Bristles shorter than the achene body 8
8 Achene pitted, but without transverse lines; tubercle gray crustaceous . . . . **R. harveyi**
8 Achene pitted and with distinct transverse lines; tubercle not crustaceous . . . . **R. globularis**

R. alba (L.) Vahl; bogs; infrequent; Jul–Oct; GA, NC.
R. caduca Ell.; wet woods and clearings; infrequent; Jul–Sep; NC.
R. capillacea Torr.; limestone seeps; infrequent; Jul–Sep; VA.
R. capitellata (Michx.) Vahl; bogs and marshes; frequent; Jul–Sep; NC, SC, TN, VA.
R. chalarocephala Fern. & Gale; bogs and wet woods; infrequent; Jul–Sep; NC, SC, VA.
R. globularis (Chapm.) Small; wet clearings; infrequent; Jul–Sep; NC, SC, VA.
R. glomerata (L.) Vahl; wet clearings; occasional; Jul–Sep; NC, SC, TN, VA.
R. gracilenta Gray; bogs; infrequent; Jul–Sep; NC, SC.
R. harveyi W. Boott; wet clearings; infrequent; Jul–Sep; NC.

Scirpus (Bulrush)

1 Involucral bract(s) absent; spikelets solitary 2
2 Stems round, smooth; perianth bristles longer than the achene . . . . . . **S. cespitosus**
2 Stems triangular, roughened; perianth bristles shorter to nearly as long as the achene . . . . **S. verecundus**
1 Involucral bract(s) present; spikelets usually more than 1 3
3 Inflorescence appearing lateral; the lowermost bract erect or nearly so and appearing as a continuation of the stem 4
4 Some or all of the spikelets pedicellate 5
5 Scales broadly ovate, about as long as the achenes . . . . **S. tabernaemontanii**
5 Scales ovate, longer than the achenes . . . . **S. acutus**
4 Spikelets sessile or subsessile 6
6 Submerged aquatics; stems nodose, about 1 mm wide; spikelets often solitary . . . . **S. subterminalis**
6 Plants emergent, only the basal portion in or near water; stems not nodose, more than 1 mm wide; spikelets more than 1 7
7 Tufted annuals; mature achene black (rarely light brown), less than 2 mm long . . . . **S. purshianus**
7 Rhizomatous perennials; mature achene brown, more than 2 mm long . . . 8
8 Spikelet scales distinctly mucronate and with a V-shaped notch at the apex; achene beak less than 0.3 mm long . . . . **S. americanus**

8 Spikelet scales only slightly mucronate, not notched at the apex; achene beak 0.4–0.8 mm long . . . . **S. torreyi**

3 Inflorescence appearing terminal; bracts 2–8, leaflike and spreading 9

9 Spikelets 1–3(–6) cm long; scales puberulent . . . . **S. robustus**

9 Spikelets less than 1 cm long; scales glabrous 10

10 Perianth bristles smooth, twisted, without teeth along the margin 11

11 Bristles much longer than the scales and giving the spikelet a woolly appearance; scales without a prominent green midrib . . . . **S. cyperinus**

11 Bristles included within or only slightly longer than the scales; scales with a prominent green midrib . . . . **S. pendulus**

10 Perianth bristles nearly straight and with teeth along the margins or completely absent 12

12 Stem leaves 10–20; spikelets reddish . . . . **S. polyphyllus**

12 Stem leaves fewer than 10; spikelets greenish-black 13

13 Lower leaf sheath purplish . . . . **S. expansus**

13 Lower leaf sheath green . . . . **S. atrovirens**

S. acutus Bigelow; marshes and swamps; infrequent; Jun–Sep; VA.

S. americanus Pers.; marshes and floodplains; occasional; Jun–Sep; TN, VA.

S. atrovirens Willd.; marshes and swamps; frequent; Jun–Sep; NC, SC, TN, VA. *S. georgianus* Harper.

S. cespitosus L.; moist slopes and rock outcrops, usually at high elevations; infrequent; Jul–Sep; GA, NC, TN.

S. cyperinus (L.) Kunth; marshes, bogs, ponds, wet meadows; frequent; Jun–Sep; NC, SC, TN, VA.

S. expansus Fern.; open wet areas; frequent; Jun–Sep; ALL.

S. pendulus Muhl.; moist circumneutral soils; occasional; Jun–Sep; TN, VA. *S. lineatus* of recent authors, not Michx.

S. polyphyllus Vahl; marshes and stream banks; frequent; Jun–Sep; ALL.

S. purshianus Fern.; open wet areas; frequent; Jun–Sep; ALL.

S. robustus Pursh; margin of saline ponds; infrequent; Jun–Sep; Saltville, VA.

S. subterminalis Torr.; ponds and streams; infrequent; Jun–Sep; VA. Our only typically submerged species.

S. tabernaemontanii K. C. Gmel.; marshes and stream banks; frequent; Jun–Sep; NC, SC, TN, VA. *S. validus* Vahl.

S. torreyi Olney; ponds; infrequent; Jun–Sep; VA.

S. verecundus Fern.; fields and woodlands; infrequent; Jun–Sep; VA.

Scleria Berg. (Nut Rush)

1 Achene body smooth 2

2 Base of achene with 8 or 9 papillose-crusty tubercles . . . . **S. oligantha**

2 Base of achene papillose-crusty, but without distinct tubercles 3

3 Achene 2–3 mm long; leaves 3–9 mm wide . . . . **S. triglomerata**

3 Achene 1–2.5 mm long; leaves less than 3 mm wide . . . . **S. minor**

1 Achene body reticulate, tuberculate, or papillose 4

4 Achene body reticulate, often pubescent; base of achene with 3 closely appressed bractlike lobes; inflorescence paniculate . . . . **S. reticularis**

4 Achene body tuberculate or papillose 5

5 Base of achene with 6 tubercles . . . . **S. pauciflora**
5 Base of achene with 3 tubercles, the tubercles sometimes slightly lobed . . **S. ciliata**

S. ciliata Michx.; wet pinelands; infrequent; Jun–Aug; SC, VA. *S. elliottii* Chapm.
S. minor (Britt.) Stone; high-elevation bogs; infrequent; Jun–Aug; NC, SC.
S. oligantha Michx.; dry woodlands; occasional; Jun–Sep; NC, SC, TN, VA.
S. pauciflora Muhl. ex Willd.; woodlands and wet meadows; occasional; Jun–Sep; NC, SC, TN, VA.
S. reticularis Michx.; wet clearings; infrequent; Jun–Sep; NC, SC, VA.
S. triglomerata Michx.; low woodlands; occasional; Jun–Sep; NC, SC, TN, VA.

DIOSCOREACEAE (Yam Family)

Dioscorea L.

1 Inflorescence arising from the axils of opposite leaves; leaves hastate, often with tubers in the axils . . . . **D. batatas**
1 Inflorescence arising from the axils of alternate or whorled leaves; leaves cordate at the base and without tubers in the axils . . . . **D. villosa**

D. batatas Dcne.*, Cinnamon Vine; waste places; frequent; Jun–Aug; NC, SC, TN, VA.
D. villosa L., Wild Yam; woodlands; common; Apr–Jun; ALL. *D. quaternata* (Walt.) J. F. Gmel.

ERIOCAULACEAE (Pipewort Family)

Kral, R. 1966. Eriocaulaceae of continental North America north of Mexico. *Sida* 2:285–332.

Eriocaulon L. (Pipewort)

1 Heads pubescent, white at maturity 2
2 Heads soft, easily flattened in pressing; sheaths of scape longer than most of the leaves . . . . **E. compressum**
2 Heads rigid, not easily flattened in pressing; sheaths of scape shorter than most of the leaves . . . . **E. decangulare**
1 Heads glabrate, gray to nearly black at maturity . . . . **E. septangulare**

E. compressum Lam.; sandy bogs, pools, and wet ditches; infrequent; Jun–Oct; NC.
E. decangulare L.; sandy wet places; infrequent; Jun–Oct; NC, SC. This and the preceding species are more abundant on the Coastal Plain.
E. septangulare With.; sandy bogs, marshes; infrequent and disjunct from the northern U.S.; Jul–Oct; NC, VA. *E. pellucidum* Michx.

HAEMODORACEAE (Bloodwort Family)

Lachnanthes Ell. (Redroot)

L. caroliniana (Lam.) Dandy; swamps and bogs; infrequent and disjunct from the Coastal Plain; Jun–Aug; VA. *L. tinctoria* (Walt.) Ell.

HYDROCHARITACEAE (Frog's-bit Family)

1 Leaves basal, up to 6 dm long and ribbonlike .... **Vallisneria**
1 Leaves cauline, opposite or whorled and usually less than 4 cm long 2
  2 Upper leaves in whorls of 3; cauline leaves less than 2 cm long .... **Elodea**
  2 Upper leaves in whorls of 4 or more; cauline leaves 2–4 cm long .... **Egeria**

Egeria Planchon (Waterweed)

E. densa Planchon*; ponds and streams; infrequent; May–Nov; NC, SC, VA. *Elodea densa* (Planchon) Casp., *Anacharis densa* (Planchon) Vict.

Elodea Michx. (Waterweed)

1 Staminate flowers sessile and free-floating at anthesis; leaves pointed at the tip and not strongly overlapping at the stem apex .... **E. nuttallii**
1 Staminate flowers distinctly pedicellate and not free-floating at anthesis; leaves rounded at the tip and strongly overlapping at the stem apex .... **E. canadensis**

E. canadensis Rich. in Michx.; lakes, ponds, and streams; occasional; Jul–Sep; NC, SC, TN, VA. *Anacharis canadensis* (Michx.) Rich.

E. nuttallii (Planchon) St. John; lakes, ponds, and streams; infrequent; Jul–Sep; NC, VA. *Anacharis nuttallii* Planchon.

Vallisneria L. (Eelgrass)

V. americana Michx.; sluggish or fast-flowing streams; infrequent; Jul–Oct; TN, VA.

IRIDACEAE (Iris Family)

1 Sepals and petals dissimilar; styles petaloid, concealing the stamens .... **Iris**
1 Sepals and petals similar; styles neither petaloid nor concealing the stamens 2
  2 Flowers orange, with dark spots, petals up to 3 cm long; rhizomatous perennials .... **Belamcanda**
  2 Flowers blue to white, petals less than 2 cm long; roots fibrous ..... **Sisyrinchium**

Belamcanda Adans. (Blackberry Lily)

B. chinensis (L.) DC.*; dry woods and bluffs, often on basic soils; occasional; Jun–Aug; NC, SC, TN, VA.

Iris L. (Iris, Flag)

1 Flowering stems less than 12 cm tall 2
  2 Sepals crested; leaves usually more than 1 cm wide .... **I. cristata**
  2 Sepals crestless; leaves usually less than 1 cm wide .... **I. verna**
1 Flowering stems more than 12 cm tall 3
  3 Flowers yellow .... **I. pseudacorus**
  3 Flowers primarily blue or violet, rarely white 4
    4 Leaves 1 cm or less wide; capsule sharply angled .... **I. prismatica**
    4 Leaves more than 1 cm wide; capsule angles rounded 5
      5 Sepal blade about 2× as wide as the claw, unspotted or with a greenish-yellow spot at the base .... **I. versicolor**

5 Sepal blade less than 2× as wide as the claw, with a yellow, pubescent spot at the base . . . . **I. virginica**

I. cristata Ait., Dwarf Crested I.; rich moist woods; common; Apr–May; ALL.

I. prismatica Pursh ex Ker-Gawl.; bogs and marshes; infrequent and more common on the Coastal Plain; May–Jun; NC.

I. pseudacorus L.*, Yellow F.; swamps, marshes, and lakeshores; occasional; May–Jun; NC, SC, TN, VA.

I. verna L., Dwarf I.; dry woods and slopes; occasional; Apr–May; NC, SC, TN, VA. Represented in our area by var. **smalliana** Fern. ex M. E. Edwards.

I. versicolor L.; bogs and marshes; infrequent; May–Jun; VA.

I. virginica L., Blue F.; marshes and stream banks; occasional; Apr–May; NC, SC, TN, VA.

Sisyrinchium L. (Blue-eyed Grass)

Contributed by K. L. Hornberger

1 Spathe valves strongly unequal, outer valve at least 1.5× longer than the inner; stems usually simple . . . . **S. mucronatum**

1 Spathe valves equal or nearly so, outer valve less than 1.5× as long as the inner; stems simple or branched 2

2 Plants drying pale green and glaucescent, yellow-green, or straw-colored 3

3 Spathes sessile and paired at top of scapes (sometimes 1 or 3 spathes present; stems with only 1 node . . . . **S. albidum**

3 Spathes pedunculate; stems with 1 or more nodes 4

4 Stems with 3–5 nodes; the branches and spathes becoming progressively smaller toward the top of the plants . . . . **S. dichotomum**

4 Stems with 1 or 2 nodes; branches and spathes not as above 5

5 Stem bases fibrous; capsules drying pale to light brown, globose to subglobose . . . . **S. nashii**

5 Stem bases never fibrous; capsules drying dark brown to black, oblong-subglobose to obovate . . . . **S. atlanticum**

2 Plants drying medium olive-green to black . . . . **S. angustifolium**

S. albidum Raf.; dry woodlands and fields; infrequent; Mar–May; NC. *S. capillare* Bickn., *S. scabrellum* Bickn.

S. angustifolium Miller; woodlands, wet meadows, and roadsides; common; Mar–Jun; ALL.

S. atlanticum Bickn.; moist clearings; occasional; Mar–Jun; NC, TN, VA. *S. mucronatum* Michx. var. *atlanticum* (Bickn.) Ahles.

S. dichotomum Bickn.; woodlands; infrequent; May–Jun; endemic to NC.

S. mucronatum Michx.; meadows and open woodlands; frequent; Apr–Jun; ALL.

S. nashii Bickn.; clearings and woodlands; occasional; Apr–Jun; ALL. *S. carolinianum* Bickn., *S. floridanum* Bickn., *S. fibrosum* Bickn.

JUNCACEAE (Rush Family)

1 Plants glabrous throughout; capsules with numerous seeds . . . . **Juncus**

1 Plants pubescent; capsules 3-seeded . . . . **Luzula**

Juncus L. (Rush)

1 Principal inflorescence bract erect and much longer than the lateral inflorescence ... 2

2 Capsule 3-angled; flowers borne singly; blade of basal leaves reduced or absent ... 3

3 Perianth about as long as the capsule; capsule flattened apically and beakless; stems densely tufted .... **J. effusus**

3 Perianth about 1/2 as long as the capsule; capsule conical and pointed at the tip; stems arising 8–10 mm apart along the rhizome .... **J. gymnocarpus**

2 Capsule subglobose; flowers borne singly or in few-flowered clusters; blade of basal leaves well developed .... **J. coriaceus**

1 Principal inflorescence bract shorter or longer than the inflorescence, but not appearing as a continuation of the stem 4

4 Plant annual; first flowers appearing below 1/2 the total plant height; auricles absent .... **J. bufonius**

4 Plants perennial; first flowers appearing in the upper 1/2 of the total plant height; auricles present 5

5 Flowers or fruits borne singly and subtended by 2, opposite bracteoles 6

6 Stem leaves present 7

7 Inflorescence simple, 1–4-flowered; auricles fringed at the summit .... **J. trifidus**

7 Inflorescence branched, many-flowered; auricles entire .... **J. gerardii**

6 Stem leaves absent, leaves basal or essentially so 8

8 Leaves terete, narrowly channeled on the upper side .... **J. dichotomus**

8 Leaves flat, sometimes inrolled in drying 9

9 Lowest involucral leaf shorter than the inflorescence; leaves mostly less than 1/3 the length of the stem 10

10 Perianth 2.5–3 mm long; flowers arising along one side of the stem (secund) .... **J. secundus**

10 Perianth 4.5–6 mm long; flowers not secund .... (**J. georgianus**)

9 Lowest involucral leaf longer than the inflorescence; leaves mostly more than 1/3 the length of the stem 11

11 Auricles of the leaf sheath delicate, prolonged 1–3 mm beyond the base of the leaf blade .... **J. tenuis**

11 Auricles of the leaf sheath firm, usually not prolonged beyond the base of the leaf blade 12

12 Auricles cartilaginous, opaque, rigid; bracteoles obtuse to acute .. .. **J. dudleyi**

12 Auricles membranous, hyaline; bracteoles acuminate to aristate .. .. **J. interior**

5 Flowers or fruits in dense clusters of glomerules of 2–many flowers; flowers not subtended by a pair of bracteoles 13

13 Leaf blades not septate 14

14 Glomerules usually more than 25/stem; stems usually more than 6 dm tall; principal leaf blades 4–6 mm wide .... **J. biflorus**

14 Glomerules usually less than 25/stem; stems usually less than 6 dm tall; principal leaf blades 1–3 mm wide .... **J. marginatus**

13 Leaf blades septate 15

15 Seeds 0.7–2.3 mm long, with slender, taillike appendages 16

16 Seed appendages constituting more than 1/2 of the total seed length .... **J. canadensis**

16 Seed appendages short, constituting less than 1/2 the total seed length 17

17 Capsules barely exceeding the perianth; glomerules 5–20-flowered .... **J. subcaudatus**

17 Capsules distinctly longer (1 mm or more) than the perianth; glomerules 2–7-flowered 18

18 Inflorescence strict, 3–6× as long as wide; petals lance-subulate .... **J. brevicaudatus**

18 Inflorescence diffuse, the branches spreading; petals obtuse to acute .... **J. brachycephalus**

15 Seeds less than 0.7 mm long, without slender, taillike appendages ... 19

19 Flowers cymose, not in dense, spherical heads 20

20 Capsule 4–5 mm long, 2× as long as the perianth .... **J. diffusissimus**

20 Capsule 2–3 mm long, less than 2× as long as the perianth .. 21

21 Capsule up to 1.5 mm longer than the perianth .... **J. debilis**

21 Capsule and perianth of equal length .... **J. acuminatus**

19 Flowers in dense, spherical heads 22

22 Blades of lower leaves laterally flattened .... **J. validus**

22 Blades of lower leaves terete 23

23 Blade of uppermost stem leaf much longer than its sheath; sepals longer than the petals .... **J. torreyi**

23 Blade of uppermost stem leaf equal to or slightly longer than its sheath; sepals and petals subequal .... **J. scirpoides**

J. acuminatus Michx.; moist open areas; common; Jun–Aug; ALL.

J. biflorus Ell.; moist open areas; occasional; Jun–Sep; NC, SC, TN, VA. Considered by some authors to be synonymous with *J. marginatus* Rostk. *J. longii* Fern.

J. brachycephalus (Engelm.) Buch.; moist open areas; infrequent; Jun–Sep; VA.

J. brevicaudatus (Engelm.) Fern.; open wet areas, often at high elevations; occasional; Jul–Sep; NC, VA. Closely related to the preceding taxon.

J. bufonius L.; sandy, wet, disturbed sites; infrequent; Jun–Nov; NC, TN, VA.

J. canadensis J. Gay ex La Harpe; open wet areas; occasional; Jun–Oct; NC, SC, TN, VA.

J. coriaceus Mackenzie; open wet areas; occasional; Jun–Sep; ALL.

J. debilis Gray; open wet areas; frequent; May–Aug; ALL.

J. dichotomus Ell.; open, damp or wet areas; occasional; Jun–Oct; NC, SC, TN, VA. *J. dichotomus* Ell. var. *platyphyllus* Wieg.

J. diffusissimus Buckl.; open wet areas; occasional; May–Sep; NC, SC, VA.

J. dudleyi Wieg.; open wet areas; infrequent; Jun–Oct; VA.

J. effusus L.; wet areas, margins of marshes and streams; common; Jun–Sep; NC, SC, TN, VA.

J. georgianus Coville; wet areas of granite outcrops; infrequent; Jun–Aug; SC.

J. gerardii Loisel.; saline marshes; infrequent; Jun–Sep; Saltville, VA.

J. gymnocarpus Coville; open or shaded wet sites, especially seepages; infrequent; Jul–Sep; NC, SC, TN.

J. interior Wieg.; open wet areas; infrequent; Jun–Sep; TN.

J. marginatus Rostk.; open wet areas; common; Jun–Sep; ALL.

J. scirpoides Lam.; open wet areas; infrequent; Jun–Sep; NC, VA.

J. secundus Beauv. ex Poir.; open, moist, often rocky sites; occasional; Jun–Sep; NC, SC, VA.

J. subcaudatus (Engelm.) Coville & Blake; wet, open, or shaded areas; frequent; Jul–Sep; NC, SC, TN, VA.

J. tenuis Willd., Path Rush; dry or moist disturbed sites; common; May–Sep; ALL.

J. torreyi Coville; wet open areas; infrequent; Jun–Sep; VA.

J. trifidus L.; rocky ledges at high elevations; infrequent; Jun–Sep; NC, TN, VA. Represented in our area by var. **monanthos** (Jacq.) Bluff & Fingerhuth.

J. validus Coville; open wet areas; infrequent; Jun–Sep; SC.

Luzula DC. (Wood Rush)

1 Flowers solitary (rarely paired) at the tips of the inflorescence rays . . . **L. acuminata**

1 Flowers few–many in glomerulate clusters . . . . **L. multiflora**

L. acuminata Raf.; woods, stream banks, and clearings; frequent; Mar–Aug; ALL. *L. acuminata* Raf. var. *carolinae* (Wats.) Fern.

L. multiflora (Retz.) Lejeune; woods, stream banks, and clearings; frequent; Mar–Aug; ALL. Plants with divergent or recurved peduncles are referred to by some authors as *L. echinata* (Small) Hermann. Individuals producing bulblets may also be segregated as *L. bulbosa* (Wood) Rydb. It is difficult to find a consistent pattern of variation in this complex, and until more data are available, a conservative treatment seems best.

LEMNACEAE (Duckweed Family)

Hellquist, C. B., and G. E. Crow. 1982. *Aquatic vascular plants of New England:* Part 5. *Araceae, Lemnaceae, Xyridaceae, Eriocaulaceae, and Pontederiaceae.* New Hampshire Agric. Exp. Sta. Bull. 523.

1 Roots present on lower side of frond; reproductive pouches 2/frond 2

  2 Roots 2 or more/frond; frond reddish-purple beneath . . . . **Spirodela**

  2 Roots 2/frond; frond green or slightly purple beneath . . . . **Lemna**

1 Roots absent; reproductive pouches 1/frond . . . . **Wolffia**

Lemna L. (Duckweed)

1 Margin of frond toothed or merely erose; fronds stalked, 6–15 mm long, usually submerged and in chainlike clones . . . . **L. trisulca**

1 Margin of frond entire; fronds sessile, less than 6 mm long, usually emersed and not in chainlike clones 2

  2 Fronds obscurely 3-nerved, elliptic to linear-oblong with mostly parallel sides . . . . **L. valdiviana**

  2 Fronds obscurely 1-nerved, broadly oblong, obovate or suborbicular with sides curved 3

    3 Root sheath with lateral wings; root tip pointed; frond with prominent apical and central papillae . . . . **L. perpusilla**

    3 Root sheath without lateral wings; root tip rounded; frond without prominent apical papillae . . . . **L. minor**

L. minor L.; infrequent; TN, VA.
L. perpusilla Torr.; infrequent; NC, VA.
L. trisulca L.; infrequent; VA.
L. valdiviana Phil.; infrequent; VA.

Spirodela Schleid. (Great Duckweed)

S. polyrhiza (L.) Schleid.*; infrequent; VA.

Wolffia Horkel (Water Meal)

W. brasiliensis Weddell; infrequent; VA. *W. papulifera* Thompson.

Lemnaceae contains the smallest known flowering plants. In our area, they occur sporadically in quiet waters of ponds, lakes, ditches, and streams. They flower infrequently, and phenology is not well documented. Identification of certain taxa is difficult to impossible with dried material.

LILIACEAE (Lily Family)
(Includes Amaryllidaceae)

1 Ovary distinctly inferior 2
  2 Perianth yellow, less than 4 cm long . . . . . **Hypoxis**
  2 Perianth white or pink, more than 4 cm long 3
    3 Stamens united into a corona; flowers 3–7, in a terminal umbel . . . **Hymenocallis**
    3 Stamens free; flowers solitary . . . . . **Zephyranthes**
1 Ovary superior or slightly inferior 4
  4 Stem leaves whorled 5
    5 Leaves in a single whorl; flowers solitary . . . . . **Trillium**
    5 Leaves in 2 or more whorls; flowers 2–many 6
      6 Flowers 2.5 cm or more long . . . . . **Lilium**
      6 Flowers less than 2.5 cm long . . . . . **Medeola**
  4 Stem leaves alternate, basal, or absent 7
    7 Flowers or flower clusters appearing axillary 8
      8 Leaves narrowly filiform, less than 1 cm wide . . . . . **Asparagus**
      8 Leaves more than 1 cm wide 9
        9 Tepals fused . . . . . **Polygonatum**
        9 Tepals separate 10
          10 Tepals recurved, pink to, rarely, whitish . . . . . **Streptopus**
          10 Tepals not recurved, yellowish . . . . . **Uvularia**
    7 Flowers solitary or in terminal inflorescences 11
      11 Perianth large and showy, more than 5 cm long . . . . . **Hemerocallis**
      11 Perianth less than 5 cm long 12
        12 Flowers solitary or in umbellate clusters of more than 3 flowers 13
          13 Flowers solitary 14
            14 Leaves basal; stem unbranched . . . . . **Erythronium**
            14 Leaves cauline; stem branched . . . . . **Uvularia**
          13 Flowers 4–many, umbellate 15
            15 Plants with an onion or garlic odor . . . . . **Allium**

15 Plants without an onion or garlic odor 16
16 Leaves linear, less than 5 mm wide . . . . **Nothoscordum**
16 Leaves oblong to obovate, 2–6 cm wide . . . . **Clintonia**
12 Flowers several–many, in pairs, spikes, racemes, panicles, or at the tips of branches 17
17 Tepals united to the middle or beyond 18
18 Flowers blue or purplish . . . . **Muscari**
18 Flowers white 19
19 Perianth smooth externally; inflorescence arising from a basal leaf sheath; ovary superior . . . . **Convallaria**
19 Perianth roughened externally; inflorescence arising from a basal rosette; ovary about 1/3 inferior . . . . **Aletris**
17 Tepals separate to the base or nearly so 20
20 Stem leaves not greatly reduced upward; basal leaves absent . . . 21
21 Leaves perfoliate . . . . **Uvularia**
21 Leaves not perfoliate 22
22 Tepals 4; leaves 2, rarely 3 . . . . **Maianthemum**
22 Tepals 6; leaves numerous 23
23 Flowers 1–3, terminating the branches . . . . **Disporum**
23 Flowers numerous, paniculate 24
24 Perianth white . . . . **Smilacina**
24 Perianth green . . . . **Veratrum**
20 Stem leaves greatly reduced upward; basal leaves present and conspicuous 25
25 Leaves 2-ranked, irislike 26
26 Style 1; filaments woolly; flowers yellow . . . **Narthecium**
26 Styles 3; filaments not woolly; flowers white . . . **Tofieldia**
25 Leaves not 2-ranked 27
27 Inflorescence paniculate 28
28 Axis of inflorescence glabrous 29
29 Tepals glandular at the base . . . . **Zigadenus**
29 Tepals without basal gland(s) . . . . . . **Stenanthium**
28 Axis of inflorescence pubescent 30
30 Tepals clawed at the base, the limb with 2 basal glands . . . . **Melanthium**
30 Tepals gradually tapering to the base, glandless . . . . **Veratrum**
27 Inflorescence spicate or racemose 31
31 Flowers predominantly blue or pink 32
32 Flowers blue; perianth segments 1.3–1.8 cm long; stamens shorter than the tepals . . . . **Camassia**
32 Flowers pink; perianth segments 5–9 mm long; stamens equal to or longer than the tepals . . . **Helonias**
31 Flowers white or creamy 33
33 Flowers mostly unisexual; bracts absent; pedicels less than 5 mm long . . . . **Chamaelirium**
33 Flowers bisexual; bracts present, at least on the

basal flowers; pedicels more than 5 mm long . . . 34
34 Stems leafless; perianth with an external green stripe and persistent in fruit . . . **Ornithogalum**
34 Stems with reduced leaves; perianth without a green stripe and deciduous in fruit 35
35 Leaves evergreen, about 2 mm wide; plants more than 1 m tall . . . . **Xerophyllum**
35 Leaves not evergreen, more than 2 mm wide; plants less than 1 m tall . . . . . **Amianthium**

Aletris L. (Colicroot)

A. farinosa L.; dry woodlands; frequent; Apr–Jun; ALL.

Allium L. (Onion)

Jones, A. G. 1979. A study of wild leek, and the recognition of *Allium burdickii* (Liliaceae). *Systematic Botany* 4:29–43.

1 Leaves appearing before the flowers and withered by anthesis, more than 2 cm wide . . 2
2 Leaves 5–8 cm wide, with a distinct petiolar base, the petioles usually pinkish; flowers 25–55/umbel . . . . **A. tricoccum**
2 Leaves 2–4 cm wide, without a distinct petiolar base, the basal portion white; flowers 10–20/umbel . . . . **A. burdickii**
1 Leaves present at anthesis, less than 2 cm wide (or wider in *A. ampeloprasum*) 3
3 Leaves round, hollow . . . . **A. vineale**
3 Leaves flat or channeled, not hollow 4
4 Lower portion of stem leafy; flowers deep purple . . . . **A. ampeloprasum**
4 Leaves all basal; flowers white or pinkish 5
5 Inflorescence nodding; bulblets absent . . . . **A. cernuum**
5 Inflorescence erect; bulblets present or absent 6
6 Bulblets present . . . . **A. canadense**
6 Bulblets absent . . . . (**A. cuthbertii**)

A. ampeloprasum L.*; waste areas; infrequent; May–Jul; NC, SC, TN, VA.
A. burdickii (Hanes) A. G. Jones, Ramps; rich, often dry woods; infrequent; Jun–Jul; NC, TN, VA.
A. canadense L., Wild O.; fields and waste places; occasional; Apr–May; NC, TN, VA.
A. cernuum Roth, Nodding O.; open woods and bluffs; common; Jul–Sep; ALL. *A. oxyphilum* Wherry.
A. cuthbertii Small; granite outcrops; infrequent; May–Jun; NC. Doubtfully distinct from *A. canadense*.
A. tricoccum Ait., Ramps; rich mountain coves; frequent; Jun–Jul; GA, NC, TN, VA.
A. vineale L.*, Field Garlic; fields and disturbed sites; frequent; May–Jun: NC, SC, TN, VA.

Amianthium Gray (Fly Poison)

A. muscaetoxicum (Walt.) Gray; mesic or dry woodlands; frequent; May–Jul; ALL.

Asparagus L. (Asparagus)

A. officinalis L.*; waste places; occasional; Apr–Jun; NC, SC, TN, VA.

Camassia Lindl. (Wild Hyacinth)

C. scilloides (Raf.) Cory; shady mesic slopes, often on basic soils; infrequent; May–Jun; VA.

Chamaelirium Willd. (Devil's Bit, Blazing Star)

C. luteum (L.) Gray; rich woods; common; Mar–May; ALL.

Clintonia Raf.

1 Perianth yellowish, the lobes 1.5 cm or more long; flowers umbellate and occasionally with lateral clusters; leaf margins glabrous to slightly ciliate .... **C. borealis**
1 Perianth white, the lobes 8–10 mm long; flowers strictly umbellate; leaf margins distinctly ciliate .... **C. umbellulata**

C. borealis (Ait.) Raf., Bluebead Lily; high elevations, particularly in spruce-fir forests; frequent; May–Jun; ALL.

C. umbellulata (Michx.) Morong, Speckled Wood Lily; rich woods at mid elevations; frequent; May–Jun; ALL.

Convallaria L. (Lily-of-the-Valley)

1 Lowermost bract 5 mm or less long; leaves 1–1.5 dm long .... **C. majalis**
1 Lowermost bract 8 mm or more long; leaves 1.5–3 dm long .... **C. montana**

C. majalis L.*; persistent after cultivation; infrequent; Apr–May; TN, VA.

C. montana Raf.; rich woods; frequent; Apr–Jun; ALL. *C. majalis* L. var. *montana* (Raf.) Ahles, *C. majuscula* Greene.

Disporum Salisb. ex D. Don (Mandarin)

1 Perianth with purple spots; fruit pubescent .... **D. maculatum**
1 Perianth unspotted; fruit glabrous .... **D. lanuginosum**

D. lanuginosum (Michx.) Nicholson, Yellow M.; rich woods; frequent; Apr–May; ALL.

D. maculatum (Buckley) Britt., Spotted M.; rich woods; infrequent; Apr–May; GA, NC, TN, VA.

Erythronium L. (Trout Lily, Dog-tooth Violet)

Parks, C. R., and J. W. Hardin. 1963. Yellow erythroniums of the eastern United States. *Brittonia* 15:245–259.

1 Perianth white .... **E. albidum**
1 Perianth yellow 2
  2 Petals auriculate at the base; fruit rounded or slightly beaked at the apex; peduncle of fruit not pendant, the fruit often held above the ground and oriented horizontally .... **E. americanum**
  2 Petals without auricles; fruit indented at the apex; peduncle of fruit arching, the fruit

resting on the ground with the apex downward .... **E. umbilicatum**

E. albidum Nutt.; rich woods, often on basic soils; infrequent; Mar–Apr; VA.

E. americanum Ker-Gawl.; rich woods; occasional; Mar–May; TN, VA. *E. americanum* Ker-Gawl. subsp. *harperi* (Wolf) Parks & Hardin.

E. umbilicatum Parks & Hardin; rich woods; frequent; Mar–May; ALL. *E. umbilicatum* Parks & Hardin subsp. *monostolum* Parks & Hardin.

Helonias L. (Swamp Pink)

H. bullata L.; bogs; infrequent; Apr–May; GA, NC, SC, VA.

Hemerocallis L. (Day Lily)

H. fulva (L.) L.*; persistent after cultivation; frequent; May–Jun; NC, SC, TN, VA.

Hymenocallis Salisb. (Spider Lily)

H. caroliniana (L.) Herb.; swamps, low woods, and stream banks; infrequent; Jun–Aug; NC, SC, TN. *H. occidentalis* (Le Conte) Kunth.

Hypoxis L. (Yellow Star Grass)

H. hirsuta (L.) Coville; woodlands and stream banks; common; Mar–Jun; ALL.

Lilium L. (Lily)

Adams, R. M., II, and W. J. Dress. 1982. Nodding *Lilium* species of eastern North America (Liliaceae). *Baileya* 21:165–188.

1 Flowers erect; tepals clawed .... **L. philadelphicum**
1 Flowers nodding; tepals not clawed 2
 2 Leaf margins papillose or scabrous 3
  3 Flowers 3–4 cm wide; tepals less than 5.5 cm long; pistil 3–4 cm long .... **L. grayi**
  3 Flowers 4.5–9 cm wide; tepals 6 cm or more long; pistil 4–6 cm long .... **L. canadense**
 2 Leaf margins smooth 4
  4 Leaves oblanceolate to obovate; inflorescence an umbel of usually 4 or fewer flowers .... **L. michauxii**
  4 Leaves elliptic or lanceolate; inflorescence a raceme of up to 30 flowers .... **L. superbum**

L. canadense L., Canada L.; open moist areas; frequent; Jun–Jul; ALL. Plants in our area with brick-red flowers may be referred to *L. canadense* L. subsp. *editorum* (Fern.) Wherry.

L. grayi Watson, Gray's L.; open moist balds and thickets; occasional; Jun–Jul; NC, TN, VA.

L. michauxii Poir., Michaux's L.; dry to moist woodlands and thickets; frequent; Jul–Aug; ALL.

L. philadelphicum L., Wood L.; meadows and woodlands; occasional; Jun–Jul; NC, VA.

L. superbum L., Turk's-cap L.; moist bogs, woods, and fields; frequent; Jun–Aug; GA, NC, TN, VA.

Maianthemum Wiggers (False Lily-of-the-Valley)

M. canadense Desf.; rich woods; frequent; Apr–Jul; GA, NC, TN, VA.

Medeola (Indian Cucumber Root)

M. virginiana L.; rich woods; common; Apr–Jun; ALL.

Melanthium L. (Bunch-flower)

1 Perianth margins entire, flat; largest leaves usually less than 2 cm wide; stamens adnate for more than 1/2 the length of the claw .... **M. virginicum**
1 Perianth margins undulate, crisped; largest leaves usually more than 2 cm wide; stamens adnate for less than 1/2 the length of the claw .... **M. hybridum**

M. hybridum Walt.; rich woods; occasional; Jul–Aug; ALL. *M. latifolium* Desr.
M. virginicum L.; bogs; occasional; Jul–Aug; NC, VA.

Muscari P. Mill. (Grape Hyacinth)

M. atlanticum Boiss. & Reut.*; persistent after cultivation; infrequent; Mar–Apr; NC, SC, TN, VA. *M. racemosum* sensu auctt., non (L.) P. Mill.

Narthecium Huds. (Bog Asphodel)

N. americanum Ker-Gawl.; bogs; infrequent; Jun–Jul; NC.

Nothoscordum Kunth (False Garlic)

N. bivalve (L.) Britt.; dry woods and fields; infrequent; Mar–May; SC, TN. *Allium bivalve* (L.) Kuntze.

Ornithogalum L. (Star-of-Bethlehem)

O. umbellatum L.*; lawns and waste areas; occasional; Mar–May; NC, SC, TN, VA.

Polygonatum P. Mill. (Solomon's Seal)

1 Leaves pubescent on the veins beneath .... **P. pubescens**
1 Leaves glabrous beneath .... **P. biflorum**

P. biflorum (Walt.) Ell.; rich woods; common; Apr–Jun; ALL. Robust tetraploid individuals have been recognized as *P. canaliculatum* of authors; *P. commutatum* (Schultes f.) A. Dietr.
P. pubescens (Willd.) Pursh; rich woods; frequent; Apr–Jun; ALL.

Smilacina Desf. (False Solomon's Seal)

1 Flowers racemose; tepals 4–6 mm long; fruit black .... **S. stellata**
1 Flowers paniculate; tepals up to 3 mm long; fruit red .... **S. racemosa**

S. racemosa (L.) Desf.; rich woods; common; Apr–Jun; ALL. *Maianthemum racemosum* (L.) Link.

S. stellata (L.) Desf.; moist sandy woods and stream banks; infrequent; Apr–Jun; TN, VA. *Maianthemum stellatum* (L.) Link.

Stenanthium (Gray) Kunth (Featherbells)

S. gramineum (Ker-Gawl.) Morong; open woods and meadows; frequent; Jul–Sep; ALL.

Streptopus Michx. (Twisted Stalk)

1 Leaf margins and nodes not coarsely ciliate . . . . **S. amplexifolius**
1 Leaf margins and nodes coarsely ciliate . . . . **S. roseus**

S. amplexifolius (L.) DC.; rich woods at high elevations; infrequent; Apr–Jul; NC, TN, VA.
S. roseus Michx.; rich woods at high elevations; occasional; Apr–Jul; GA, NC, TN, VA.

Tofieldia Huds. (False Asphodel)

1 Perianth equal to or longer than the capsule; seeds with tails ½ or less as long as the body . . . . **T. racemosa**
1 Perianth shorter than the capsule; seeds with at least 1 tail equal to or longer than the body . . . . **T. glutinosa**

T. glutinosa (Michx.) Pers.; bogs; infrequent; Jul–Aug; NC. *T. racemosa* (Walt.) BSP. var. *glutinosa* (Michx.) Ahles.
T. racemosa (Walt.) BSP.; wet woods and seeps; infrequent; Jun–Aug; VA.

Trillium L. (Trillium)

Contributed by Thomas S. Patrick

1 Flowers sessile; leaves mottled 2
  2 Stem decumbent, minutely pubescent near the leaves . . . . **T. decumbens**
  2 Stem erect, glabrous 3
    3 Stamens 2× as high as pistil or more; petals spatulate, pale ivory-yellow with green to maroon claws . . . . **T. discolor**
    3 Stamens 1.5× as high as pistil or less; petals otherwise 4
      4 Stamens ½ as long as petals; anthers prolonged into a beak, 1–5 mm long . . . . **T. sessile**
      4 Stamens ⅓ as long as petals; anthers blunt, extended less than 1.5 mm 5
        5 Flower lemon-scented; young ovary greenish-white; petals yellow . . . . **T. luteum**
        5 Flower more spicy, banana-scented; young ovary purplish; petals maroon, green, bronze, rarely yellow . . . . **T. cuneatum**
1 Flowers pedicellate; leaves not mottled 6
  6 Petals delicate, wavy-margined, from white to deep pink (if white, becoming pink or transparent with age); stigmas thin, of uniform diameter, weakly recurved, somewhat fused at base 7
    7 Petals white with red blaze, rarely entirely white; anther sacs lavender to white, dehiscence extrorse . . . . **T. undulatum**
    7 Petals white to deep pink; anther sacs yellow, dehiscence introrse 8
      8 Anthers irregularly twisted outward; pollen egg-yolk yellow; pedicel declined below leaves, rarely otherwise . . . . **T. catesbaei**

8 Anthers straight, erect; pollen light yellow; pedicel inclined above leaves, rarely otherwise 9

9 Sepals obtuse, as broad as petals or broader; anther connectives purplish-green . . . . . **T. pusillum**

9 Sepals acute-acuminate, narrower than petals; anther connectives white to greenish-white 10

10 Petals obovate, tightly rolled basally, flaring outward distally; leaves acuminate, broadly elliptic; style minute, less than 1 mm long . . . . . **T. grandiflorum**

10 Petals elliptic, loose, gradually separating; leaves acute, ovate; style conspicuous, more than 1.5 mm long . . . . . **T. persistens**

6 Petals veiny, straight-margined, white, creamy, yellow, green, pink, maroon, or mixed (if white, turning brown with age); stigmas coarse, thicker toward base, strongly coiled, distinct 11

11 Pedicel abruptly declined below leaves; leaves petiolate; petals recurved between sepals; anthers often darker than filaments, or stamens far exceeding ovary in length 12

12 Stamens far exceeding ovary; ovary globose, shorter than filaments . . . . . **T. vaseyi**

12 Stamens 1.5× height of ovary or shorter; ovary ovoid, longer than filaments 13

13 Anthers more than 7 mm long, longer than filaments . . . . . **T. rugelii**

13 Anthers less than 7.5 mm long, often shorter than filaments . . . . . **T. cernuum**

11 Pedicel declined, more often horizontal to erect; leaves sessile to subsessile; petals variable; anthers and filaments similar in color 14

14 Ovary commonly white to pink, broader near base; anthers creamy, more than twice as long as the filaments; plants of calcareous soils . . . . . **T. flexipes**

14 Ovary commonly maroon to black, subglobose; anthers creamy, yellow, or purplish, less than 3× as long as the filaments; plants of circumneutral to acidic soils 15

15 Petals lanceolate, spreading from base, exposing ovary in side view; leaves often rhombic; sepals 0.5–0.8× length of pedicel . . **T. erectum**

15 Petals ovate, forming more of a cup-shaped base, concealing ovary in side view; leaves elliptic-obovate; sepals variable, often less than 0.5× length of pedicel 16

16 Sepals often suffused with purple, 0.2–0.4× length of pedicel; stamens 0.9–1.6× height of pistil; petals apically recurved, appearing barely longer than sepals . . . . . **T. sulcatum**

16 Sepals green, 0.4–0.7× length of pedicel; stamens 1.2–1.8× pistil height; petals not apically recurved, much longer than sepals . . . . . **T. simile**

T. catesbaei Ell., Catesby's T.; oak and oak-pine woods; occasional; Apr–May; GA, NC, SC, TN. *T. nervosum* Ell., *T. stylosum* Nutt.

T. cernuum L., Nodding T.; damp woods with black ash and American elm; infrequent; May; VA. *T. cernuum* L. var. *macranthum* Eames & Wieg.

T. cuneatum Raf., Sweet Betsy, Purple Toadshade; rich woods; occasional, more common southward; Apr–May; GA, NC, SC, TN. *T. hugeri* Small.

T. decumbens Harbison, Decumbent T.; rich woods; infrequent; Apr–May; GA, TN.

T. discolor Wray ex Hook., Pale Yellow T.; rich woods; infrequent; May; GA, NC, SC. A species restricted to Savannah River drainage.

T. erectum L., Red/Purple T., Stinking Willie/Benjamin; rich coves and spruce-fir forests; common; Apr–May; GA, NC, TN, VA. A variable species exhibiting a wide array of color forms. Plants with white petals resemble depauperate *T. simile*.

T. flexipes Raf., Bent/White T.; infrequent in rich calcareous woods of adjacent Ridge and Valley Province, perhaps extirpated from southern Blue Ridge Province, but easily confused with *T. rugelii*; Apr–May; NC, TN. *T. gleasonii* Fern.

T. grandiflorum (Michx.) Salisb., Large-flowered T., White Wake Robin; rich woods; frequent, becoming local southward; Apr–May; ALL.

T. luteum (Muhl.) Harbison, Yellow T.; rich woods, common at low elevations; Apr–May; GA, NC, TN. *T. cuneatum* Raf. var. *luteum* (Muhl.) Ahles.

T. persistens Duncan, Persistent T.; acid woods with hemlock and heaths; extremely rare; Apr; GA, SC. A federally endangered species restricted to a five-mile stretch of the Tallulah-Tugaloo river system.

T. pusillum Michx., Dwarf/Least T.; oak woods; Apr; NC. An elusive bloomer resembling a miniature *T. grandiflorum*.

T. rugelii Rendle, Southern Nodding T.; rich woods, often with rosebay rhododendron; occasional; Apr–May; GA, NC, SC, TN. Considered by some to be conspecific with *T. cernuum*. *T. affine* Rendle.

T. sessile L., Sessile T./Toadshade; rich woods; infrequent; Apr–May; VA.

T. simile Gleason, Sweet White T.; rich woods; occasional; Apr–May; GA, NC, SC, TN. Considered by some as conspecific with *T. erectum* but displaying features as unique as other taxa within the complex; often conspicuous at low elevations as trilliums with broad, white petals and a dark ovary.

T. sulcatum Patrick, Southern Red/Barksdale's T.; rich woods, often with heath understory; occasional; Apr–May; NC, TN, VA. In our area, primarily found within the New River drainage.

T. undulatum Willd., Painted T.; acid woods; occasional; Apr–May; ALL.

T. vaseyi Harbison, Vasey's T.; rich woods; frequent; Apr–Jun; GA, NC, SC, TN. One of the last trilliums to bloom; the largest of the eastern North American species. *T. erectum* L. var. *vaseyi* (Harbison) Ahles.

Some color variation is expected in all trilliums—even the painted trillium rarely has petals without red blazes. In the southern Blue Ridge, morphological features also seem to defy stability. Intermediates may be encountered in this region between these members of the *T. erectum* complex: *T. erectum*, *T. flexipes*, *T. simile*, *T. sulcatum*, *T. rugelii* and *T. vaseyi*.

Uvularia L. (Bellwort)

Wilbur, R. L. 1963. A revision of the North American genus *Uvularia* (Liliaceae). *Rhodora* 65:158–188.

1 Leaves perfoliate 2
  2 Tepals glabrous within; leaves pubescent beneath . . . . **U. grandiflora**
  2 Tepals papillose within; leaves glabrous beneath . . . . **U. perfoliata**
1 Leaves sessile, not perfoliate 3

3 Undivided style as long as or slightly longer than the stigmatic lobes; rhizome short; leaves shiny . . . . **U. puberula**

3 Undivided style 3–5× longer than the stigmatic lobes; rhizome elongate; leaves not shiny . . . . **U. sessilifolia**

U. grandiflora Smith; rich woods; frequent; Apr–May; NC, TN, VA.

U. perfoliata L.; rich woods and lowlands; common; Apr–May; ALL.

U. puberula Michx.; dry or moist woods; frequent; Mar–May; ALL. *U. pudica* (Walt.) Fern.

U. sessilifolia L.; rich woods and lowlands; occasional; Mar–May; ALL.

Veratrum L. (False Hellebore)

1 Perianth segments glabrous, less than 3 mm wide; leaves predominantly basal . . . . **V. parviflorum**

1 Perianth segments pubescent, 3–5 mm wide; leaves gradually reduced upward . . . . **V. viride**

V. parviflorum Michx.; rich woods; occasional; Jul–Sep; ALL.

V. viride Ait.; open woods, bogs, and meadows; occasional; Jun–Aug; GA, NC, TN, VA.

Xerophyllum Michx. (Turkey Beard)

X. asphodeloides (L.) Nutt.; dry open woods; occasional; May–Jun; ALL.

Zephyranthes Herb. (Atamasco Lily)

Z. atamasco (L.) Herb.; damp woods and meadows; infrequent; Mar–Apr; NC, SC.

Zigadenus Michx.

1 Tepals 3–4 mm long, each with a single gland at the base; plants not glaucous . . . . **Z. leimanthoides**

1 Tepals 8–12 mm long, each with a lobed gland at the base; plants glaucous . . . . **Z. glaucus**

Z. glaucus (Nutt.) Nutt., White Camass; calcareous slopes; infrequent; Jul–Aug; NC, TN, VA.

Z. leimanthoides Gray; wet woods; infrequent; Jul–Aug; NC, VA.

NAJADACEAE (Naiad Family)

Contributed by David H. Webb

Haynes, R. R. 1979. A revision of North and Central American *Najas* (Najadaceae). *Sida* 8:34–56.

Najas L.

1 Sheath of leaf base with truncate to auriculate shoulders; seeds asymmetrical at apex or slightly curved 2

2 Leaf blades spreading to ascending, not conspicuously toothed to the unaided eye; areolae of seed coat longer than wide, in more than 20 longitudinal rows .... **N. gracillima**

2 Leaf blades sometimes spreading to ascending when young, usually recurved with age, conspicuously toothed to the unaided eye; areolae of seed coat wider than long, in less than 20 longitudinal rows .... **N. minor**

1 Sheath of leaf base with rounded, tapering shoulders; seeds symmetrical at apex, not recurved 3

3 Leaves acute or rounded, not sharp-pointed; seeds brown to purplish, pitted, fusiform; areolae of seed coat in about 20 longitudinal rows .... **N. guadalupensis**

3 Leaves tapered to a sharp point; seeds brown to yellowish, smooth, obovate; areolae of seed coat in about 50 longitudinal rows .... **N. flexilis**

N. flexilis (Willd.) Rostk. & Schmidt; rivers and lakes; infrequent; Jul–Aug; VA.

N. gracillima (A. Braun) Magnus; ponds and lakes; occasional; Jul–Oct; NC, TN.

N. guadalupensis (Spreng.) Magnus, Southern Naiad; ponds, lakes and reservoirs; occasional; Jul–Oct; NC, TN, VA.

N. minor All.*, Spinyleaf Naiad; lakes and reservoirs; infrequent; Jul–Oct; TN, VA. A weedy species spreading into our area.

ORCHIDACEAE (Orchid Family)

Luer, C. A. 1975. *The native orchids of the United States and Canada, excluding Florida.* Bronx: New York Botanical Garden.

1 Leaves absent or withering at anthesis 2

2 Flowers solitary (rarely 2) .... **Arethusa**

2 Flowers in spikes or racemes 3

3 Base of lip with a conspicuous spur .... **Tipularia**

3 Base of lip without a spur 4

4 Inflorescence spicate, spiraled; flowers predominately white ..... **Spiranthes**

4 Inflorescence racemose, not spiraled; flowers not white or with a whitish lip only 5

5 Leaf solitary, basal, withering at anthesis .... **Aplectrum**

5 Leaves absent; plants saprophytic 6

6 Perianth more than 1 cm long .... **Hexalectris**

6 Perianth less than 1 cm long .... **Corallorhiza**

1 Leaves present at anthesis 7

7 Leaf solitary 8

8 Leaf basal 9

9 Leaf broad, elliptic or ovate .... **Platanthera**

9 Leaf narrow, grasslike 10

10 Flowers solitary (rarely 2) .... **Arethusa**

10 Flowers numerous .... **Calopogon**

8 Leaf cauline 11

11 Flowers solitary (rarely 2) 12

12 Sepals pink to rose; lip 2.5 cm or less long, with a fringed crest .... **Pogonia**

12 Sepals brownish-purple; lip 3 cm or more long, without a fringed crest .... **Cleistes**

11 Inflorescence a multiflowered spike 13
13 Lip spurred .... **Platanthera**
13 Lip not spurred .... **Malaxis**
7 Leaves 2–many 14
14 Lip inflated, pouchlike, or saccate 15
15 Leaves uniformly green, plicate; flowers 1–3 .... **Cypripedium**
15 Leaves white-reticulate above, not plicate; flowers more than 3 .... **Goodyera**
14 Lip not inflated, pouchlike, or saccate 16
16 Base of lip prolonged into an elongate spur or, rarely, short and less than 3 mm long 17
17 Lip white, entire, the remaining perianth purplish .... **Galearis**
17 Lip the same color as other perianth parts, entire, lacerate, or 3-parted 18
18 Spur as long as or longer than the lip .... **Platanthera**
18 Spur shorter than the lip .... **Coeloglossum**
16 Base of lip not prolonged into a spur 19
19 Stem leaves opposite or whorled 20
20 Stem leaves opposite .... **Listera**
20 Stem leaves whorled .... **Isotria**
19 Stem leaves alternate or the leaves predominately basal 21
21 Flowers inverted, the lip uppermost and bearded ...... **Calopogon**
21 Flowers not inverted, the lip lowermost and not bearded 22
22 Leaves cauline, flowers 1–6 .... **Triphora**
22 Leaves mostly basal, flowers numerous 23
23 Leaves narrow, less than 3 cm wide; flowers spiraled .... **Spiranthes**
23 Leaves elliptic to ovate, more than 3 cm wide; flowers racemose .... **Liparis**

Aplectrum (Nutt.) Torr. (Putty Root, Adam-and-Eve)

A. hyemale (Muhl. ex Willd.) Nutt.; rich woods; frequent; May–Jun; ALL.

Arethusa L. (Bog-rose)

A. bulbosa L.; bogs and wet meadows; infrequent; May–Jun; NC, SC, VA.

Calopogon R. Br. (Grass-pink)

C. tuberosus (L.) BSP.; bogs and seeps; occasional; Apr–Jul; ALL. *C. pulchellus* (Salisb.) R. Br.

Cleistes L. C. Rich. ex Lindl. (Spreading Pogonia)

C. divaricata (L.) Ames; open, mesic, or dry woods, wet meadows; occasional; May–Jul; ALL.

Coeloglossum Hartman (Long-bracted Orchid)

C. viride (L.) Hartman; balds, meadows, and rich woods; infrequent; Apr–Jun; NC, TN, VA. Represented in our area by var. **virescens** (Muhl. ex Willd.) Luer. *Habenaria viridis* (L.) R. Br. var. *bracteatum* (Muhl. ex Willd.) Gray.

Corallorhiza Gagnebin (Coral-root)

1 Lip 3-lobed or with a pair of lateral lobes 2
  2 Plants more than 15 cm tall; lip 5–9 mm long .... **C. maculata**
  2 Plants 10–15 cm tall; lip 3–5 mm long .... **C. trifida**
1 Lip entire, not lobed 3
  3 Perianth 5 mm or more long; spring flowering .... **C. wisteriana**
  3 Perianth less than 5 mm long; fall flowering .... **C. odontorhiza**

C. maculata (Raf.) Raf., Spotted C. R.; rich woods and stream banks; frequent; Jul–Aug; GA, NC, TN, VA.

C. odontorhiza (Willd.) Nutt., Autumn C. R.; woodlands; common; Aug–Oct; ALL.

C. trifida (L.) Chatelain, Early/Northern C. R.; hemlock and mixed hardwoods; infrequent; May–Jul; VA.

C. wisteriana Conrad, Spring C. R.; low woods; occasional; Apr–May; NC, SC, TN, VA.

Cypripedium L. (Lady's Slipper, Moccasin Flower)

1 Leaves basal; pouch with a longitudinal groove .... **C. acaule**
1 Leaves cauline; pouch without a longitudinal groove 2
  2 Pouch yellow; petals twisted, acute at the apex 3
    3 Pouch 3–5 cm long .... **C. calceolus** var. **pubescens**
    3 Pouch 2–2.5 cm long .... **C. calceolus** var. **parviflorum**
  2 Pouch roseate, rarely white; petals not twisted, obtuse at the apex ..... **C. reginae**

C. acaule Ait., Pink L. S.; pine or pine-heath woodlands; frequent; Apr–Jun; ALL.

C. calceolus L. var. **parviflorum** (Salisb.) Fern., Small Yellow L. S.; rich deciduous woods; infrequent; Apr–Jun; GA, NC, TN, VA.

C. calceolus L. var. **pubescens** (Willd.) Correll, Large Yellow L. S.; rich deciduous woods; occasional; Apr–Jun; ALL.

C. reginae Walt., Queen L. S.; bogs and seeps; infrequent; May–Jun; NC, TN, VA.

Galearis Raf. (Showy Orchis)

G. spectabilis (L.) Raf.; rich woods and stream banks; frequent; Apr–Jun; ALL. *Orchis spectabilis* L.

Goodyera R. Br. (Rattlesnake Plantain)

1 Inflorescence cylindric, densely flowered (up to 80) .... **G. pubescens**
1 Inflorescence 1-sided or spiraled, few-flowered (up to 20) .... **G. repens**

G. pubescens (Willd.) R. Br., Downy R. P.; damp or dry woodlands; common; Jun–Aug; ALL.

G. repens (L.) R. Br., Lesser R. P.; damp rich woods; occasional; Jun–Sep; NC, TN, VA.

Hexalectris Raf. (Crested Coral-root)

H. spicata (Walt.) Barnhart; dry woodlands, especially over calcareous rocks; occasional; Jul–Aug; ALL.

Isotria Raf. (Whorled Pogonia)

1 Stem purplish; leaves erect at early anthesis; flowers sessile or subsessile .... **I. verticillata**

1 Stem greenish-white; leaves drooping at early anthesis; flowers with a pedicel 5–15 mm long . . . . **I. medeoloides**

I. medeoloides (Pursh) Raf., Small W. P.; open pine or mixed pine-hardwoods, stream banks; infrequent; May–Jun; GA, NC, SC.

I. verticillata (Muhl. ex Willd.) Raf., Large W. P.; rich woods and stream banks; occasional; Apr–Jul; ALL.

Liparis L. C. Rich.

1 Lip purplish, more than 8 mm long . . . . **L. lilifolia**
1 Lip yellow-green, less than 5 mm long . . . . **L. loeselii**

L. lilifolia (L.) L. C. Rich. ex Lindl., Lily-leaved Twayblade; moist woods and stream banks; frequent; May–Jul; ALL.

L. loeselii (L.) L. C. Rich., Fen Orchid; bogs and seeps; infrequent; May–Jul; NC, TN.

Listera R. Br. (Twayblade)

1 Lip deeply divided into linear-lanceolate lobes and without a tooth in the sinus 2
  2 Lip 3–6 mm long, with a pair of lateral horns at the base . . . . **L. cordata**
  2 Lip 6–12 mm long, without a pair of lateral horns at the base . . . . **L. australis**
1 Lip shallowly cleft at the apex, the lobes obovate-cuneate and with a tooth in the sinus . . . . **L. smallii**

L. australis Lindl., Southern T.; damp woods under rhododendron; infrequent; Apr–Jul; NC.

L. cordata (L.) R. Br., Heart-leaved T.; in moss under evergreens; infrequent and possibly extirpated from our area; Jun–Jul; NC.

L. smallii Wiegand, Appalachian T.; damp shady woods, often under rhododendron; occasional; Jun–Jul; ALL.

Malaxis Soland. ex Swartz (Green Adder's Mouth)

M. unifolia Michx.; moist woods and stream banks; frequent; Jun–Aug; ALL.

Platanthera L. C. Rich.

1 Lip lacerate, fringed, or distinctly 3-parted 2
  2 Lip lacerate or fringed, but not 3-parted 3
    3 Perianth white . . . . **P. blephariglottis**
    3 Perianth yellow to orange 4
      4 Spur longer than the ovary and pedicel . . . . **P. ciliaris**
      4 Spur equal to or shorter than the ovary and pedicel . . . . **P. cristata**
  2 Lip 3-parted 5
    5 Perianth white, creamy, or greenish-white 6
      6 Perianth white or creamy; petals broad and crenate . . . . **P. leucophaea**
      6 Perianth greenish-white; petals narrow and obscurely crenulate . . . . **P. lacera**
    5 Perianth purplish or white in albino forms 7
      7 Lip scarcely fringed to nearly entire . . . . **P. peramoena**
      7 Lip distinctly fringed 8
        8 Entrance to the nectary dumbbell-shaped; lip fringed less than 1/3 its length; racemes 8–15 cm long . . . . **P. psycodes**
        8 Entrance to the nectary round; lip fringed more than 1/3 its length; racemes up to 25 cm long . . . . **P. grandiflora**

1 Lip entire, merely erose, or shallowly lobed at the base or apex, but not lacerate, fringed, or distinctly 3-parted 9
9 Leaves basal, 1 or 2(3); stem with a single bract near the middle . . . . **P. orbiculata**
9 Leaves cauline or subbasal; stem leafy or conspicuously bracteate 10
10 Flowers white . . . . **P. integrilabia**
10 Flowers yellow-green 11
11 Lip lobed or angled at the base and with a small tubercle near center of base; floral bracts 10–15 mm long . . . . **P. flava**
11 Lip neither lobed at base nor with a tubercle; floral bracts up to 6 mm long . . . . **P. clavellata**

P. blephariglottis (Willd.) Lindl., Large White Fringed Orchid; wet meadows; infrequent; Jul–Aug; GA, VA. *Habenaria blephariglottis* (Willd.) Hook.

P. ciliaris (L.) Lindl., Yellow Fringed Orchid; bogs, meadows, and low woods; frequent; Jul–Sep; ALL. *Habenaria ciliaris* (L.) R. Br.

P. clavellata (Michx.) Luer, Small Green Wood Orchid; bogs and low woods; frequent; Jun–Sep; ALL. *Habenaria clavellata* (Michx.) Sprengel.

P. cristata (Michx.) Lindl., Crested Fringed Orchid; bogs and low woods; infrequent; Jun–Sep; NC, SC, TN, VA. *Habenaria cristata* (Michx.) R. Br.

P. flava (L.) Lindl., Tubercled Rein Orchid; low woods and meadows; occasional; Mar–Sep; NC, TN, VA. Represented in our area by var. **herbiola** (R. Br.) Luer. *Habenaria flava* (L.) R. Br. var. *herbiola* (R. Br.) Ames & Correll.

P. grandiflora (Bigelow) Lindl., Large Purple Fringed Orchid; thickets and meadows, often at high elevations; occasional; Jun–Aug; NC, VA. *Habenaria psycodes* (L.) Sprengel var. *grandiflora* (Bigelow) Gray, *H. fimbriata* (Dryander) R. Br. in Ait.

P. integrilabia (Correll) Luer, Monkey Face; low woods and bogs; infrequent; Jul–Sep; GA, NC, SC, TN. *Habenaria blephariglottis* (Willd.) Hook. var. *integrilabia* Correll.

P. lacera (Michx.) G. Don, Green Fringed Orchid; thickets and wet meadows; occasional; Jun–Aug; NC, SC, TN, VA. *Habenaria lacera* (Michx.) R. Br.

P. leucophaea (Nutt.) Lindl., Prairie Fringed Orchid; damp meadows and bogs; infrequent; May–Jul; VA. *Habenaria leucophaea* (Nutt.) Gray.

P. orbiculata (Pursh) Lindl., Large Round-leaved Orchid; deep woods; infrequent; Jun–Sep; GA, NC, TN, VA. *Habenaria orbiculata* (Pursh) Torr.

P. peramoena (Gray) Gray, Purple Fringeless Orchid; meadows and low woods; occasional; Jun–Oct; NC, SC, TN, VA. *Habenaria peramoena* Gray.

P. psycodes (L.) Lindl., Small Purple Fringed Orchid; low woods and meadows; occasional; Jun–Oct; ALL. *Habenaria psycodes* (L.) Sprengel.

Pogonia Juss. (Rose Pogonia)

P. ophioglossoides (L.) Juss.; open bogs and seeps; infrequent and more common on the Coastal Plain; May–Jun; NC, SC, TN, VA.

Spiranthes Richard (Ladies' Tresses)

Contributed by Thomas S. Patrick

1 Plants flowering in late spring 2
2 Lip white with bright yellow center; leaves elliptic-lanceolate . . . . **S. lucida**
2 Lip white to pale green, sometimes green-veined; leaves linear . . . . **S. praecox**
1 Plants flowering in midsummer to frost 3

3 Lip less than 4 mm long; spike once-spiraled, slender 4
4 Lip white with green center 5
5 Inflorescence minutely pubescent, loosely spiraled; leaves present at flowering .... **S. lacera** var. **lacera**
5 Inflorescence glabrous, tightly spiraled; leaves absent or withering at flowering .... **S. lacera** var. **gracilis**
4 Lip white .... **S. tuberosa**
3 Lip 5–15 mm long; spike once- or, more commonly, 2–3×-spiraled, coarse 6
6 Lip 5 mm long; leaves basal, sheathing, oblanceolate, up to 15 cm long .... **S. ovalis**
6 Lip 7–12 mm long; leaves basal and cauline, linear to narrowly oblanceolate, the longer ones more than 20 cm long, rarely absent during flowering 7
7 Inflorescence with pointed hairs; lip 5–8 mm long; plants flowering Jun–early Sep .... **S. vernalis**
7 Inflorescence with knobbed hairs; lip 7–15 mm long; plants flowering Sep–Nov 8
8 Lip white with creamy center; lower bracts about as long as the flowers .... **S. cernua**
8 Lip creamy with dull yellowish center; lower bracts longer than the flowers .... **S. ochroleuca**

S. cernua (L.) Richard, Nodding L. T.; moist openings and meadows; common; Sep–Nov; ALL.

S. lacera (Raf.) Raf. var. **gracilis** (Bigel.) Luer; openings and woodland margins; frequent; Jul–Aug; ALL. *S. gracilis* (Bigel.) Beck. More common south of our area.

S. lacera (Raf.) Raf. var. **lacera**, Slender L. T.; upland margins, barrens, and old fields; frequent; Jul–Aug; TN, VA. More common north of our area.

S. lucida (H. H. Eat.) Ames, Shining L. T.; openings in calcareous seeps; infrequent; May–Jun; TN, VA.

S. ochroleuca (Rydb.) Rydb., Yellow Nodding L. T.; moist open slopes at higher elevations; infrequent; Oct–Nov; NC, TN, VA. *S. cernua* (L.) Richard var. *ochroleuca* (Rydb. ex Britt.) Ames. A northern member of the *S. cernua* group, this robust species has only recently been recognized from the southern Appalachians.

S. ovalis Lindl., Lesser/Oval L. T.; disturbed woods; infrequent; Sep–Oct; GA, SC, TN, VA.

S. praecox (Walt.) S. Wats., Giant L. T.; damp meadows; very rare or possibly extirpated from our area; May–Jun; VA. More common on the Coastal Plain.

S. tuberosa Raf., Little L. T.; dry barrens and upland woods; occasional; Jul–Aug; ALL. *S. beckii* Gray, *S. grayi* Ames. Our smallest species and easily overlooked.

S. vernalis Engelm. & Gray, Spring L. T.; gravelly fields and clearings; frequent; Jun–early Sep; NC, TN, VA. This robust species blooms much earlier on the Coastal Plain, hence its common name.

Tipularia Nutt. (Crane-fly Orchid)

T. discolor (Pursh) Nutt.; woodlands; common; Jul–Sep; ALL.

Triphora Nutt. (Three Birds Orchid)

T. trianthophora (Swartz) Rydb.; rich woods; occasional; Jul–Sep; ALL.

POACEAE (Grass Family)

Contributed by W. T. Batson (except *Panicum*)

1 Plants woody . . . . **Arundinaria**
1 Plants herbaceous 2
2 Large grass; inflorescence a spike, spikelets unisexual; pistillate portion below and breaking up into 1-seeded joints . . . . **Tripsacum**
2 Not as above 3
3 Glumes wanting; spikelets 1-flowered and strongly flattened . . . . **Leersia**
3 Not as above 4
4 Spikelets more or less laterally flattened; sterile florets, when present, above the fertile; articulation above the glumes 5
5 Sterile lemma(s) present 6
6 Sterile lemmas small and unlike fertile; panicle dense 7
7 Sterile lemmas awned; plants fragrant . . . . **Anthoxanthum**
7 Sterile lemmas awnless; plants not fragrant; panicle very dense and spikelike . . . . **Phalaris**
6 Sterile lemmas similar to fertile; panicles very narrow, or diffuse . . . . **Chasmanthium**
5 Sterile lemma(s) absent 8
8 Spikelets pediceled; panicles open or contracted 9
9 Spikelets 1-flowered . . . . **Key A**
9 Spikelets 2- or more-flowered 10
10 Glumes shorter than the lowest lemma . . . . **Key B**
10 Glumes longer than the lowest lemma . . . . **Key C**
8 Spikelets sessile; inflorescence of 1 or more spikes 11
11 Spikes with spikelets alternating on opposite side of rachis . . . **Key D**
11 Spikes with spikelets on only 1 side of the rachis . . . . **Key E**
4 Spikelets more or less dorsally flattened and with 1 perfect and 1 sterile or incomplete floret; articulation below glumes 12
12 Glumes tough; lemma and palea membranous; spikelets in pairs, lower sessile and perfect, upper pediceled, smaller, sterile, and sometimes abortive . . . . **Key F**
12 Glumes membranous; lemma and palea thick, hard, and glossy . . . **Key G**

KEY A

1 Inflorescence a very dense, uninterrupted, spikelike panicle . . . . **Phleum**
1 Not as above 2
2 Lemma(s) awned 3
3 Lemma 3-awned . . . . **Aristida**
3 Lemma(s) 1-awned from summit 4
4 Awns 3–5 cm long and twisted . . . . **Stipa**
4 Awns shorter to nearly absent 5
5 Lower glume well developed (obsolete in *Muhlenbergia schreberi*) 6
6 Lemmas with long silky hairs at base . . . . **Calamagrostis**
6 Lemmas not as above 7
7 Glumes shorter than lemma . . . . **Muhlenbergia**

7 Glumes as long as or longer than lemma; lemma at maturity nearly black . . . . **Oryzopsis**

5 Lower glume very small to obsolete, the upper slightly larger . . . . **Brachyelytrum**

2 Lemma(s) awnless, or awned from back 8

8 Spikelets dorsally compressed; lemmas hard and glossy (resembling *Panicum*) . . . . **Milium**

8 Not as above 9

9 Erect plants to 1.5 m tall; panicle drooping; articulation above the glumes, spikelet falling entire; upper lemma 3-nerved and may bear minute awn from just below summit . . . . **Cinna**

9 Plants smaller; glumes persistent 10

10 Glumes shorter than lemma . . . . **Sporobolus**

10 Glumes as long as or longer than lemma . . . . **Agrostis**

KEY B

1 Tall stout reeds with plumose panicles . . . . **Arundo**

1 Plants usually less than 1.5 m tall 2

2 Spikelets in dense, somewhat 1-sided clusters . . . . **Dactylis**

2 Not as above 3

3 Lowest lemmas sterile . . . . **Chasmanthium**

3 Lowest lemmas fertile 4

4 Upper florets very short and sterile . . . . **Melica**

4 Upper florets similar to lower 5

5 Leaf sheaths closed, at least below 6

6 Lemmas at least 6 mm long . . . . **Bromus**

6 Lemmas less than 6 mm long . . . . **Glyceria**

5 Leaf sheaths open 7

7 Lemmas awned from summit, some obscurely so . . . . **Festuca**

7 Lemmas awnless 8

8 Lemmas 5-nerved, sometimes obscurely . . . . **Poa**

8 Lemmas 3-nerved 9

9 Lemmas glabrous 10

10 Blades 1–2 cm wide; lemmas acuminate but not awned; grains 6–10 mm long and pointed, with partial exposure at maturity . . . . **Diarrhena**

10 Blades and grains smaller . . . . **Eragrostis**

9 Lemmas conspicuously pubescent on veins . . . . **Tridens**

KEY C

1 Spikelets with 2 or more florets, all fertile 2

2 Glumes similar 3

3 Lemmas awned from near tip, or awnless 4

4 Glumes with 7 or more nerves . . . . **Avena**

4 Glumes with 1–5 nerves . . . . **Danthonia**

3 Lemmas awned from near middle or base 5

5 Awn arising from near base .... **Deschampsia**
5 Awn arising from near middle .... **Trisetum**
2 Glumes dissimilar, second widening toward tip .... **Sphenopholis**
1 Spikelets with 1 fertile and 1 sterile floret 6
6 Lower floret sterile and awned .... **Arrhenatherum**
6 Lower floret fertile and awnless .... **Holcus**

KEY D

1 Spikelets solitary at each joint of rachis 2
2 Spikelets borne edgewise to rachis .... **Lolium**
2 Spikelets borne flatwise to rachis .... **Agropyron**
1 Spikelets 2 or more at each joint of rachis 3
3 Spikelets 2 at each joint of rachis 4
4 Spikelets appressed to slightly spreading .... **Elymus**
4 Spikelets widely spreading to horizontal .... **Hystrix**
3 Spikelets 3 at each joint of rachis .... **Hordeum**

KEY E

1 Spikes or spikelike racemes borne digitately or nearly so 2
2 Spikelets with 2 or 3 perfect florets; annuals .... **Eleusine**
2 Spikelets with 1 perfect floret .... **Cynodon**
1 Spikes or spikelike racemes borne racemosely 3
3 Spikelets appressed and borne remotely along triangular rachis; leaves short and broad .... **Gymnopogon**
3 Not as above 4
4 Annuals with very slender racemes and 2 or more perfect florets/spikelet .... **Leptochloa**
4 Perennials with only 1 perfect floret/spikelet 5
5 One perfect and 1 or more sterile florets/spikelet .... **Bouteloua**
5 Spikelet 1-flowered .... **Spartina**

KEY F

1 Both spikelets of each pair similar and perfect 2
2 Straggling annual; leaves less than 10 cm long .... **Microstegium**
2 Tall perennials; leaves more than 10 cm long 3
3 Inflorescence of slender, spreading, spikelike racemes .... **Miscanthus**
3 Inflorescence a rather narrow, silky panicle .... **Erianthus**
1 Only sessile spikelet perfect 4
4 Pediceled spikelet much smaller and sterile 5
5 Inflorescence of 1–3 racemes; rachis and pedicels of sterile spikelets more or less silky .... **Andropogon**
5 Inflorescence a rather dense, large-seeded panicle .... **Sorghum**
4 Pediceled spikelet wanting; pedicel present or wanting 6
6 Pedicel present; leaves more than 10 cm long; plant erect .... **Sorghastrum**
6 Pedicel wanting, except sometimes at lower nodes of inflorescence; leaves ovate and less than 5 cm long; plant creeping; awned or awnless .... **Arthraxon**

KEY G

1 Spikelet surrounded by a prickly bur or subtended by bristles 2
  2 Spikelet surrounded by a prickly bur . . . . **Cenchrus**
  2 Spikelet subtended by bristles . . . . **Setaria**
1 Not as above 3
  3 Inflorescence paniculate 4
    4 Second glume inflated or saclike at base and strongly nerved . . . . . . **Sacciolepis**
    4 Not as above 5
      5 Panicle of short racemose branches; spikelets crowded . . . . **Eriochloa**
      5 Panicle spreading . . . . **Panicum**
  3 Inflorescence of 1-sided spikelike racemes; first glume minute or wanting 6
    6 Spikelets flat to plano-convex, broad and blunt-tipped . . . . **Paspalum**
    6 Spikelets little compressed, lanceolate; margins of fertile lemma hyaline and partly enclosing palea; first glume minute or wanting 7
      7 Fertile lemma leathery and usually dark . . . . **Digitaria**
      7 Fertile lemma papery . . . . **Axonopus**

Agropyron Gaertn.

1 Plant from or forming rhizomes 2
  2 Glumes membranous and abruptly awn-pointed; lemmas awned or awnless . . . . **A. repens**
  2 Glumes rigid and only gradually awn-pointed; plants glaucous . . . . **A. smithii**
1 Plants cespitose . . . . **A. trachycaulum**

A. repens (L.) Beauv.*, Quack Grass; moist fields, roadsides, and waste areas; occasional; Jul–Aug; NC, VA.

A. smithii Rydb.; infrequent along transportation routes; Jul–Aug; GA. A western species perhaps not native in our area.

A. trachycaulum (Link) Malte ex H. F. Lewis; pinewoods; an infrequent northern species; Jul–Aug; NC, VA.

Agrostis L. (Bent Grass)

1 Spikelet of 4 bracts, palea at least ½ as long as lemma 2
  2 Ligule of culm leaves 2 mm long . . . . **A. tenuis**
  2 Ligule of culm leaves 6 mm long 3
    3 Culms erect, to 1 m or more tall; rhizomes present . . . . **A. gigantea**
    3 Culms decumbent at base, shorter; rhizomes absent . . . . **A. stolonifera**
1 Spikelet of 3 bracts, palea obsolete or wanting 4
  4 Plants annual; lemmas awned . . . . **A. elliottiana**
  4 Plants perennial; lemmas awned or awnless 5
    5 Lemmas awned 6
      6 Spikelet about 1.5 mm long . . . . **A. hiemalis**
      6 Spikelet about 2.5 mm long . . . . **A. mertensii**
    5 Lemmas awnless . . . . **A. perennans**

A. elliottiana Schultes; roadsides and fields; infrequent; Jun–Oct; NC, TN.

A. gigantea Roth*, Common Redtop; escaped to margins and waste places; infrequent; Jun–Sep; GA.

A. hiemalis (Walt.) BSP.; pastures, roadsides, and meadows; common; May–Oct; ALL.

A. mertensii Trin.; mountain summits, balds; infrequent; Jul–Aug; NC, VA. *A. borealis* Hartman.

A. perennans (Walt.) Tuckerman; woods, margins, and meadows; common; Aug–Oct; ALL.

A. stolonifera L.*, Redtop; fields, margins, and waste places; common; Jun–Oct; ALL. *A. alba* of authors, *A. palustris* Huds.

A. tenuis Sibthorp; meadows and roadsides; infrequent; Jun–Aug; NC, TN, VA.

Andropogon L. (Broom Grass, Broomsedge, Broomstraw, Bluestem)

1 Raceme 1 on each peduncle .... **A. scoparius**
1 Racemes 2 or more on each peduncle 2
  2 Pediceled spikelet about as large as sessile spikelet but awnless; inflorescence exserted .... **A. gerardii**
  2 Pediceled spikelet much reduced or wanting 3
    3 Inflorescence a large, heavy, leafy glom .... **A. glomeratus**
    3 Inflorescence not as above 4
      4 Upper leaf sheaths conspicuously inflated .... **A. gyrans**
      4 Upper leaf sheaths not as above 5
        5 Upper culm and inflorescence leafy; racemes partly hidden .... **A. virginicus**
        5 Upper culm and inflorescence leaves remote; racemes exserted .... **A. ternarius**

A. gerardii Vitman, Big Bluestem; dry woods and rock outcrops; infrequent; Aug–Oct; ALL.

A. glomeratus (Walt.) BSP.; wet or moist open places; frequent; Jul–Oct; ALL.

A. gyrans Ashe; dry open woods, edges, and abandoned fields; infrequent; Jul–Oct; ALL. *A. elliottii* Chapman.

A. scoparius Michx., Little Bluestem; dry woods and margins; common; Aug–Oct; ALL. *Schizachyrium scoparium* (Michx.) Nash.

A. ternarius Michx.; moist margins and abandoned fields; common; Aug–Oct; ALL.

A. virginicus L., Broomstraw; old fields, roadsides, and waste places; common; Aug–Oct; ALL. Once a common source of brooms.

Anthoxanthum L. (Sweet Vernal Grass)

A. odoratum L.*; meadows, margins, and waste areas; common; Apr–Jun; ALL.

Aristida L. (Triple Awn Grass)

Allred, K. W. 1986. Studies in the *Aristida* (Gramineae) of the southeastern United States. IV. Key and conspectus. *Rhodora* 88:367–387.

1 First glume 3–7-nerved .... **A. oligantha**
1 First glume 1–2-nerved 2
  2 Central awn spirally coiled and corkscrew at the base 3
    3 First glume essentially equal to the second glume; lemma sparsely appressed-pubescent, 3–8 mm long .... **A. dichotoma** var. **dichotoma**
    3 First glume 1/2–2/3 as long as the second glume; lemma glabrous to scaberulous, 6–11 mm long .... **A. dichotoma** var. **curtissii**
  2 Central awn straight, curved, or contorted, but not corkscrewlike at the base 4

4 Lateral awns reduced, 1/3 or less the length of the central awn .... **A. longespica**
4 Lateral awns well developed, at least 1/2 the length of the central awn 5
5 Plant annual .... **A. longespica**
5 Plant perennial .... **A. purpurascens**

A. dichotoma Michx. var. **curtissii** Gray; margins and waste areas; infrequent; Aug–Sep; VA. *A. curtissii* (Gray) Nash.

A. dichotoma Michx. var. **dichotoma**; dry open ground and old fields; common; Aug–Oct; ALL.

A. longespica Poir.; old fields, roadsides, and waste places; frequent; Aug–Oct; ALL.

A. oligantha Michx.; dry roadsides and old fields; occasional; Aug–Sep; ALL.

A. purpurascens Poir.; dry margins and waste places; frequent; Aug–Oct; ALL.

Arrhenatherum Beauv. (Oat Grass)

A. elatius (L.) Beauv. ex J. & C. Presl.*; moist woods and margins; occasional; May–Jun; NC, SC, TN, VA.

Arthraxon Beauv.

A. hispidus (Thunb.) Makino*; ditches and wet margins; frequent; Sep–Oct; ALL. Represented in our area by var. **cryptherus** (Hack.) Honda.

Arundinaria Michx. (Cane)

A. gigantea (Walt.) Muhl.; low or moist woods and stream banks; frequent; Apr–Jul; ALL. Our only native woody grass, seldom flowering.

Avena L. (Oats)

A. sativa L.*; escaped to field margins and waste places; occasional; May–Jun; ALL.

Axonopus Beauv. (Carpet Grass)

A. affinis Chase; yards, margins, and roadsides; infrequent; Jun–Oct; SC. Sometimes used as a lawn grass.

Bothriochloa Kuntze (Silver Beard Grass)

B. saccharoides (Swartz) Rydb.; dry margins; infrequent; Aug–Oct; TN. More common west of our area. *Andropogon saccharoides* (Sw.) Rydb.

Bouteloua Lag. (Side-oats Grama)

B. curtipendula (Michx.) Torr.; dry margins; infrequent; Jul–Sep; VA.

Brachyelytrum Beauv.

B. erectum (Schreb.) Beauv.; rich, moist wooded slopes; frequent; Jun–Aug; ALL.

Bromus L. (Brome Grass)

1 Spikelets very strongly laterally compressed; lemmas keeled .... **B. catharticus**
1 Not as above 2
2 Plants perennial 3
3 Creeping rhizomes present .... **B. inermis**
3 No creeping rhizomes present 4

4 Lemmas glabrous . . . . **B. pubescens**
4 Lemmas pubescent . . . . **B. ciliatus**
2 Plants annual 5
5 Leaf sheaths glabrous . . . . **B. secalinus**
5 Leaf sheaths pubescent 6
6 Spikelets pubescent; awns varying in length from 2 mm for lowest in spikelet to 8–12 mm for the uppermost; panicle branches much longer than spikelets, flexuous, and drooping . . . . **B. japonicus**
6 Not as above 7
7 Awns less than 1.4 cm long 8
8 Spikelets narrow and drooping; first glume 1-nerved . . . . . . **B. tectorum**
8 Spikelets broader, not drooping; first glume 3-nerved . . . . **B. racemosus**
7 Awns 2–3 cm long . . . . **B. sterilis**

B. catharticus Vahl*; fields, margins, and waste places; infrequent; Apr–Jun; NC, SC, VA.

B. ciliatus L.; wet areas; infrequent; Jul–Aug; NC, TN, VA.

B. inermis Leysser*; roadsides and waste places; occasional; Jun–Jul; NC, TN, VA.

B. japonicus Thunb. ex Murray*; waste places; common; May–Jun; NC, SC, TN, VA.

B. pubescens Willd.; moist waste areas; occasional; Jun–Aug; NC, TN, VA. *B. purgans* L.

B. racemosus L.*; fields and roadsides; frequent; May–Jun; NC, SC, TN, VA. *B. commutatus* Schrad.

B. secalinus L.*; roadsides and waste ground; frequent; May–Jun; NC, SC, TN, VA.

B. sterilis L.*; dry margins; infrequent; May–Jun; VA.

B. tectorum L.*; fields and margins; occasional; Apr–Jun; NC, SC, TN, VA.

Calamagrostis Adans. (Reed Grass)

1 Awn sharply bent 2
2 Plant tufted; leaves 1–2 mm wide and soon involute . . . . **C. cainii**
2 Rhizomes present; leaves flat and 4–8 mm wide . . . . **C. porteri**
1 Awn straight 3
3 Panicle loose, spreading . . . . **C. canadensis**
3 Panicle dense, erect . . . . **C. cinnoides**

C. cainii Hitchc.; moist cliffs and landslides; infrequent; Jul–Sep; a strict endemic to Mt. LeConte, TN.

C. canadensis (Michx.) Beauv.; mountain summits; infrequent; Aug–Sep; NC, VA.

C. cinnoides (Muhl.) Barton; ditches and wet openings; frequent; Jul–Oct; ALL.

C. porteri Gray; openings and rocky places; occasional; Jul–Aug; NC, TN, VA.

Cenchrus L. (Sandspurs)

1 Body of bur between the spines distinctly woolly; spines 5–9 mm long . . . . **C. tribuloides**
1 Body of bur not woolly; spines 1–4 mm long . . . . **C. longispinus**

C. longispinus (Hackel) Fern.; dry margins; frequent; Jun–Oct; NC, SC, VA.

C. tribuloides L.; dry fields and margins; infrequent; Aug–Oct; VA.

Chasmanthium Link

1 Spikelets very flat, ovate, and 8–20-flowered .... **C. latifolium**
1 Spikelets with 5 or fewer flowers 2
  2 Sheaths glabrous or nearly so .... **C. laxum**
  2 Sheaths pubescent, at least in most cases .... **C. sessiliflorum**

C. latifolium (Michx.) Yates, River Oats; damp woods and stream margins; common; Jun–Oct; ALL. *Uniola latifolia* Michx.

C. laxum (L.) Yates; low woods and margins; common; Jun–Oct; ALL. *Uniola laxa* (L.) BSP.

C. sessiliflorum (Poir.) Yates; rich woods; frequent; Aug–Oct; GA, SC, TN. *Uniola sessiliflora* Poir.

Cinna L.

1 Panicle dense; spikelets 5 mm long .... **C. arundinacea**
1 Panicle loose, spreading; spikelets 3–4 mm long .... **C. latifolia**

C. arundinacea L., Wood Reed; moist woods and openings; frequent; Aug–Oct; NC, SC, TN, VA.

C. latifolia (Trev. ex Goepp.) Grisebach; moist woods at high elevations; occasional; Jun–Aug; NC, TN, VA.

Cynodon L. C. Rich. (Bermuda Grass)

C. dactylon (L.) Pers.*; lawns, pastures, playgrounds, cultivated areas; common; May–Oct; ALL.

Dactylis L. (Orchard Grass)

D. glomerata L.*; fields and margins; common; May–Oct; ALL.

Danthonia Lam. & DC.

1 Sheaths long pubescent .... **D. sericea**
1 Sheaths glabrous or nearly so 2
  2 Blades mostly 15 cm or more long .... **D. compressa**
  2 Blades mostly less than 15 cm long and curled .... **D. spicata**

D. compressa Austin; mountain woods and summits; frequent; Jun–Aug; ALL.

D. sericea Nutt.; dry woods and margins; very common; Apr–Jun; ALL.

D. spicata (L.) Beauv. ex R. & S., Poverty Oat Grass; dry openings and balds; frequent; May–Jul; ALL.

Deschampsia Beauv. (Hair Grass)

1 Blades flat or folded .... **D. caespitosa**
1 Blades inrolled .... **D. flexuosa**

D. caespitosa (L.) Beauv.; dry open woods; infrequent; Jun–Jul; NC, VA.

D. flexuosa (L.) Trin.; rocky outcrops and balds; frequent; Apr–Jul; GA, NC, SC, VA.

Diarrhena Beauv.

D. americana Beauv.; rich woods; infrequent; Jul–Aug; TN, VA.

Digitaria Heist. ex Fabr. (Crab Grass)

1 Culms freely rooting at nodes; rachis winged 2
  2 Fertile lemmas pale brown or straw-colored . . . . **D. sanguinalis**
  2 Fertile lemmas dark brown . . . . **D. ischaemum**
1 Culms erect; rachis slender . . . . **D. filiformis**

D. filiformis (L.) Koeler; fields and roadsides; frequent; Sep–Oct; ALL.
D. ischaemum (Schreb. ex Schweig.) Schreb. ex Muhl.*; fields and roadsides; frequent; Jul–Oct; ALL.
D. sanguinalis (L.) Scop.*; fields, roadsides, waste places, lawns, gardens, and pastures; a far too common pernicious species; Jul–Oct; ALL.

Echinochloa Beauv. (Barnyard Grass)

E. crusgalli (L.) Beauv.*; low open places; frequent; Jul–Oct; ALL. A variable species.

Eleusine Gaertner (Goose Grass)

E. indica (L.) Gaertn.*; fields, margins, and disturbed ground; common; Jun–Oct; NC, SC, TN, VA.

Elymus L. (Wild Rye)

1 Awns divergent . . . . **E. canadensis**
1 Awns straight 2
  2 Glumes 4–5-nerved . . . . **E. virginicus**
  2 Glumes 1–3-nerved 3
    3 Blades hirsute above . . . . **E. villosus**
    3 Blades glabrous or only scabrous . . . . **E. riparius**

E. canadensis L.; rich, dry or moist soil, woods, and openings; infrequent; Jul–Oct; NC, VA.
E. riparius Wiegand; moist to wet woods; infrequent; Jul–Aug; NC, TN, VA.
E. villosus Muhl. ex Willd.; moist woods and margins; occasional; Jul–Aug; NC, TN, VA.
E. virginicus L.; mixed woods, wet banks, and openings; frequent; Jun–Oct; ALL. A rather variable species.

Eragrostis von Wolf (Love Grass)

1 Plants perennial 2
  2 Sheaths glabrous . . . . **E. lugens**
  2 Sheaths long pubescent 3
    3 Panicle purple and about as long as broad; spikelets with 6 or more florets . . . . **E. spectabilis**
    3 Panicle not purple; spikelets mostly with fewer than 6 florets . . . . **E. hirsuta**
1 Plants annual 4
  4 Plants rooting at the nodes, forming mats . . . . **E. hypnoides**
  4 Not as above 5
    5 Spikelets mostly with 2–5 florets 6
      6 Culms erect; panicle long and diffuse . . . . **E. capillaris**
      6 Culms spreading . . . . **E. frankii**
    5 Spikelets mostly more than 5-flowered 7

7 Glumes and lemmas glandular on keel 8
8 Lemmas 2–2.5 mm long . . . . **E. cilianensis**
8 Lemmas less than 2 mm long . . . . **E. poaeoides**
7 Glumes and lemmas not glandular on keel 9
9 Spikelets about 1 mm wide . . . . **E. pilosa**
9 Spikelets about 1.5 mm wide . . . . **E. pectinacea**

E. capillaris (L.) Nees; abandoned fields and margins; frequent; Jul–Oct; NC, SC, TN, VA.

E. cilianensis (All.) Mosher*; old fields and margins; frequent; Jul–Oct; NC, SC, TN, VA.

E. frankii C. A. Mey. ex Steud.; waste areas and roadsides; infrequent; Aug–Sep; VA.

E. hirsuta (Michx.) Nees; dry open woods, old fields, and margins; occasional; Jul–Oct; NC, SC, TN, VA.

E. hypnoides (Lam.) BSP.; low to wet mostly open areas; occasional; Jun–Sep; NC, TN, VA.

E. lugens Nees; low old fields, ditches, and other wet places; infrequent; Jun–Oct; SC.

E. pectinacea (Michx.) Nees; moist open ground and cultivated areas; infrequent; Jul–Aug; GA, TN.

E. pilosa (L.) Beauv.*; old fields and margins; common; Jul–Oct; NC, SC, TN, VA.

E. poaeoides Beauv. ex R. & S.*; margins and waste places; occasional; Jul–Aug; NC, TN, VA.

E. spectabilis (Pursh) Steud.; sandy old fields and waste places; frequent; Aug–Oct; ALL.

Erianthus Michx.

1 Awn twisted at base 2
2 Panicle densely woolly . . . . **E. alopecuroides**
2 Panicle not densely woolly . . . . **E. contortus**
1 Awn straight; panicle very densely woolly . . . . **E. giganteus**

E. alopecuroides (L.) Ell., Beard Grass; low, wet to moist borders; occasional; Sep–Oct; NC, SC, TN, VA.

E. contortus Baldw., Beard Grass; wet woods and openings; occasional; Jul–Oct; NC, SC, TN.

E. giganteus (Walt.) Muhl., Plume Grass; wet open places, pond margins, and ditches; Sep–Oct; NC, SC, TN, VA.

Festuca L. (Fescue)

1 Plants annual 2
2 Spikelets 6–10-flowered; awns of lemma 1–4 mm long . . . . **F. octoflora**
2 Spikelets 4–5-flowered; awns of lemma 8–14 mm long . . . . **F. myuros**
1 Plants perennial 3
3 Blades flat and 3 mm or more wide 4
4 Spikelets 8–10-flowered; plants robust . . . . **F. elatior**
4 Spikelets with 5 or fewer flowers 5
5 Spikelets crowded, overlapping one another toward ends of panicle branches . . . . **F. paradoxa**
5 Spikelets not crowded as above . . . . **F. obtusa**

3 Blades less than 3 mm wide and sometimes involute 6
6 Basal sheaths soon deteriorating to reddish fibers . . . . **F. rubra**
6 Basal sheaths persisting 7
7 Lemmas awnless . . . . **F. tenuifolia**
7 Lemmas awned . . . . **F. ovina**

F. elatior L.*; fields, margins, and pastures; frequent; May–Jul; NC, SC, TN, VA. A variable species and valuable pasture plant.
F. myuros L.*; dry fields and roadsides; frequent; May–Jun; NC, SC, TN, VA. *Vulpia myuros* (L.) Gmel.
F. obtusa Biehler; moist woods and margins; common; May–Jul; ALL.
F. octoflora Walt.; dry fields and margins; common; Apr–Jul; ALL. *Vulpia octoflora* (Walt.) Rydb.
F. ovina L.*; meadows, margins, and summits; occasional; May–Jul; NC, SC, VA.
F. paradoxa Desv.; rich woods; frequent; May–Jul; GA, NC, SC.
F. rubra L.; margins and openings in high woods and summits; occasional; May–Jul; NC, TN, VA.
F. tenuifolia Sibthorp*; rock outcrops; infrequent; May–Jun; NC, VA.

Glyceria R. Br. (Manna Grass)

1 Spikelets 1 cm or more long, and narrow 2
2 Lemmas 6–8 mm long and long-pointed; paleas longer . . . . **G. acutiflora**
2 Lemmas 4 mm long and rounded . . . . **G. septentrionalis**
1 Spikelets shorter 3
3 Leaf sheaths open . . . . **G. pallida**
3 Sheaths, at least the upper, closed 4
4 Panicle narrow 5
5 Panicle dense and heavily fruited; lemmas 3–4 mm long . . . . **G. obtusa**
5 Panicle not heavily fruited; lemmas 2–2.5 mm long . . . . **G. melicaria**
4 Panicle open, spreading 6
6 Palea showing conspicuously . . . . **G. canadensis**
6 Palea mostly hidden by the lemma 7
7 Lemmas 6–10 mm long . . . . **G. nubigena**
7 Lemmas mostly 2–4 mm long . . . . **G. striata**

G. acutiflora Torr.; wet openings; infrequent; Jun–Jul; VA.
G. canadensis (Michx.) Trin.; wet boggy places at high elevations; infrequent; Jun–Jul; NC, TN, VA. A somewhat taller variety with only 3–5-flowered spikelets may be segregated as var. *laxa* (Schribn.) Hitchc.
G. melicaria (Michx.) Hubbard; wet slopes and ditches; occasional; Jul–Aug; GA, NC, TN, VA.
G. nubigena W. A. Anderson; wet, semiopen slopes, stream sides, and ditches; infrequent; Jun–Jul; endemic to Great Smoky Mountains, NC, TN.
G. obtusa (Muhl.) Trin.; swamps, bogs, and ditches; infrequent; Jun–Sep; VA.
G. pallida (Torr.) Trin.; boggy places; infrequent; Jun–Sep; NC.
G. septentrionalis Hitchc.; wet, mostly open places; infrequent; Jun–Sep; NC, VA.
G. striata (Lam.) Hitchc.; meadows and ditch banks; frequent; Apr–Jun; ALL.

Gymnopogon Beauv.

G. ambiguus (Michx.) BSP.; dry open woods and abandoned fields; frequent; Aug–Oct; NC, SC, TN, VA.

Holcus L. (Velvet Grass)

H. lanatus L.*; roadsides, waste places, pastures; frequent; May–Oct; ALL.

Hordeum L. (Barley)

1 Glume (including awns) up to 15 mm long .... **H. pusillum**
1 Glume (including awns) up to 7 cm long .... **H. jubatum**

H. jubatum L., Foxtail B.; meadows and waste areas; infrequent; May–Aug; VA.
H. pusillum Nutt., Little B.; roadsides and waste places; frequent; Apr–Jun; NC, SC, TN, VA.

Hystrix Moench (Bottlebrush Grass)

H. patula Moench; moist woods and margins; frequent; May–Jul; NC, SC, TN, VA.

Leersia Swartz (Cut Grass)

1 Panicle well developed; lower branches opposite or whorled .... **L. oryzoides**
1 Panicle weakly developed; branches arising singly .... **L. virginica**

L. oryzoides (L.) Swartz; low open areas and ditches; frequent; Aug–Oct; ALL.
L. virginica Willd.; ditches, marshes, and other wet semiopen areas; frequent; Aug–Oct; ALL.

Leptochloa Beauv. (Sprangletop)

L. filiformis (Lam.) Beauv.; fields, gardens, and other moist open places; infrequent; Jun–Oct; NC, SC, TN.

Lolium L. (Rye Grass)

L. perenne L.*; waste ground and margins; Apr–Jun; NC, SC, TN, VA.

Melica L. (Melic Grass)

1 Spikelets 2-flowered .... **M. mutica**
1 Spikelets 3-flowered .... **M. nitens**

M. mutica Walt.; dry woods; occasional; Apr–May; NC, SC, TN, VA.
M. nitens (Schribn.) Nutt. ex Piper; mountain forests; infrequent; May–Jun; NC.

Microstegium Nees

M. vimineum (Trin.) Camus*; a rapidly spreading pernicious invader on moist ground; too common; Sep–Oct; ALL.

Miscanthus Andersson

M. sinensis Andersson*; frequently spreading to moist woodland margins and roadsides; Sep–Nov; ALL.

Milium L.

M. effusum L.; rich moist mountain woods; infrequent; May–Jul; NC, TN, VA. More common north of our area.

Muhlenbergia Schreb.

1 Culms more or less tufted; rhizomes absent 2
  2 Culms decumbent, rooting at the nodes; panicle narrow . . . . **M. schreberi**
  2 Culms erect, not rooting at the nodes; 6–10 dm tall; panicle very diffuse . . . . **M. capillaris**
1 Rhizomes present 3
  3 Panicle densely flowered 4
    4 Glumes long-awned, awn length exceeding that of glume body; lemmas awnless . . . . **M. glomerata**
    4 Glumes with awns shorter than body of glume 5
      5 Lemmas long-awned . . . . **M. sylvatica**
      5 Lemmas awnless or nearly so 6
        6 Culms glabrous below the nodes . . . . **M. frondosa**
        6 Culms pubescent below the nodes . . . . **M. mexicana**
  3 Panicle not densely flowered 7
    7 Culms weakly erect and freely branching; panicle very slender . . . . . . **M. bushii**
    7 Culms erect and little-branched 8
      8 Lemmas long-awned . . . . **M. tenuiflora**
      8 Lemmas awnless or with only very short awns . . . . **M. sobolifera**

M. bushii Pohl; low moist woods; infrequent; Aug–Sep; GA.

M. capillaris (Lam.) Trin.; moist to dry old fields and borders; occasional; Aug–Sep; ALL.

M. frondosa (Poir.) Fern.; moist woods and margins; occasional; Sep–Oct; GA, NC, TN, VA.

M. glomerata (Willd.) Trin.; bogs and swamps; infrequent; Sep–Oct; NC, VA.

M. mexicana (L.) Trin.; roadsides and disturbed areas; infrequent; Sep–Oct; NC, VA.

M. schreberi J. F. Gmel.; moist and shaded places; frequent; Aug–Oct; ALL.

M. sobolifera (Willd.) Trin.; moist woods; infrequent; Jul–Sep; NC, TN, VA.

M. sylvatica (Torr.) Torr. ex Gray; moist woods and banks; occasional; Sep–Oct; GA, NC, TN, VA.

M. tenuiflora (Willd.) BSP.; moist woods, borders, and pastures; frequent; Aug–Oct; ALL.

Oryzopsis Michx. (Rice Grass)

O. racemosa (Sm.) Ricker; dry woods and borders; infrequent; Jul–Aug; VA.

Panicum L. (Panic Grass)

Contributed by M. G. Lelong

Gould, F. W., and C. A. Clark. 1978. *Dichanthelium* (Poaceae) in the United States and Canada. *Ann. Missouri Bot. Gard.* 65:1088–1132.

1 Plants annual or perennial, lacking an overwintering basal rosette of leaves or a dense overwintering basal cushion of leaves (Subgenus *Panicum*) . . . . **Key A**

1 Plants perennial, producing an overwintering rosette of leaves with short, broad blades or a dense overwintering basal cushion of leaves (Subgenus *Dichanthelium*) ... **Key B**

KEY A

1 Plants annual ... 2
  2 Spikelets verrucose or tuberculate ... **P. verrucosum**
  2 Spikelets not verrucose nor tuberculate ... 3
    3 Sheaths glabrous ... **P. dichotomiflorum**
    3 Sheaths more or less densely pubescent ... 4
      4 Spikelets 1.7–2.5 mm long; primary panicle usually more than 0.5× as broad as long ... **P. capillare**
      4 Spikelets 2.7–3.5 mm long; primary panicle about 0.5× as broad as long ... **P. flexile**
1 Plants perennial ... 5
  5 Terminal panicle narrow, usually less than 2 cm broad with few subsessile spikelets ... **P. hemitomon**
  5 Terminal panicle more than 2 cm broad with numerous pedicellate spikelets ... 6
    6 Sterile paleas greatly inflated and indurate at maturity; culms slender, densely tufted ... **P. hians**
    6 Sterile paleas not inflated at maturity; culms robust, tufted, clumped, or solitary ... 7
      7 Plant with short, hard, knotty bases or caudexes; lower sheaths laterally compressed, keeled, subequitant ... **P. rigidulum**
      7 Plants with stout, elongate, scaly rhizomes; lower sheaths terete, not keeled nor subequitant ... 8
        8 Spikelets subsecund, often gaping and falcate, on short, appressed pedicels ... **P. anceps**
        8 Spikelets not as above, usually on long flexuous pedicels ... **P. virgatum**

KEY B

1 Plants producing dense basal cushion of leaves by extensive axillary branching from lower nodes and relative lack of elongation of lower internodes ... 2
  2 Blades elongate, linear, fairly rigid, up to 20 cm long and 5 mm wide, usually more than 20× as long as wide ... 3
    3 Spikelets 2.8–4 mm long, beaked or acuminate ... **P. depauperatum**
    3 Spikelets 2.3–2.9 mm long, obtuse or acute ... **P. linearifolium**
  2 Blades lanceolate, thin, up to 15 cm long and 12 mm wide, usually much less than 20× as long as wide ... 4
    4 Sheaths with fine, long, spreading or retrorse hairs; spikelets 1.7–2.3 mm long, pustulose-pubescent ... **P. laxiflorum**
    4 Sheaths glabrous; spikelets 1.1–1.6 mm long, glabrous ... **P. strigosum**
1 Plants producing a distinct overwintering rosette of short, broad blades usually unlike cauline blades; eventually branching at least somewhat from upper nodes ... 5
  5 Spikelets 0.8–1.9 mm long ... 6
    6 Ligules actually or apparently 1–5 mm long ... 7

7 Blades small, 1.5–3.5 cm long and 1–4 mm wide; ligules up to 1.5 mm long . . . . . **P. ensifolium**

7 Blades up to 10 cm long and 10 mm wide; ligules up to 5 mm long (Key C, key to varieties) . . . . . **P. acuminatum**

6 Ligules less than 1 mm long or obsolete 8

8 Spikelets subspherical to broadly ellipsoid, puberulent; blades 4–25 mm wide, broadly cordate or subcordate at base 9

9 Cauline blades usually less than 4, 4.5–10 cm long and 5–14 mm wide . . . . . **P. sphaerocarpon**

9 Cauline blades more than 4, 10–23 cm long and 14–25 mm wide . . . . . **P. polyanthes**

8 Spikelets usually ellipsoid or obovoid, glabrous or pubescent; blades 3–14 mm wide, tapering, constricted, or rounded at base 10

10 Blades thin, 4–14 cm long and 3–14 mm wide; spikelets 1.5–2.6 mm long, glabrous or sparsely pubescent (Key D, key to varieties) . . . . . **P. dichotomum**

10 Blades usually stiffly ascending, 2.5 cm long and 1.5–6 mm wide, conspicuously white-margined; spikelets 1.3–1.6 mm long, puberulent . . . . . **P. tenue**

5 Spikelets 1.9–4.7 mm long 11

11 Spikelets 1.9–3.2 mm long 12

12 Cauline blades cordate at base, often more than 11 mm wide 13

13 Plant densely and softly pubescent throughout; nodes densely bearded with lustrous glandular rings below them . . . . . **P. scoparium**

13 Plants mostly glabrous, sparsely pubescent, or puberulent; nodes glabrous or sparsely pubescent 14

14 Culms robust, often more than 7 dm tall; sheaths, at least the lower or axillary ones, papillose-hirsute or papillose; blades usually more than 10 cm long . . . . . **P. clandestinum**

14 Culms usually less than 7 dm tall; sheaths glabrous, sparsely pubescent, or puberulent; blades usually less than 10 cm long . . . . . **P. commutatum**

12 Cauline blades rounded or tapering at base, seldom more than 11 mm wide (except occasionally in *P. scabriusculum*) 15

15 Leaf blades elongate, linear, usually more than 14× as long as wide, often ascending 16

16 Blades thick, often striate above and pleated beneath, usually pubescent at least beneath 17

17 Plant densely grayish-pubescent throughout; nodes bearded . . . . . **P. consanguineum**

17 Plant glabrous or sparsely papillose-pilose; nodes not bearded . . . . . **P. angustifolium**

16 Blades thin, not striate or pleated beneath, essentially glabrous . . . . . **P. bicknellii**

15 Leaf blades wider, often lanceolate, usually 10× as long as wide or less 18

18 Ligules ciliate, 1.5–5 mm long 19

19 Spikelets 1.1–1.9 mm long, seldom longer (Key C, key to varieties) .... **P. acuminatum**

19 Spikelets 2–4.2 mm long 20

20 Sheaths pilose, with spreading or ascending hairs up to 4 mm long; spikelets 2–3 mm long .... **P. ovale** var. **villosum**

20 Sheaths glabrous or papillose-hispid, with ascending hairs less than 2 mm long; spikelets 2.7–4.2 mm long .... **P. oligosanthes**

18 Ligules ciliate or membranous, less than 1.5 mm long, or obsolete 21

21 Culms robust, 7–14 dm tall; blades large, 12–25 cm long .... **P. scabriusculum**

21 Culms usually less than 7 dm tall or, if more, with bearded nodes; blades usually less than 12 cm long 22

22 Culms sparsely ascending-pilose or strigose; blades mostly ascending, often with whitish, scaberulous margins .... **P. ovale** var. **pseudopubescens**

22 Culms glabrous or puberulent; blades often spreading to reflexed or ascending, usually without whitish margins 23

23 Spikelets 2.7–3.5 mm long, obovoid; sheaths papillose-hispid or papillose-pilose .... **P. oligosanthes**

23 Spikelets 1.5–2.7 mm long, ellipsoid or asymmetrically obovoid; sheaths glabrous or puberulent 24

24 Culms wiry, often purple and puberulent, up to 5 dm tall; spikelets asymmetrically pyriform, usually puberulent .... **P. portoricense**

24 Culms robust or weak, usually glabrous, up to 10 dm tall; spikelets mostly ellipsoid or obovoid, usually glabrous (Key D, key to varieties) .... **P. dichotomum**

11 Spikelets 3.2–4.7 mm long 25

25 Spikelets more than 3.7 mm long 26

26 Blades broadly cordate at base, 10–35 mm wide 27

27 Sheaths glabrous, pilose, or puberulent; nodes bearded with long, retrorse hairs .... **P. boscii**

27 Sheaths densely papillose-hispid or papillose-pilose; nodes bearded with short, tangled hairs .... **P. ravenelii**

26 Blades tapering at base or subcordate, 5–15 mm wide .... **P. oligosanthes**

25 Spikelets less than 3.7 mm long 28

28 Ligules ciliate, 1–3 mm long; blades usually less than 10 mm wide .... **P. oligosanthes**

28 Ligules ciliate, erose, or obsolete, less than 1 mm long; blades usually 10–35 mm wide 29

29 Sheaths, at least the lowermost and the axillary ones, papillose-hispid or papillose-pilose .... **P. clandestinum**

29 Sheaths glabrous, sparsely pilose, or puberulent 30

30 Spikelets 3.2–3.8 mm long .... **P. latifolium**
30 Spikelets usually less than 3.2 mm long ...... **P. commutatum**

KEY C (Varieties of *P. acuminatum*)

1 Culms and sheaths densely and variously pubescent 2
2 Culms and sheaths densely spreading-villous and occasionally also inconspicuously puberulent; blades softly pubescent beneath .... var. **acuminatum**
2 Culms and sheaths ascending to spreading papillose-pilose or densely puberulent with scattered ascending trichomes; blades densely to sparsely appressed-pilose or puberulent beneath 3
3 Sheaths and culms ascending to spreading papillose-pilose, occasionally also puberulent underneath 4
4 Blades usually more than 6 mm wide, spreading, short-pilose to nearly glabrous above; spikelets 1.5–1.8 mm long .... var. **fasciculatum**
4 Blades usually less than 6 mm wide, ascending to reflexed, long-pilose; spikelets 1.3–1.6 mm long .... var. **unciphyllum**
3 Sheaths and culms densely puberulent, often with scattered long trichomes intermixed .... var. **columbianum**
1 Culms and sheaths usually glabrous, or at least lowermost sparsely pubescent 5
5 Blades often yellowish-green, conspicuously papillose-ciliate at base; spikelets 1.3–1.6 mm long, usually obovoid .... var. **lindheimeri**
5 Blades often dark green or purplish, less conspicuously ciliate at base; spikelets 1.1–1.5 mm long, usually ellipsoid .... var. **longiligulatum**

KEY D (Varieties of *P. dichotomum*)

1 Blades at midculm seldom over 7 mm wide, often narrowed or constricted at base .. 2
2 Culms stiff, erect, terete, often purplish .... var. **dichotomum**
2 Culms weak, reclining or sprawling, usually flattened, pale green .. var. **lucidum**
1 Blades at midculm 7–14 mm wide, usually not constricted at base or subcordate ... 3
3 Nodes densely bearded with retrorse hairs; spikelets pubescent or glabrous, obtuse or subacute 4
4 Spikelets 1.5–1.8 mm long, glabrous .... var. **ramulosum**
4 Spikelets 1.8–2.5 mm long, pubescent 5
5 Sheaths and blades essentially glabrous .... var. **nitidum**
5 Sheaths and blades more or less densely covered with soft or velvety pubescence .... var. **mattamuskeetense**
3 Nodes glabrous or sparsely pubescent; spikelets glabrous, acute or faintly beaked .... var. **yadkinense**

P. acuminatum Swartz var. **acuminatum**; moist, open sandy areas; infrequent; May–Sep; GA, SC. *P. lanuginosum* Ell. var. *lanuginosum*, *Dichanthelium acuminatum* (Swartz) Gould & Clark var. *acuminatum*, in part. This species is probably the most polymorphic and troublesome one in the genus. The varieties recognized are usually separable, but occasional intergrading specimens occur.

P. acuminatum Swartz var. **columbianum** (Scribner) Lelong; dry, open sandy areas; occasional; May–Sep; NC, TN, VA. *P. columbianum* Nash, *P. tsugetorum* Nash, *Di-*

chanthelium sabulorum (Lam.) Gould & Clark var. *patulum* (Scribn. & Merr.) Gould & Clark, in part, *D. sabulorum* (Lam.) Gould & Clark var. *thinium* (Hitchc. & Chase) Gould & Clark, in part.

P. acuminatum Swartz var. **fasciculatum** (Torr.) Lelong; various habitats, especially dry open woods, roadsides, and disturbed areas; common; May–Sep; ALL. *P. glutinoscabrum* Fern., *P. huachucae* Ashe, *P. tennesseense* (Ashe) Gleason, *P. lanuginosum* Ell. var. *fasciculatum* (Torr.) Fern., *Dichanthelium acuminatum* (Swartz) Gould & Clark var. *acuminatum*, in part. This variety is possibly the most common, variable, and ubiquitous one in this perplexing complex.

P. acuminatum Swartz var. **lindheimeri** (Nash) Lelong; moist to dry open sites, often in clay or calcareous soil; occasional; May–Sep; NC, TN, VA. *P. lindheimeri* Nash, *P. lanuginosum* Ell. var. *lindheimeri* (Nash) Fern., *Dichanthelium acuminatum* (Swartz) Gould & Clark var. *lindheimeri* (Nash) Gould & Clark.

P. acuminatum Swartz var. **longiligulatum** (Nash) Lelong; moist to wet, sandy open areas, bogs; infrequent and more common on the Coastal Plain; May–Sep; GA. *P. longiligulatum* Nash, *Dichanthelium acuminatum* (Swartz) Gould & Clark var. *longiligulatum* (Nash) Gould & Clark.

P. acuminatum Swartz var. **unciphyllum** (Trinius) Lelong; dry rocky or sandy woodlands, alluvial woods; occasional; May–Sep; ALL. *P. albemarlense* Ashe, *P. implicatum* Scribn., *P. meridionale* Ashe, *P. oricola* Hitchc. & Chase, *P. lanuginosum* Ell. var. *implicatum* (Scribn.) Fern., *Dichanthelium acuminatum* (Swartz) Gould & Clark var. *implicatum* (Scribn.) Gould & Clark, in part.

P. anceps Michx.; moist to dry woodlands, roadsides, waste places, often on sandy soils; common; Jun–Oct; ALL.

P. angustifolium Ell.; sandy open pine-oak woods, roadsides; infrequent; May–Sep; GA, NC, SC, TN. *Dichanthelium aciculare* (Desv. ex Poir.) Gould & Clark, in part.

P. bicknellii Nash; dry open woods; infrequent; May–Sep; NC, VA. Included by Gould & Clark in *Dichanthelium boreale* (Nash) Freckmann. This dubious and poorly defined species resembles *P. dichotomum* var. *yadkinense*, *P. laxiflorum* and subglabrous forms of *P. linearifolium* treated by some as *P. werneri* Scribn.

P. boscii Poir.; mesic to dry, often rocky woods; common; May–Sep; ALL. *P. boscii* Poir. var. *molle* (Vasey) Hitchc. & Chase, *Dichanthelium boscii* (Poir.) Gould & Clark. Most specimens of this widespread and variable species are more or less densely pubescent.

P. capillare L. var. **campestre** Gattinger; moist to dry open habitats, often on disturbed ground, limestone, sandstone and granite outcrops; frequent; Sep–Nov; NC, SC, TN, VA. *P. gattingeri* Nash. Blades 5–10 mm wide; terminal panicles usually green and less than 0.5× the length of the culms; spikelets numerous, green, 1.9–2.5 mm long, acute.

P. capillare L. var. **capillare**; habitats similar to preceding; occasional; Sep–Nov; NC, SC, VA. Blades 6–15 mm wide; terminal panicles often purple and 0.5× or more the length of the culms; spikelets often purple and long-acuminate.

P. capillare L. var. **sylvaticum** Torr.; habitats similar to preceding; frequent; Sep–Nov; GA, NC, TN, VA. *P. philadelphicum* Bernh., *P. lithophilum* Swallen. Blades 2–7 mm wide, ascending; terminal panicles with few spikelets often paired near branch tips; spikelets purplish, 1.7–2 mm long, acute.

P. clandestinum L.; moist, often sandy open woods, thickets, roadsides; common; May–

Sep; ALL. *Dichanthelium clandestinum* (L.) Gould. Some specimens of this common species closely resemble *P. latifolium,* which is usually less robust, has somewhat larger spikelets, and occurs in well-drained mesic woods.

P. commutatum Schultes var. **ashei** Fern.; often on drier, more open, disturbed sites than the following; common; May–Sep; ALL. *P. ashei* Pearson. Culms and sheaths usually densely puberulent, blades thickish, less than 10 mm wide, spikelets often less than 2.7 mm long.

P. commutatum Schultes var. **commutatum**; moist or dry woods, thickets, roadsides; common; May–Sep; ALL. *P. joorii* Vasey, *P. mutabile* Scribn. & Smith, *P. commutatum* Schultes var. *joorii* (Vasey) Fern., *Dichanthelium commutatum* (Schultes) Gould.

P. consanguineum Kunth; low sandy pinelands, savannahs, bogs; infrequent and more common on the Coastal Plain; May–Sep; VA. *Dichanthelium consanguineum* (Kunth) Gould & Clark.

P. depauperatum Muhl.; dry, sandy, or rocky open woodlands, waste ground; occasional; May–Sep; GA, NC, SC, VA. *Dichanthelium depauperatum* (Muhl.) Gould.

P. dichotomiflorum Michx.; moist, open, mostly disturbed areas, shores, fields, roadsides; common; Jun–Oct; ALL. This species varies greatly in size; robust intergrading specimens with wide blades have been treated by some as var. *geniculatum* (Wood) Fern.

P. dichotomum L. var. **dichotomum**; mesic to dry woods; common; May–Sep; ALL. *P. barbulatum* Michx., *P. dichotomum* L. var. *barbulatum* (Michx.) Wood, *Dichanthelium dichotomum* (L.) Gould var. *dichotomum*. This species is almost as variable and troublesome taxonomically as *P. acuminatum*. The six varieties recognized are usually separable, but occasional specimens intergrade or exhibit features of other species.

P. dichotomum L. var. **lucidum** (Ashe) Lelong; wet woods, sphagnum bogs; infrequent; May–Sep; GA, NC, VA. *P. lucidum* Ashe, *P. sphagnicola* Nash, *P. lucidum* Ashe var. *opacum* Fern., *Dichanthelium dichotomum* (L.) Gould var. *dichotomum*, in small part.

P. dichotomum L. var. **mattamuskeetense** (Ashe) Lelong; moist to wet sites, bogs; infrequent; May–Sep; VA. *P. annulum* Ashe, *P. clutei* Nash, *P. mattamuskeetense* Ashe, *P. mattamuskeetense* Ashe var. *clutei* (Nash) Fern., *Dichanthelium dichotomum* (L.) Gould var. *dichotomum*, in small part. This dubious variety is very similar to the next two and intergrades with them.

P. dichotomum L. var. **nitidum** (Lam.) Woods; moist to wet areas, swamps; infrequent; May–Sep; GA, VA. *P. nitidum* Lam., *Dichanthelium dichotomum* (L.) Gould var. *dichotomum*, in small part.

P. dichotomum L. var. **ramulosum** (Torr.) Lelong; moist to wet woods, swamps, thickets; common; May–Sep; ALL. *P. microcarpon* Muhl., *P. nitidum* Lam. var. *ramulosum* Torr., *Dichanthelium dichotomum* (L.) Gould var. *dichotomum*, in part.

P. dichotomum L. var. **yadkinense** (Ashe) Lelong; moist to wet woods, thickets; infrequent; May–Sep; GA, NC, VA. *P. yadkinense* Ashe, *Dichanthelium dichotomum* (L.) Gould var. *dichotomum*, in small part. This uncommon variety grades into var. *ramulosum* and var. *dichotomum*. It also exhibits features of *P. commutatum* and *P. laxiflorum*.

P. ensifolium Bald. ex. Ell. var. **curtifolium** (Nash) Lelong; moist to wet places such as bogs, pine savannahs; rare outside of Coastal Plain; May–Sep; NC. *P. curtifolium* Nash, *Dichanthelium acuminatum* (Swartz) Gould & Clark var. *implicatum* (Scribn.)

Gould & Clark, in small part. Sheaths sparsely spreading-pilose; blades often sparsely pilose or glabrous on both surfaces.

P. ensifolium Bald. ex Ell. var. **ensifolium**; habitats similar to preceding; May–Sep; NC. *P. vernale* Hitchc. & Chase, *Dichanthelium dichotomum* (L.) Gould var. *ensifolium* (Bald.) Gould & Clark. Sheaths glabrous, blades usually puberulent beneath and glabrous above.

P. flexile (Gattinger) Scribner; moist to dry open areas, woodlands, cedar glades, limestone outcrops, meadows; infrequent; Jul–Oct; GA, NC, TN, VA. This species often resembles *P. capillare* var. *sylvaticum* from which it can be distinguished by its longer, acuminate spikelets on long pedicels.

P. hemitomon Schultes; marshy shores of lakes, ponds, and streams; infrequent and more common on the Coastal Plain; Jul–Aug; VA.

P. hians Ell.; wet open areas; infrequent and more common on the Coastal Plain; Jul–Sep; TN.

P. latifolium L.; mesic to dryish, sandy or rocky woods, thickets; frequent, especially in the northern part of our area; May–Sep; GA, NC, TN, VA. *Dichanthelium latifolium* (L.) Gould & Clark. Occasional specimens grade into *P. boscii*, others into *P. clandestinum*.

P. laxiflorum Lam.; moist or dryish woods, usually in shade; occasional; May–Sep; ALL. *P. xalapense* HBK., *Dichanthelium laxiflorum* (Lam.) Gould.

P. linearifolium Scribner; dry sandy or rocky woods, glades; occasional; May–Sep; NC, VA. *P. werneri* Scribner, *Dichanthelium linearifolium* (Scribner) Gould. This species intergrades somewhat with *P. depauperatum* and possibly also with narrow-blade forms of *P. laxiflorum*, occasionally approaching the dubious *P. bicknellii*.

P. oligosanthes Schultes; dry open sandy or clayey woods, glades, prairies, waste places; infrequent; May–Sep; VA. Represented in our area by var. **scribnerianum** (Nash) Fern. *P. scribnerianum* Nash, *Dichanthelium oligosanthes* (Schult.) Gould var. *scribnerianum* (Nash) Gould.

P. ovale Ell. var. **pseudopubescens** (Nash) Lelong; dry, open sandy woods, openings; occasional; May–Sep; GA, NC, TN, VA. *P. addisonii* Nash, *P. commonsianum* Ashe, *P. pseudopubescens* Nash, *P. commonsianum* Ashe var. *addisonii* (Nash) Fern., *P. villosissimum* Nash var. *pseudopubescens* (Nash) Fern., *Dichanthelium ovale* (Ell.) Gould & Clark var. *addisonii* (Nash) Gould & Clark.

P. ovale Ell. var. **villosum** (A. Gray) Lelong; same habitats as the preceding; frequent; May–Sep; ALL. *P. villosissimum* Nash, *Dichanthelium acuminatum* (Swartz) Gould & Clark var. *villosum* (A. Gray) Gould & Clark.

P. polyanthes Schultes; low woods, ditches, openings; common; May–Sep; ALL. *Dichanthelium sphaerocarpon* (Ell.) Gould var. *isophyllum* (Scribn.) Gould & Clark. This species is closely related to *P. sphaerocarpon*, and occasional intergrading specimens occur.

P. portoricense Desv. ex Hamilt.; sandy, often moist woods; infrequent and more common on the Coastal Plain; May–Sep; GA. Represented in our area by var. **nashianum** (Scribner) Lelong. *P. lancearium* Trin., *P. patulum* (Scribn. & Merr.) Hitchc., *P. lancearium* Trin. var. *patulum* (Scribn. & Merr.) Fern., *Dichanthelium sabulorum* (Lam.) Gould & Clark var. *patulum* (Scribn. & Merr.) Gould & Clark.

P. ravenelii Scribner & Merrill; dry sandy woods, openings; infrequent; May–Sep; GA, SC, VA. *Dichanthelium ravenelii* (Scribn. & Merr.) Gould.

P. rigidulum Bosc ex Nees var. **elongatum** (Scribner) Lelong; marshes, low woods, stream banks, and other moist to wet habitats; occasional; Jul–Oct; ALL. *P. stipitatum* Nash, *P. agrostoides* Spreng. var. *elongatum* (Pursh) Scribner. Blades and ligules similar to the typical variety; spikelets 2.4–3 mm long, less than 0.6 mm wide, usually purple, stipitate.

P. rigidulum Bosc ex Nees var. **pubescens** (Vasey) Lelong; moist, often sandy areas; infrequent; Jul–Oct; GA, TN, VA. *P. longifolium* Torrey, *P. longifolium* Torrey var. *pubescens* (Vasey) Fern. Blades 2–7 mm wide, often pilose above at base; ligules usually fimbriate, 0.5–3 mm long; spikelets 2–2.7 mm long, often obliquely set on pedicels.

P. rigidulum Bosc ex Nees var. **rigidulum**; habitats similar to var. *elongatum*; infrequent; Jul–Oct; GA, NC, VA. *P. agrostoides* Spreng., *P. condensum* Nash. Blades 5–12 mm wide, mostly glabrous; ligules membranous, 0.2–1 mm long; spikelets 1.6–2.5 mm long, usually more than 0.6 mm wide, green or purple-tinged.

P. scabriusculum Ell.; wet, sandy open sites, bogs; infrequent and more common on the Coastal Plain; May–Sep; GA. *Dichanthelium scabriusculum* (Ell.) Gould & Clark.

P. scoparium Lam.; moist to wet open sandy areas; infrequent; May–Sep; GA, NC, TN, VA. *Dichanthelium scoparium* (Lam.) Gould. This robust, predominantly Coastal Plain grass is perhaps the most distinctive species of the subgenus.

P. sphaerocarpon Ell.; dry, open sandy woods, openings; frequent; May–Sep; ALL. *Dichanthelium sphaerocarpon* (Ell.) Gould var. *sphaerocarpon*. This species is closely related to *P. polyanthes*, and occasional specimens are intermediate between these two taxa.

P. strigosum Muhl.; moist sandy woods, bogs; infrequent and more common on the Coastal Plain; May–Sep; VA. *Dichanthelium leucoblepharis* (Trin.) Gould & Clark var. *pubescens* (Vasey) Gould & Clark, *D. strigosum* (Muhl.) Freckmann var. *strigosum*.

P. tenue Muhl.; moist to dry sandy areas, bogs; infrequent; May–Sep; GA, TN. *P. albomarginatum* Nash, *P. trifolium* Nash, *Dichanthelium dichotomum* (L.) Gould var. *tenue* (Muhl.) Gould & Clark. This predominantly Coastal Plain species exhibits characteristics of *P. dichotomum*, *P. ensifolium*, and *P. sphaerocarpon*.

P. verrucosum Muhl.; moist to wet sandy, open areas; occasional; Sep–Oct; ALL.

P. virgatum L.; moist to dry open areas such as marshes, prairies, open woods; occasional; Jul–Oct; GA, NC, VA.

Paspalum L.

1 Racemes mostly 2 and close together in inflorescence .... **P. distichum**
1 Racemes several and alternate, or if 1, then terminal 2
  2 Rather succulent annual; fertile lemmas dark brown .... **P. boscianum**
  2 Perennials with rhizomes 3
    3 Spikelets with long silky hairs on margins .... **P. dilatatum**
    3 Not as above 4
      4 Culms robust, to 2 m tall; spikelets about 4 mm long; plant bluish-green .... **P. floridanum**
      4 Not as above 5
        5 Spikelets arising singly .... **P. laeve**
        5 Spikelets arising in pairs 6

6 Spikelets about 1.5 mm long .... **P. setaceum**
6 Spikelets about 3 mm long .... **P. pubiflorum**

P. boscianum Flugge; moist to wet open places; occasional; Jul–Oct; GA, SC, TN, VA.

P. dilatatum Poir.*, Dallas Grass; moist open areas; frequent; May–Oct; NC, SC, TN, VA.

P. distichum L.; wet or swampy places; infrequent; Jun–Aug; NC.

P. floridanum Michx.; low margins and cleared rights-of-way; occasional; Aug–Oct; GA, NC, SC, VA.

P. laeve Michx.; old fields, roadsides, and openings; frequent; Jun–Aug; ALL.

P. pubiflorum Rupr. ex Fourn.; moist roadsides; infrequent; Sep–Oct; NC, TN, VA.

P. setaceum Michx.; old fields, margins, and waste places; frequent; Jun–Sep; ALL.

Phalaris L. (Canary Grass)

P. arundinacea L., Reed C. G.; low fields, meadows, and roadsides; occasional; Jun–Aug; NC, TN, VA.

Phleum L. (Timothy)

P. pratense L.*; fields, margins, and meadows, often planted for hay; frequent; Jun–Oct; ALL.

Poa L. (Blue Grass)

1 Annuals 2
2 Lemmas without cobwebby hairs at base .... **P. annua**
2 Lemmas with cobwebby hairs at base .... **P. chapmaniana**
1 Perennials 3
3 Culms strongly flattened throughout .... **P. compressa**
3 Culms terete 4
4 Rhizomes present 5
5 Lowest panicle branches commonly in whorls of 5 .... **P. pratensis**
5 Lowest panicle branches commonly opposite .... **P. cuspidata**
4 Rhizomes absent 6
6 Lemmas with pubescence other than woolly hairs at base 7
7 Lemmas with pubescence only on midveins toward base .... **P. alsodes**
7 Lemmas with pubescence on both lateral and midveins, especially toward base 8
8 Lemmas about 4 mm long 9
9 Leaf blades 2–3 mm wide .... **P. autumnalis**
9 Leaf blades narrower .... **P. wolfii**
8 Lemmas 2–3.5 mm long 10
10 Culms decumbent, flattened, and purplish at base; ligules 3.5 mm long .... **P. palustris**
10 Culms erect; ligules 1 mm long .... **P. sylvestris**
6 Lemmas glabrous 11
11 Sheaths glabrous 12
12 Lemmas pointed at tips .... **P. saltuensis**
12 Lemmas rounded at tips .... **P. languida**
11 Sheaths scabrous; culms decumbent at base .... **P. trivialis**

P. alsodes Gray; high woods and summits; infrequent; May–Jun; NC, TN, VA.
P. annua L.*; fields, roadsides, margins, and yards; common; Mar–May; NC, SC, TN, VA.
P. autumnalis Muhl. ex Ell.; moist wooded slopes; frequent; Apr–May; NC, SC, TN, VA. An inappropriate specific epithet for a spring-flowering plant.
P. chapmaniana Scribn.; margins and waste areas; frequent; Apr–May; NC, SC, TN, VA.
P. compressa L.*; roadsides, meadows, and grasslands; occasional; May–Aug; ALL.
P. cuspidata Nutt.; rocky and moist wooded slopes; occasional; Mar–Apr; ALL.
P. languida Hitchc.; dry open woods; infrequent; Apr–May; VA.
P. palustris L.; moist banks and meadows; infrequent; Jun–Jul; NC, TN, VA.
P. pratensis L., Kentucky Blue Grass; borders, meadows, and lawns; common; Apr–Aug; ALL.
P. saltuensis Fern. & Wiegand; open woodlands; infrequent; Jul–Aug; NC, TN?, VA.
P. sylvestris Gray; rich moist woods; occasional; Apr–May; NC, SC, TN, VA.
P. trivialis L.*; moist woods and clearings; infrequent; Apr–Jun; NC, TN, VA.
P. wolfii Scribn.; moist woods and borders; infrequent; Jun–Jul; VA.

Sacciolepis Nash

S. striata (L.) Nash; wet open margins; occasional; Jul–Oct; NC, SC, TN.

Setaria Beauv. (Foxtail Grass)

1 Annuals 2
  2 Panicles heavy and lobed .... **S. italica**
  2 Panicles smaller and cylindric 3
    3 Bristles 1–3 below each spikelet 4
      4 Spikelets 2.5–3 mm long .... **S. faberi**
      4 Spikelets 1.8–2 mm long .... **S. viridis**
    3 Bristles 5 or more below each spikelet .... **S. glauca**
1 Perennial from short, knotty rhizomes .... **S. geniculata**

S. faberi W. Herrm.*; margins and waste areas; frequent; Jul–Oct; NC, SC, TN, VA.
S. geniculata (Lam.) Beauv.; fields, margins, and roadsides; frequent; Jun–Oct; NC, SC, TN, VA.
S. glauca (L.) Beauv.*; fields and margins; frequent; Jul–Oct; NC, SC, TN, VA.
S. italica (L.) Beauv.*, Foxtail Millet; field margins; infrequent; Jul–Aug; NC, VA.
S. viridis (L.) Beauv.*; margins and waste areas; occasional; Jul–Oct; NC, SC, TN, VA.

Sorghastrum Nash (Indian Grass)

1 Awn brown and more than 2 cm long .... **S. elliottii**
1 Awn yellow and shorter .... **S. nutans**

S. elliottii (Mohr) Nash; dry woods and clearings; occasional; Sep–Oct; NC, SC, TN.
S. nutans (L.) Nash; dry woods and clearings; occasional; Sep–Oct; ALL.

Sorghum Moench (Johnson Grass)

S. halepense (L.) Pers.*; a too common weed in fields and waste areas; May– Oct; ALL.

Spartina Schreb. (Cord Grass)

S. pectinata Link, Slough Grass; wet open banks and swamps; infrequent; Jul–Sep; NC, VA.

Sphenopholis Scribn. (Oat Grass)

1 Panicle dense and spikelike . . . . **S. obtusata**
1 Panicle loose and weak or nodding 2
  2 Second lemma awnless or nearly so; spikelets 3–3.5 mm long; sheaths usually pubescent . . . . **S. nitida**
  2 Second lemma awned; spikelets 5–7 mm long; sheaths glabrous or slightly scabrous . . . . **S. pensylvanica**

S. nitida (Biehler) Scribn.; moist woods; frequent; Apr–Jun; NC, SC, TN, VA.

S. obtusata (Michx.) Scribn.; moist open margins; occasional; Apr–May; NC, SC, TN, VA. *S. intermedia* (Rydb.) Rydb.

S. pensylvanica (L.) Hitchc.; moist banks and woodlands; frequent; Apr–Jun; ALL. *Trisetum pensylvanicum* (L.) Beauv. ex R. & S.

Sporobolus R. Br. (Dropseed Grass)

1 Annuals; panicles mostly included in leaf sheaths 2
  2 Lemmas pubescent . . . . **S. vaginiflorus**
  2 Lemmas glabrous . . . . **S. neglectus**
1 Perennials 3
  3 Panicles mostly included in leaf sheaths . . . . **S. clandestinus**
  3 Panicles well exserted 4
    4 Panicles narrow and spikelike . . . . **S. indicus**
    4 Panicles open . . . . **S. heterolepis**

S. clandestinus (Biehler) Hitchc.; dry open woods and borders; occasional; Sep–Oct; VA.

S. heterolepis (Gray) Gray; dry pinewoods; infrequent; Aug–Sep; NC.

S. indicus (L.) R. Br.*, Smut Grass; yards, margins, and waste places; frequent; Jul–Oct; NC, TN, VA. *S. poiretii* (R. & S.) Hitchc.

S. neglectus Nash; dry or sandy, mostly open areas; occasional; Aug–Sep; TN, VA.

S. vaginiflorus (Torr. ex Gray) Wood; margins and clearings; occasional; Sep–Oct; NC, TN, VA.

Stipa L. (Needle Grass)

S. avenacea L.; dry woods; frequent; Apr–Jun; ALL.

Tridens R. & S. (Purpletop)

T. flavus (L.) Hitchc.; margins and openings; common; Jul–Oct; NC, SC, TN, VA.

Tripsacum L. (Gamma Grass)

T. dactyloides (L.) L.; moist banks, open woods, and margins; occasional; Jun–Nov; NC, SC, TN, VA.

Trisetum Pers.

T. triflorum (Bigelow) Löve & Löve; high mountain outcrops; infrequent; Jul–Aug; Roan Mt., NC. Represented in our area by subsp. **molle** (Michx.) Löve & Löve. *T. spicatum* (L.) Richter var. *molle* (Michx.) Beal.

PONTEDERIACEAE (Pickerelweed Family)

1 Stamens 3, all exserted . . . . **Heteranthera**
1 Stamens 6, 3 exserted and 3 included . . . . **Pontederia**

Heteranthera R. & P. (Mud Plantain)

1 Petals yellow; leaves linear; flowers solitary . . . . **H. dubia**
1 Petals blue; leaves reniform; flowers 3–10/inflorescence . . . . **H. reniformis**

H. dubia (Jacq.) Macm., Water Star Grass; shallow streams; infrequent; Aug–Sep; NC, TN, VA.
H. reniformis R. & P.; shallow water and mud flats; infrequent; Jun–Oct; TN, VA.

Pontederia L. (Pickerelweed)

P. cordata L.; shorelines; occasional; May–Oct; NC, SC, TN, VA.

POTAMOGETONACEAE (Pondweed Family)

Potamogeton L. (Pondweed)

Contributed by David H. Webb

Haynes, R. R. 1974. A revision of North American *Potamogeton* subsection *Pusilli* (Potamogetonaceae). *Rhodora* 76:564–649.
Hellquist, C. B., and G. E. Crow. 1980. *Aquatic vascular plants of New England:* Part 1. *Zosteraceae, Potamogetonaceae, Zannichelliaceae, Najadaceae.* New Hampshire Agric. Exp. Sta. Bull. 515.

1 Submersed leaves nonlinear, 1–7.5 cm wide 2
2 Stem flattened; leaf margins undulate with small but conspicuous teeth; fruit with a beak 2–3 mm long . . . . **P. crispus**
2 Stem round; leaf margins sometimes undulate but never toothed; beak of fruit absent or less than 2 mm long 3
3 Submersed leaves arched, nerves more than 21 . . . . **P. amplifolius**
3 Submersed leaves not arched or weakly so, nerves 21 or less 4
4 Stems black-spotted; floating leaves with cordate to rounded bases; submersed leaves with undulate margins . . . . **P. pulcher**
4 Stems not black-spotted; floating leaves with rounded to tapered bases; submersed leaves with flattened margins 5
5 Submersed leaves generally sessile or having petioles less than 2 cm long, rarely 4 cm long; apex of submersed leaves acuminate to mucronate; fruits less than 3.5 mm long . . . . **P. illinoensis**
5 Submersed leaves with petioles 2–13 cm long; apex of submersed leaves acute, but not acuminate or mucronate; fruits 3.5–4.3 mm long . . **P. nodosus**

1 Submersed leaves linear, 0.1–10 mm wide 6
6 Leaf base adnate to the stipule for about 1/2 to almost the entire length of the stipule 7
7 Floating leaves never present; submersed leaves setaceous; adnation of stipule and leaf exceeding 10 mm; rhizomes often bearing tubers .... **P. pectinatus**
7 Floating leaves frequently present; submersed leaves flattened and narrowly linear; adnation of stipule and leaf less than 10 mm; rhizomes not bearing tubers .... **P. diversifolius**
6 Leaf base not adnate to the stipule or, if so, only at the very base of the stipule 8
8 Floating leaves absent, never formed by any plants of the population; peduncles filiform to slightly thickened; fruiting spike subcapitate or interrupted 9
9 Fruit with undulate to dentate dorsal keel to 0.4 mm high; nodal glands absent .... **P. foliosus**
9 Fruit with rounded dorsal surface, keel absent; nodal glands usually present .. 10
10 Stipule delicate, whitish, green, or brownish, disintegrating with age; peduncles cylindric .... **P. pusillus**
10 Stipule more or less coarsely fibrous, whitish, shredding with age; peduncle clavate .... **P. strictifolius**
8 Floating leaves present in some of the population; peduncles generally thickened; fruiting spike cylindric, not interrupted 11
11 Submersed leaves ribbonlike, 2–10 mm broad, with prominent bands of lacunae along both sides of the midvein .... **P. epihydrus**
11 Submersed leaves narrower, 0.2–2 mm broad, lacking prominent bands of lacunae 12
12 Submersed leaves adnate at base of stipule .... **P. tennesseensis**
12 Submersed leaves free of stipule 13
13 Submersed leaves 0.8–2 mm wide; blades of floating leaves 2.5–6.5 cm wide, with cordate bases; fruits 2.5–3.5 mm broad, scarcely keeled .... **P. natans**
13 Submersed leaves 0.3–1 mm wide; blades of floating leaves 1–3 cm wide, with rounded to tapering bases; fruits 1.6–2.4 mm broad, with prominent keel .... **P. oakesianus**

P. amplifolius Tuckerman; rivers and streams; infrequent; Jul–Sep; TN, VA.

P. crispus L.*, Curly P.; streams, rivers, and lakes; occasional; Apr–Jun; NC, TN, VA.

P. diversifolius Raf.; streams, lakes, and ponds; common; Jun–Sep; ALL.

P. epihydrus Raf.; streams, rivers, and lakes; occasional; Jun–Sep; NC, TN, VA.

P. foliosus Raf.; streams, lakes, and ponds; occasional; May–Oct; NC, TN, VA.

P. illinoensis Morong; streams, lakes, and ponds; infrequent; May–Sep; VA.

P. natans L.*; lakes; infrequent; Jul–Sep; Buncombe Co., NC, where apparently a short-lived population.

P. nodosus Poir., American P.; streams, rivers, and lakes; occasional; May–Sep; NC, VA.

P. oakesianus Robbins; ponds; infrequent; Jul–Sep; VA.

P. pectinatus L., Sago P.; streams and rivers; infrequent; Jun–Sep; VA. A preferred food for waterfowl.

P. pulcher Tuckerman; slow streams and ponds; occasional; Jun–Sep; ALL.

P. pusillus L.; streams, lakes, and ponds; infrequent; Jun–Sep; NC, VA. Plants in our area are referrable to var. **tenuissimus** Mert. & Koch. *P. berchtoldii* Fieber.

P. strictifolius A. Bennett; rivers, lakes, and ponds; infrequent; Jul–Sep; VA. Disjunct from New England and Great Lakes region.

P. tennesseensis Fern.; streams and rivers; infrequent; May–Sep; TN, VA.

To accurately identify pondweeds, especially the narrow-leaved species, effort should be made to collect flowering and fruiting specimens and to determine the presence or absence of floating leaves in the population.

SMILACACEAE (Catbrier Family)

Smilax L.

1 Stems herbaceous; peduncles usually more than 4 cm long; stems without prickles; ovules 2/carpel (Carrion Flowers) 2
  2 Plants viny, up to 3 m high; tendrils numerous 3
    3 Leaves pubescent beneath . . . . . **S. pulverulenta**
    3 Leaves glabrous and glaucous beneath 4
      4 Fruiting pedicels 2× or more as long as the fruit; larger leaves broadly ovate . . . . . **S. herbacea**
      4 Fruiting pedicels less than 2× as long as the fruit; larger leaves hastate . . . . . **S. pseudo-china**
  2 Plants erect, 1 m or less high; tendrils absent or rudimentary 5
    5 Leaves glabrous and glaucous beneath . . . . . **S. biltmoreana**
    5 Leaves pubescent beneath . . . . . **S. hugeri**
1 Stems woody; peduncles usually less than 3 cm long; stems with prickles; ovules 1/carpel (Cat-, Saw-, Greenbriers) 6
  6 Leaves glaucous beneath . . . . . **S. glauca**
  6 Leaves green beneath 7
    7 Leaves evergreen; the lower 1/3 of the midrib elevated and much more prominent than the lateral veins . . . . . **S. laurifolia**
    7 Leaves deciduous, or if evergreen, the midrib similar to the lateral veins 8
      8 Peduncles 1.5× or more longer than the subtending leaf petiole; fruits usually with 1 seed 9
        9 Stems with stiff, broad-based, pale or dark-tipped prickles; leaf margin thickened, entire or spinulose-ciliate . . . . . **S. bona-nox**
        9 Stems with bristly, dark prickles; leaf margin not thickened, minutely serrulate . . . . . **S. tamnoides**
      8 Peduncles less than 1.5× as long as the subtending leaf petiole, or if peduncles longer, the stem without dark bristles or leaves without a marginal rib; fruits 1–3-seeded 10
        10 Leaves lanceolate to elliptic-lanceolate, the base broadly cuneate . . . . . **S. smallii**
        10 Leaves elliptic, ovate, or suborbicular, the base rounded to cordate 11
          11 Fruits dark blue to black; plants usually in dry habitats . . **S. rotundifolia**
          11 Fruits red; plants of wet habitats . . . . . **S. walteri**

S. biltmoreana (Small) J. B. S. Norton ex Pennell; woodlands; occasional; May–Jun; GA, NC, SC. *S. ecirrata* (Engelm. ex Kunth) S. Wats. var. *biltmoreana* (Small) Ahles.

S. bona-nox L.; thickets and roadsides; occasional; Apr–May; ALL.
S. glauca Walt.; woodlands and thickets; common; Apr–Jun; ALL.
S. herbacea L.; open woods and clearings; frequent; May–Jun; ALL.
S. hugeri (Small) J. B. S. Norton ex Pennell; rich woods; infrequent; Mar–Apr; NC, TN. *S. ecirrata* (Engelm. ex Kunth) S. Wats. var. *hugeri* (Small) Ahles.
S. laurifolia L.; moist thickets; infrequent; Jul–Aug; ALL.
S. pseudo-china L.; damp woods; infrequent; May–Jun; GA, VA. *S. tamnifolia* Michx.
S. pulverulenta Michx.; damp woodlands; occasional; Apr–Jun; NC, SC, TN, VA. *S. herbacea* L. var. *pulverulenta* (Michx.) Gray.
S. rotundifolia L.; woodlands and dry thickets; common; Apr–May; ALL. The terminal branches are often quadrangular, and this variant may be recognized as var. *quadrangularis* (Muhl.) Wood.
S. smallii Morong; damp woods and thickets; infrequent and more common on the Coastal Plain; Jun–Jul; VA. The leaves of this taxon are often used in decorations.
S. tamnoides L.; woodlands and stream banks; frequent; Apr–May; ALL. *S. hispida* Muhl.
S. walteri Pursh; damp woods and thickets; infrequent; Apr–May; GA, SC.

SPARGANIACEAE (Bur-reed Family)

Sparganium L. (Bur-reed)

Beal, E. O. 1960. *Sparganium* (Sparganiaceae) in the southeastern United States. *Brittonia* 12:176–181.
Crow, G. E., and C. B. Hellquist. 1981. *Aquatic vascular plants of New England:* Part 2. *Typhaceae and Sparganiaceae*. New Hampshire Agric. Exp. Sta. Bull. 517.

1 Stigmas 2; fruits truncate at the apex, 4–8 mm broad .... **S. eurycarpum**
1 Stigma 1; fruits tapering to a beak, less than 4 mm broad 2
  2 Pistillate heads (at least some) borne above the axils of leaves or bracts .... **S. chlorocarpum**
  2 Pistillate heads borne directly in the axils of leaves or bracts 3
    3 Mature fruit dull; fruiting heads 1.5–2.5 cm wide .... **S. americanum**
    3 Mature fruit shiny (rarely dull); fruiting heads 2.5–3.5 cm wide .... **S. androcladum**

S. americanum Nutt.; shallow streams and shorelines; common; May–Sep; ALL.
S. androcladum (Engelm.) Morong; ponds and streams; infrequent; May–Sep; TN.
S. chlorocarpum Rydb.; ponds and streams; infrequent; May–Sep; NC, VA.
S. eurycarpum Engelm.; ponds and streams; infrequent; May–Sep; VA.

TYPHACEAE (Cat-tail Family)

Typha L. (Cat-tail)

1 Staminate and pistillate flowers continuous along the spike; leaves flat on back, usually 8 mm or more wide .... **T. latifolia**
1 Staminate and pistillate flowers discontinuous along the spike; leaves convex on back, usually less than 8 mm wide .... **T. angustifolia**

T. angustifolia L., Narrowleaf C-t.; shallow margins of lakes, ponds, marshes; infrequent and more common on the Coastal Plain; May–Jul; TN, VA.
T. latifolia L., Common C-t.; wet ditches, shorelines, marshes; frequent; May–Jul; ALL.

XYRIDACEAE (Yellow-eyed Grass Family)

Xyris L. (Yellow-eyed Grass)

Kral, R. 1966. *Xyris* (Xyridaceae) of the continental United States and Canada. *Sida* 2:177–260.

1 Keel of lateral sepals ciliate or fimbriate 2
  2 Leaves ascending, strongly twisted; spikes ovoid . . . . **X. torta**
  2 Leaves spreading, scarcely twisted; spikes lance-ovoid to ellipsoid . . . **X. ambigua**
1 Keel of lateral sepals lacerate, thin, rarely entire 3
  3 Lateral sepals extended beyond the tips of subtending bract . . . . **X. smalliana**
  3 Lateral sepals hidden by the subtending bract 4
    4 Plant base pink to maroon (when fresh); spike ovoid to subglobose; leaves spreading . . . . **X. difformis**
    4 Plant base greenish-brown; spike narrowly ovoid to oblong; leaves ascending . . . . **X. jupicai**

X. ambigua Bey. ex Kunth; wet woods and roadside ditches; infrequent; Jun–Aug; VA.
X. difformis Chapm.; stream and pond margins; wet ditches; infrequent; Aug–Oct; NC, SC, TN.
X. jupicai L. C. Rich.; disturbed wet areas; infrequent; Jul–Sep; VA.
X. smalliana Nash; wet marshes and ditches; infrequent; Jul–Aug; SC.
X. torta J. E. Smith; river banks and wet woods; occasional; Jun–Aug; ALL.

All of our species, except *X. torta*, are Coastal Plain disjuncts.

ZANNICHELLIACEAE (Horned Pondweed Family)

Zannichellia L. (Horned Pondweed)

Z. palustris L.; ponds and pools; infrequent; Feb–Oct; TN, VA.

Dicot Families

ACANTHACEAE (Acanthus Family)

1 Corolla zygomorphic; stamens 2 . . . . **Justicia**
1 Corolla nearly actinomorphic; stamens 4 . . . . **Ruellia**

Justicia L. (Water Willow)

J. americana (L.) Vahl; river bars and sandy stream banks; occasional; Jun–Sep; NC, TN, VA.

Ruellia L. (Wild Petunia)

1 Flowers solitary, axillary, usually at lower nodes 2
  2 Calyx lobes lanceolate, 2 mm or more wide . . . . **R. strepens**
  2 Calyx lobes linear, less than 2 mm wide . . . . **R. purshiana**
1 Flowers in cymes or glomerules, usually at upper nodes 3
  3 Calyx lobes lanceolate, 2 mm or more wide . . . . **R. strepens**
  3 Calyx lobes linear, less than 2 mm wide 4
    4 Leaves sessile or subsessile . . . . **R. humilis**
    4 Leaves petiolate . . . . **R. caroliniensis**

R. caroliniensis (J. F. Gmel.) Steudel; dry rocky woods; frequent; May–Sep; NC, SC, TN, VA. *R.ciliosa* Pursh.
R. humilis Nutt.; dry open woods; infrequent; May–Sep; TN, VA.
R. purshiana Fern.; dry woods and bluffs; infrequent; May–Sep; SC, TN, VA.
R. strepens L.; dry woods; infrequent; May–Sep; TN, VA.

ACERACEAE (Maple Family)

Acer L. (Maple)

1 Leaves compound, usually appearing with the flowers . . . . **A. negundo**
1 Leaves simple, sometimes undeveloped at flowering 2
  2 Flowers in terminal panicles or racemes; shrubs or small trees 3
    3 Panicles erect; leaves coarsely serrate; bark not striped . . . . **A. spicatum**
    3 Racemes drooping; leaves finely serrate; bark striped . . . . **A. pensylvanicum**
  2 Flowers in corymbs, umbels, or fascicles; medium to large trees 4
    4 Leaf petiole and young twigs with a milky sap; introduced . . . . . **A. platanoides**
    4 Leaf petiole and young twigs without a milky sap; native 5
      5 Flowers appearing before the leaves 6
        6 Leaves shallowly 3–5-lobed, pale green to whitish beneath; fruits less than 3 cm long; flowers reddish . . . . **A. rubrum**

6 Leaves deeply 5-lobed, white to silvery beneath; fruits more than 4 cm long; flowers yellowish . . . . **A. saccharinum**
5 Flowers appearing with or after the leaves 7
7 Leaves essentially glabrous beneath . . . . **A. saccharum**
7 Leaves soft pubescent beneath at maturity 8
8 Leaf blades less than 8 cm long, yellow-green above; bark nearly white, mostly smooth . . . . **A. leucoderme**
8 Leaf blades more than 8 cm long, dark green above and often drooping; bark dark and roughened . . . . **A. nigrum**

A. leucoderme Small, Chalk M.; moist woods along rivers and ravines; infrequent; Mar–Apr; GA, NC, SC, TN.
A. negundo L., Box Elder; lowlands along rivers and streams; common; Mar–Apr; ALL.
A. nigrum Michx. f., Black M.; rich lowlands; infrequent; May–Jun; NC, VA.
A. pensylvanicum L., Striped M.; rich mountain woods; frequent; May–Jun; ALL.
A. platanoides L.*, Norway M.; infrequently escaping from cultivation; Mar–Apr; VA.
A. rubrum L., Red M.; low woods, coves, and stream banks; common; Feb–Mar; ALL.
A. saccharinum L., Silver M.; flood plains, streams, and river bottoms; frequent; Feb–Apr; ALL.
A. saccharum Marsh., Sugar M.; rich woods; common; Apr–Jun; ALL.
A. spicatum Lam., Mountain M.; rich rocky woods at high elevations; frequent; May–Jul; GA, NC, TN, VA.

AMARANTHACEAE (Amaranth Family)

1 Leaves alternate . . . . **Amaranthus**
1 Leaves opposite . . . . **Froelichia**

Amaranthus L. (Pigweed)

1 Plants dioecious . . . . **A. tuberculatus**
1 Plants monoecious, or occasionally with perfect flowers 2
2 Stipular spines present at base of most of the leaves . . . . **A. spinosus**
2 Stipular spines absent 3
3 Stamens 2 or 3; plants decumbent . . . . **A. albus**
3 Stamens 5; plants with erect stems 4
4 Bracts of pistillate flowers 2–3× as long as the sepals and firm . . . . **A. retroflexus**
4 Bracts of pistillate flowers as long as or slightly longer than the sepals, not firm . . . . **A. hybridus**

A. albus L., Tumbleweed; waste places, roadsides, and railroads; occasional; Jul–Oct; VA.
A. hybridus L.*; waste places, fields, and pastures; common; Jul–Oct; NC, SC, TN, VA.
A. retroflexus L.*, Green P.; fields and disturbed sites; occasional; Jul–Oct; NC, VA.
A. spinosus L.*, Thorny P.; fields and disturbed sites; frequent; Jul–Oct; NC, SC, TN. VA.
A. tuberculatus (Moq.) Sauer, Water Hemp; waste places; infrequent; Jul–Oct; NC.

Froelichia Moench (Cottonweed)

F. gracilis (Hook.) Moq.; sandy soil, especially along railroads; infrequent; Jul–Sep; NC, TN, VA.

ANACARDIACEAE (Cashew Family)

1 Leaflets 3 2
  2 Flowers appearing after the leaves in axillary panicles; fruit white; plant with contact poison .... **Toxicodendron**
  2 Flowers appearing before or with the leaves in terminal panicles; fruit red; plants without contact poison .... **Rhus**
1 Leaflets 5 or more 3
  3 Inflorescence axillary; fruit white; plant with contact poison .... **Toxicodendron**
  3 Inflorescence terminal; fruit red; plant without contact poison .... **Rhus**

Rhus L. (Sumac)

1 Leaflets 3; low shrubs .... **R. aromatica**
1 Leaflets 5 or more 2
  2 Petiole winged .... **R. copallina**
  2 Petiole smooth, not winged 3
    3 Branches velvety pubescent and rounded .... **R. typhina**
    3 Branches glabrous and somewhat angled .... **R. glabra**

R. aromatica Ait., Fragrant S.; bluffs and open rocky woods; frequent; Feb–Apr; NC, TN, VA.

R. copallina L., Winged S.; woodlands, old fields, road banks; common; Jul–Sep; ALL.

R. glabra L., Smooth S.; woodlands, thickets, old fields; common; Jun–Aug; ALL.

R. typhina L., Staghorn S.; woodlands, meadows, thickets; common; May–Jun; ALL.

Toxicodendron Mill.

1 Leaflets 7 or more .... **T. vernix**
1 Leaflets 3 2
  2 Stems mostly erect; leaves thick, obtuse, usually with 2–4 lobes on each side .... **T. toxicarium**
  2 Stems mostly trailing or climbing; leaves thin, acute to acuminate, mostly entire .... **T. radicans**

T. radicans (L.) Kuntze, Poison Ivy; thickets and disturbed sites; common; May–Jun; ALL. *Rhus radicans* L.

T. toxicarium (Salisb.) Gillis, Poison Oak; usually dry disturbed sites; occasional; May–Jun; GA, SC, VA. Scarcely distinct from *T. radicans*. *Rhus toxicodendron* L.

T. vernix (L.) Kuntze, Poison Sumac; swamps and low woods; occasional; May–Jun; ALL. *Rhus vernix* L.

ANNONACEAE (Custard Apple Family)

Asimina Adans. (Pawpaw)

1 Flowers 1–2 cm wide; pedicels 1–8 mm long; shrubs, less than 1.5 m tall .... **A. parviflora**
1 Flowers more than 2 cm wide; pedicels more than 8 mm long; small trees, 3–12 m tall .... **A. triloba**

A. parviflora (Michx.) Dunal, Dwarf P.; sandy dry woodlands; infrequent; Apr–May; GA, NC, SC.

A. triloba (L.) Dunal; rich woods and stream banks; common; Apr–May; ALL.

APIACEAE (Carrot Family)

1 Leaves all simple 2
2 Petals yellow; leaves perfoliate . . . . **Bupleurum**
2 Petals white, greenish, or bluish; leaves not perfoliate 3
3 Flowers in simple umbels of 2–10 flowers; leaves nearly round . . . **Hydrocotyle**
3 Flowers in dense heads of more than 10 flowers; leaves various but not rounded . . . . **Eryngium**
1 At least some leaves compound or deeply dissected 4
4 Ovary and fruit with dense bristles or prickles 5
5 Stems glabrous; leaves palmately divided; heads with both bisexual and staminate flowers . . . . **Sanicula**
5 Stems pubescent; leaves pinnately divided; all flowers bisexual 6
6 Bracts pinnately divided; bristles of fruit in simple rows . . . . **Daucus**
6 Bracts simple, linear or occasionally absent; bristles of fruit not in simple rows . . . . **Torilis**
4 Ovary and fruit glabrous or pubescent, but not bristly or prickly 7
7 Leaflets or leaf segments narrowly filiform, less than 1 mm wide 8
8 Flowers white . . . . **Ptilimnium**
8 Flowers yellow . . . . **Foeniculum**
7 Leaflets or leaf segments more than 1 mm wide 9
9 Fruits 2.5× or more longer than wide and not winged 10
10 Leaves trifoliolate . . . . **Cryptotaenia**
10 Leaves ternately or ternate-pinnately decompound 11
11 Fruits with stiff, appressed hairs; ultimate leaf segments more than 1 cm wide; mature fruits 1.5–2.5 cm long . . . . **Osmorhiza**
11 Fruits glabrous or soft pubescent; ultimate leaf segments less than 1 cm wide; mature fruits less than 1.5 cm long . . . . **Chaerophyllum**
9 Fruits less than 2.5× as long as wide or winged 12
12 Petals yellow, rarely maroon 13
13 Leaflets entire 14
14 Fruit flattened laterally (perpendicular to the septum); leaves with a celerylike fragrance . . . . **Taenidia**
14 Fruit flattened dorsally (parallel to the septum); leaves with an aniselike fragrance . . . . **Pseudotaenidia**
13 Leaflets toothed 15
15 Leaves once-pinnate; coarse biennials up to 1.5 m tall . . . . **Pastinaca**
15 Leaves ternately compound; perennials less than 1 m tall 16
16 Central flower of each umbel sessile; fruits not winged . . . . **Zizia**
16 All flowers pedicellate; fruits winged . . . . **Thaspium**
12 Petals white or greenish-white 17
17 Plants less than 15 cm tall, delicate, from small underground tubers; flowering stem leafless . . . . **Erigenia**
17 Plants more than 15 cm tall; flowering stem leafy 18
18 Stem leaves much dissected, the ultimate divisions usually less than 1 cm wide 19

19 Stem red-spotted; fruit tuberculate . . . . **Conium**
19 Stem without red spots; fruit not tuberculate 20
20 Petioles broadly winged . . . . **Conioselinum**
20 Petioles not winged . . . . **Thaspium**
18 Stem leaves divided into distinct, uniform leaflets usually more than 1.5 cm wide 21
21 Leaves once-ternate or once-pinnate 22
22 Upper petioles 1 cm or more wide; fruit 10 mm wide, with 4 dark lines on each side . . . . **Heracleum**
22 Upper petioles not dilated, less than 1 cm wide; fruit 3–6 mm wide, without dark lines 23
23 Leaf margins serrate with numerous teeth . . . . **Sium**
23 Leaf margins entire or with fewer than 5 coarse teeth on each side . . . . **Oxypolis**
21 Leaves pinnately decompound 24
24 Upper petioles 1 cm or more wide when flattened; mature fruits glabrous or pubescent and prominently winged . . . . **Angelica**
24 Upper petioles less than 1 cm wide; fruits glabrous, wings absent or narrow 25
25 Lateral leaf veins terminating at the sinuses between teeth; fruit ribbed; upper branches of stem not whorled . . **Cicuta**
25 Lateral leaf veins usually anastomosing and terminating at a tooth; fruit narrowly winged; upper branches of stem whorled . . . . **Ligusticum**

Angelica L.

1 Umbels densely pubescent . . . . **A. venenosa**
1 Umbels glabrous or nearly so 2
2 Leaves with a narrow, pale, glabrous margin . . . . **A. atropurpurea**
2 Leaf margins not differentiated, somewhat ciliate . . . . **A. triquinata**

A. atropurpurea L.; wet areas; infrequent; May–Jun; NC, TN(?).

A. triquinata Michx.; rocky slopes and balds; occasional; Aug–Sep; ALL.

A. venenosa (Greenway) Fern.; dry open areas, especially road banks; frequent; Jun–Aug; ALL.

Bupleurum L. (Thorough Wax)

B. rotundifolium L.*; waste places; infrequent; Jun–Jul; TN, VA.

Chaerophyllum L. (Wild Chervil)

1 Plants pubescent; stem mostly erect . . . . **C. tainturieri**
1 Plants glabrous; stem decumbent . . . . **C. procumbens**

C. procumbens (L.) Crantz; alluvial woods; occasional; Apr–May; TN, VA.

C. tainturieri Hook.; waste places; frequent; Mar–Apr; ALL.

Cicuta L. (Water Hemlock)

C. maculata L.; swamps and stream banks; frequent; May–Aug; ALL. Highly poisonous to both humans and livestock. *C. mexicana* Coult. & Rose, a dubious taxon with stouter stems and broader leaves, has been attributed to our area by some authors.

Conioselinum Hoffm. (Hemlock Parsley)

C. chinense (L.) BSP.; mesic woodlands; infrequent; Jul–Sep; NC. Not seen in our area in recent years.

Conium L. (Poison Hemlock)

C. maculatum L.*; disturbed sites and stream banks; occasional; May–Jun; NC, VA. Highly poisonous to humans and livestock.

Cryptotaenia DC. (Honewort)

C. canadensis (L.) DC.; mesic woods and stream banks; common; May–Jun; ALL.

Daucus L. (Carrot)

1 Rays 3–7.5 cm long; teeth of fruit not prominently barbed at the tip; stem freely branching . . . . **D. carota**

1 Rays 0.4–4 cm long; teeth of fruit barbed at the tip; stem usually unbranched . . . . **D. pusillus**

D. carota L.*, Wild C./Queen Anne's Lace; waste places; common; May–Sep; ALL.

D. pusillus Michx.; sandy waste places; infrequent; Apr–May; SC.

Erigenia Nutt. (Harbinger-of-Spring)

E. bulbosa (Michx.) Nutt.; rich woods, often on basic soils; infrequent; Mar–Apr; TN, VA. One of our earliest spring wildflowers.

Eryngium L.

1 Plants prostrate, often mat-forming; heads borne singly from the leaf axils . . . . **E. prostratum**

1 Plants erect; inflorescence branched 2

  2 Leaves parallel-veined; flowers white or greenish-white . . . . **E. yuccifolium**

  2 Leaves net-veined; flowers bluish . . . . **E. integrifolium**

E. integrifolium Walt.; mesic woodlands; occasional; Aug–Oct; GA, NC, SC.

E. prostratum Nutt.; swamps, wet ditches, and shorelines; infrequent; May–Oct; NC, SC, TN.

E. yuccifolium Michx., Rattlesnake Master; dry woodlands; occasional; Jun–Oct; GA, NC, SC, VA.

Foeniculum P. Mill. (Fennel)

F. vulgare Mill.*; disturbed sites; occasional; Jun–Aug; NC, TN, VA.

Heracleum L. (Cow Parsnip)

H. lanatum Michx.; meadows and alluvial woods; occasional; May–Jul; GA, NC, TN, VA. *H. maximum* Bartram.

Hydrocotyle L. (Pennywort)

1 Umbels sessile or nearly so; flowering stems mostly ascending ...... **H. americana**
1 Umbels pedunculate; flowering stems horizontal .... **H. ranunculoides**

H. americana L.; seeps and waterfalls; occasional and easily overlooked; Jun–Sep; NC, SC, TN, VA.
H. ranunculoides L. f.; seeps and shorelines; infrequent; Jun–Aug; VA.

Ligusticum L. (American Lovage)

L. canadense (L.) Britt.; rich woodland margins and stream banks; common; Jun–Jul; ALL.

Osmorhiza Raf. (Sweet Cicely)

Lowry, P. P., II, and A. G. Jones. 1979. Biosystematic investigations and taxonomy of *Osmorhiza* Raf. sect. *Osmorhiza* (Apiaceae) in North America. *Amer. Midl. Naturalist* 101:21–27.

1 Style and stylopodium less than 1 mm long at anthesis, shorter than the petals .... **O. claytonii**
1 Style and stylopodium 1.5–2.5 mm long at anthesis, longer than the petals .... **O. longistylis**

O. claytonii (Michx.) Clarke; rich woods; frequent; Apr–May; ALL.
O. longistylis (Torr.) DC.; rich woods; occasional; Apr–May; NC, SC, TN, VA.

Oxypolis Raf. (Cowbane)

O. rigidior (L.) Raf.; stream banks and open wet areas; frequent; Aug–Oct; ALL.

Pastinaca L. (Wild Parsnip)

P. sativa L.*; waste places; occasional; Jun–Jul; NC, SC, TN, VA.

Pseudotaenidia Mackenzie (Mountain Pimpernel)

Cronquist, A. 1982. Reduction of *Pseudotaenidia* to *Taenidia* (Apiaceae). *Brittonia* 34:365–367.

P. montana Mackenzie; dry shaley woods; infrequent; May–Jun; VA. *Taenidia montana* (Mackenzie) Cronquist.

Ptilimnium Raf.

P. capillaceum (Michx.) Raf.; low wet areas; infrequent; May–Jul; SC.

Sanicula L. (Snakeroot)

Phillippe, L. R. 1978. A biosystematic study of *Sanicula* sect. *Sanicula*. Ph.D. diss., Univ. of Tennessee, Knoxville.

1 Styles recurved, longer than the bristles of the fruit 2
  2 Staminate flowers greenish-yellow; calyx lobes ovate, 0.3–0.6 mm long; fruit pedicellate .... **S. gregaria**

2 Staminate flowers greenish-white; calyx lobes lanceolate, 0.7–1.5 mm long; fruit sessile or nearly so .... **S. marilandica**

1 Styles not recurved, shorter than the bristles of the fruit 3

3 Sepals beaklike and exceeding the bristles .... **S. trifoliata**

3 Sepals hidden by the bristles of the fruit 4

4 Styles equal to or longer than the calyx; fruit sessile or subsessile; roots thick and cordlike .... **S. smallii**

4 Styles shorter than the calyx; fruit pedicellate; roots fibrous .... **S. canadensis**

S. canadensis L.; woodlands; common; Apr–Jun; ALL.

S. gregaria Bickn.; moist woods and stream banks; frequent; Mar–Jun; ALL.

S. marilandica L.; woodlands; frequent; Mar–Jul; ALL.

S. smallii Bickn.; dry woods; occasional; Apr–Jun; GA, NC, SC, TN.

S. trifoliata Bickn.; rich woods; infrequent; Jun–Jul; NC, SC, TN, VA.

Sium L. (Water Parsnip)

S. suave Walt.; swamps and low areas; infrequent; Jun–Aug; NC, VA.

Taenidia (T. & G.) Drude (Yellow Pimpernel)

T. integerrima (L.) Drude; dry woodlands and road banks; frequent; Apr–Jun; NC, SC, TN, VA.

Thaspium Nutt. (Meadow Parsnip)

1 Stem leaves ternately decompound, the ultimate segments narrowly lanceolate to linear; leaf sheaths essentially without a hyaline margin; flowers creamy .... **T. pinnatifidum**

1 Stem leaves trifoliolate, 1- or 2-pinnately compound, or ternately decompound, the ultimate segments ovate to lanceolate; leaf sheaths with a hyaline margin; flowers yellow, maroon, or, rarely, creamy 2

2 Basal leaves 2-pinnately compound or ternately decompound; stem leaves without a hyaline margin .... **T. barbinode**

2 Basal leaves simple or ternately compound; stem leaves with a hyaline margin 3

3 Petals purple-maroon .... **T. trifoliatum** var. **trifoliatum**

3 Petals yellow .... **T. trifoliatum** var. **flavum**

T. barbinode (Michx.) Nutt.; rich woods; common; Apr–May; ALL.

T. pinnatifidum (Buckley) Gray; rich limestone slopes; infrequent; May–Jun; NC, SC, TN.

T. trifoliatum (L.) Gray var. **flavum** Blake; rich woods and stream banks; frequent; Apr–May; ALL. More common than the following variety.

T. trifoliatum (L.) Gray var. **trifoliatum**; primarily mountain woods; occasional; Apr–May; NC, SC, TN.

Torilis Adanson (Hedge Parsley)

T. arvensis (Hudson) Link*; disturbed sites; occasional; May–Jun; NC, TN, VA. *T. japonica* (Houtt.) DC.

Zizia Koch (Golden Alexander)

1 Basal leaves simple, cordate .... **Z. aptera**

1 Basal leaves 1- or 2-ternately compound 2

2 Leaf margins finely serrate, 5–10 teeth/cm .... **Z. aurea**
2 Leaf margins coarsely serrate, 2 or 3 teeth/cm .... **Z. trifoliata**

Z. aptera (Gray) Fern.; woodlands and roadsides; frequent; Apr–May; ALL.
Z. aurea (L.) Koch; meadows and wet woodlands; frequent; Apr–May; ALL.
Z. trifoliata (Michx.) Fern.; mesic woods and stream banks; frequent; Apr–May; ALL.

APOCYNACEAE (Dogbane Family)

1 Leaves alternate .... **Amsonia**
1 Leaves opposite 2
2 Trailing vine; flowers blue, rarely pink or white .... **Vinca**
2 Erect herb; flowers white to pink .... **Apocynum**

Amsonia Walt. (Blue Star)

A. tabernaemontana Walt.; deciduous woodlands, river banks; occasional; Apr–May; NC, SC, TN. *A. tabernaemontana* Walt. var. *salicifolia* (Pursh) Woodson.

Apocynum L. (Indian Hemp)

1 Corolla 4–10 mm long, pinkish 2
2 Leaves drooping; corolla 3× the length of the calyx tube ... **A. androsaemifolium**
2 Leaves spreading; corolla 2× the length of the calyx tube .... **A. medium**
1 Corolla 2–4 mm long, white to greenish-white 3
3 Leaves petiolate; follicles 10–15 cm long .... **A. cannabinum**
3 Leaves sessile or subsessile; follicles 5–9 cm long .... **A. sibiricum**

A. androsaemifolium L., Dogbane; open woodlands and roadbanks; common; Jun–Aug; GA, NC, TN, VA.
A. cannabinum L.; open woodlands and disturbed sites; common; Jun–Aug; NC, SC, TN, VA.
A. medium Greene; open woodlands and disturbed sites; common; Jun–Aug; GA, NC, TN, VA. Considered to be a hybrid between the two preceding species.
A. sibiricum Jacq.; dry sandy soils; infrequent; Jul–Sep; VA.

Vinca L. (Periwinkle)

1 Leaves ciliate, 3 cm or more long .... **V. major**
1 Leaves entire, up to 3 cm long .... **V. minor**

V. major L.*, Large P.; rarely escaping from cultivation; Apr–May; GA, NC, SC, VA.
V. minor L.*, Common P.; frequently escaping from cultivation; Apr–May; ALL.

AQUIFOLIACEAE (Holly Family)

Contributed by R. C. Clark

1 Petals separate, linear; stamens free from the corolla .... **Nemopanthus**
1 Petals fused at the base, oblong-ovate; stamens fused to the corolla .... **Ilex**

Ilex L. (Holly)

1 Leaves evergreen .... **I. opaca**
1 Leaves deciduous 2
  2 Nutlets smooth; peduncles of staminate flowers more than 0.5 mm long; corolla lobes of pistillate flowers not ciliate .... **I. verticillata**
  2 Nutlets with striate ridges; peduncles of staminate flowers less than 0.5 mm long; corolla lobes of pistillate flowers ciliate 3
    3 Leaves elliptic to broadly ovate, the apex abruptly to gradually acuminate; petioles (mature leaves) usually less than 1 cm long .... **I. ambigua** var. **ambigua**
    3 Leaves narrowly to broadly ovate, the apex gradually acuminate to attenuate; petiole (mature leaves) usually more than 1 cm long .. **I. ambigua** var. **monticola**

I. ambigua (Michx.) Torr. var. **ambigua**, Carolina H.; upland woods; occasional; May–Jun; GA, NC, SC, TN. *I. beadlei* Ashe.

I. ambigua (Michx.) Torr. var. **monticola** (Gray) Wunderlin & Poppleton, Mountain H.; mesic to dry woods; common; May–Jun; ALL. *I. montana* T. & G., *I. monticola* Gray, *I. ambigua* (Michx.) Torr. var. *montana* (T. & G.) Ahles. The *I. ambigua* complex is complicated in our area by numerous intermediates and local variants. Leaf shape and degree of pubescence are highly variable.

I. opaca Ait., American H.; rich woods, often cultivated; common; May–Jun; ALL.

I. verticillata (L.) Gray, Winterberry; wet woods and stream banks; frequent; Apr–May; ALL.

Nemopanthus Raf.

N. collinus (Alexander) Clark; wet woods and seeps; infrequent; May–Jun; NC, VA. *Ilex collina* Alexander.

ARALIACEAE (Ginseng Family)

1 Leaves simple, evergreen; woody vine .... **Hedera**
1 Leaves compound, deciduous; herb or shrub 2
  2 Leaves odd-pinnate, alternate .... **Aralia**
  2 Leaves palmate, whorled .... **Panax**

Aralia L.

1 Stem woody, with stout thorns .... **A. spinosa**
1 Stem herbaceous or somewhat woody at base, thornless or bristly 2
  2 Plant scapose; leaf solitary .... **A. nudicaulis**
  2 Plant not scapose; leaves several 3
    3 Umbels 2–10 in a loose cluster; stem bristly at base .... **A. hispida**
    3 Umbels more than 10 in a large compound panicle; stem not bristly at base .... **A. racemosa**

A. hispida Vent., Bristly Sarsaparilla; dry rocky or sandy soils; infrequent; Jun–Jul; VA.

A. nudicaulis L., Wild Sarsaparilla; woodlands; frequent; May–Jul; ALL.

A. racemosa L., Spikenard; rich woods; common; Jun–Aug; ALL.

A. spinosa L., Hercules Club, Devil's Walking Stick; rich open woodlands, roadsides; common; Jun–Aug; ALL.

Hedera L. (English Ivy)

H. helix L.*; persistent and often escaping from cultivation; Jun–Jul; ALL.

Panax L. (Ginseng)

1 Leaflets 3 (rarely 5), sessile .... **P. trifolius**
1 Leaflets 5, stalked .... **P. quinquefolius**

P. quinquefolius L., Ginseng/Sang; rich woods; occasional; May–Jun; ALL.
P. trifolius L., Dwarf G.; rich woods; infrequent; Apr–May; GA, NC, TN, VA.

ARISTOLOCHIACEAE (Birthwort Family)

1 Leafy erect herbs or woody vines; stamens 6 .... **Aristolochia**
1 Stemless herbs; stamens 12 2
  2 Leaves deciduous, pubescent .... **Asarum**
  2 Leaves evergreen, glabrous .... **Hexastylis**

Aristolochia L.

1 Woody vine; leaves 8 cm or more wide .... **A. macrophylla**
1 Erect herb; leaves less than 8 cm wide .... **A. serpentaria**

A. macrophylla Lam., Dutchman's Pipe; rich woods; frequent; Apr–Jun; ALL.
A. serpentaria L.; rich woods; occasional and easily overlooked; May–Jun; ALL.

Asarum L. (Wild Ginger)

A. canadense L.; rich woods; frequent; Apr–May; ALL. Extremely variable in length and position of the calyx lobes.

Hexastylis Raf. (Heartleaf, Wild Ginger)

Contributed by L. L. Gaddy

Blomquist, H. L. 1957. A revision of *Hexastylis* of North America. *Brittonia* 8:255–281.

1 Leaf blades triangular to ovate-sagittate or subhastate; style extension bifid to stigma 2
  2 Calyx lobes spreading, 2.5–8 mm long .... **H. arifolia** var. **arifolia**
  2 Calyx lobes erect, 2–4 mm long .... **H. arifolia** var. **ruthii**
1 Leaf blades cordate to orbicular- or triangular-cordate or reniform; style extension notched or divided at apex, not bifid to stigma 3
  3 Calyx broadly flask-shaped, rhombic-ovate to rhombic-urceolate 4
    4 Calyx broadest just above the base, internal reticulation of low relief (less than 1 mm) .... **H. contracta**
    4 Calyx broadest near the middle, internal reticulation of high relief (more than 1.5 mm) .... **H. rhombiformis**
  3 Calyx cylindrical, urceolate-campanulate, or urceolate 5
    5 Calyx cylindrical with or without a flare in the tube, or urceolate, 10 mm or less broad 6

6 Calyx lobes 3–4 mm long, internal reticulation a close network of high relief (more than 1.5 mm) . . . . **H. virginica**
6 Calyx lobes 6–10 mm long, internal reticulation an open network of low relief (less than 1 mm) . . . . **H. heterophylla**
5 Calyx urceolate-campanulate, 15 mm or more broad . . . . **H. shuttleworthii**

H. arifolia (Michx.) Small var. **arifolia**, Little Brown Jug; woodlands; occasional; Apr–May; ALL.

H. arifolia (Michx.) Small var. **ruthii** (Ashe) Blomquist, Little Brown Jug; woodlands; occasional; Apr–May; ALL.

H. contracta Blomquist; rich, acidic *Kalmia-Rhododendron* woodlands; infrequent; Apr–May; NC.

H. heterophylla (Ashe) Small; *Kalmia-Rhododendron* woodlands; occasional; Mar–Jun; ALL.

H. rhombiformis Gaddy; sandy, acidic *Kalmia-Rhododendron* slopes; infrequent; Apr–May; NC.

H. shuttleworthii (B. & B.) Small; stream banks and bog edges under *Rhododendron maximum*; occasional; May–Jul; ALL.

H. virginica (L.) Small; woodlands; occasional; Apr–May; NC, TN, VA.

H. minor (Ashe) Blomquist, which has a broader calyx and finer reticulation within the calyx than the closely related *H. heterophylla*, is found in the Piedmont of Patrick Co., VA, and may occur in the Piedmont of other VA and some NC counties along the Blue Ridge escarpment.

H. naniflora Blomquist, which can be separated from *H. minor* and *H. heterophylla* by its narrower calyx and overall smaller size, occurs in the Piedmont of Greenville Co., SC, and Rutherford Co., NC.

ASCLEPIADACEAE (Milkweed Family)

1 Erect herb . . . . **Asclepias**
1 Twining vine 2
2 Corolla white, sometimes tinged with green; stems glabrous or with nonglandular hairs . . . . **Cynanchum**
2 Corolla maroon, sometimes tinged with green or yellow; stems glandular-pubescent . . . . **Matelea**

Asclepias L. (Milkweed)

1 Leaves alternate; sap not milky . . . . **A. tuberosa**
1 Leaves opposite or whorled; sap milky 2
2 Leaves linear, at least 10× as long as wide . . . . **A. verticillata**
2 Leaves lanceolate to ovate, less than 10× as long as wide 3
3 Midstem leaves whorled . . . . **A. quadrifolia**
3 All leaves opposite or subopposite 4
4 Leaves sessile, the margin undulate . . . . **A. amplexicaulis**
4 Leaves petiolate, the margin not undulate 5
5 Leaves glabrous beneath . . . . **A. exaltata**
5 Leaves variously pubescent beneath 6

6 Petals predominantly white or green 7
7 Inflorescence sessile or nearly so; hood without a horn . . . . **A. viridiflora**
7 Inflorescence pedunculate; hood with a horn . . . . **A. variegata**
6 Petals predominantly purple or sometimes tinged with green 8
8 Peduncles 1–2 cm long; leaves primarily lanceolate 9
9 Stems weakly pubescent to glabrous . . . . **A. incarnata** subsp. **incarnata**
9 Stems distinctly pubescent . . . . **A. incarnata** subsp. **pulchra**
8 Peduncles 2–3.5 cm long; leaves widely elliptic to ovate 10
10 Corolla greenish-purple; hood margin with a median lobe; fruit spiny . . . . **A. syriaca**
10 Corolla deep maroon; hood margin without a median lobe; fruit smooth . . . . **A. purpurascens**

A. amplexicaulis Smith; sandy clearings; frequent; May–Jul; ALL.
A. exaltata L., Poke M.; clearings by forest margins; frequent; Jun–Jul; ALL.
A. incarnata L. subsp. **incarnata**, Swamp M.; moist areas; occasional; Jun–Aug; NC, TN, VA.
A. incarnata L. subsp. **pulchra** (Ehrhart ex Willd.) Woodson, Swamp M.; open swamps and wet ditches; common; Jun–Aug; ALL.
A. purpurascens L.; dry open sites; occasional; Jun–Jul; NC, VA.
A. quadrifolia Jacq., Whorled M.; woodlands; frequent; May–Jun; ALL.
A. syriaca L.; fields and waste areas; frequent; May–Jun; NC, SC, TN, VA.
A. tuberosa L., Butterfly Weed; fields and waste areas; common; May–Jul; ALL.
A. variegata L.; woodland margins; frequent; May–Jun; ALL.
A. verticillata L.; dry open clearings; occasional; Jun–Aug; ALL.
A. viridiflora Raf.; dry sandy barrens; occasional; Jul–Aug; GA, SC, VA.

Cynanchum L.

C. laeve (Michx.) Pers.; thickets and disturbed sites; occasional; Jun–Aug; SC, TN, VA.

Matelea Aublet

1 Pedicels and lower surface of corolla lobes glabrous; fruit smooth and angled 2
2 Corolla glabrous on both surfaces . . . . **M. gonocarpa**
2 Corolla glabrous below, pubescent above . . . . **M. suberosa**
1 Pedicels and lower surface of corolla lobes glandular-pubescent; fruit prickly and round 3
3 Petals 4× or more as long as wide . . . . **M. obliqua**
3 Petals less than 4× as long as wide . . . . **M. carolinensis**

M. carolinensis (Jacq.) Woodson; mesic thickets and woodlands; infrequent; Jun–Jul; NC, SC, VA.
M. gonocarpa (Walt.) Shinners; open woodlands; infrequent; Jun–Jul; TN.
M. obliqua (Jacq.) Woodson; open calcareous sites; occasional; Jun–Jul; NC, TN, VA.
M. suberosa (L.) Shinners; alluvial woods; infrequent; Jun–Jul; NC, SC.

ASTERACEAE (Sunflower Family)

Contributed by E. E. Schilling (except *Aster*)

Adapted primarily from A. Cronquist. *Vascular flora of the southeastern United States.* Vol. I, *Asteraceae*. (Chapel Hill: University of North Carolina Press, 1980).

1 Heads with ray flowers only; juice milky .... **Key A**
1 Heads with disk flowers; ray flowers present or absent; juice watery 2
  2 Heads with both ray and disk flowers 3
    3 Rays yellow or orange .... **Key B**
    3 Rays white, pink, or purplish, not yellow or orange .... **Key C**
  2 Heads with only disk flowers (the outermost disk flowers sometimes enlarged and different from the inner ones) .... **Key D**

KEY A

1 Pappus of scales or none, no bristles present 2
  2 Corollas blue to pink or white .... **Cichorium**
  2 Corollas yellow or orange 3
    3 Involucral bracts in a single series; pappus usually present .... **Krigia**
    3 Involucre with a basal series of short bracts and an inner series of long, narrow bracts; pappus none .... **Lapsana**
1 Pappus of bristles, at least in part 4
  4 Pappus bristles simple, not plumose 5
    5 Achenes flattened 6
      6 Flowers blue or cream .... **Lactuca**
      6 Flowers yellow 7
        7 Heads with more than 80 flowers; achenes beakless .... **Sonchus**
        7 Heads with fewer than 80 flowers; achenes usually with a beak .. **Lactuca**
    5 Achenes terete or prismatic, not flattened 8
      8 Flowers white, pink, or cream; heads usually nodding .... **Prenanthes**
      8 Flowers bright yellow to orange or orange-red; heads held erect 9
        9 Achenes beakless or with a short, stout beak less than 1/2 as long as the body of the achene 10
          10 Involucral bracts in a single series .... **Krigia**
          10 Involucral bracts in more than one series, the outer series frequently very short 11
            11 Plant tap-rooted annual or biennial; pappus bristles whitish .... **Crepis**
            11 Plant fibrous-rooted perennial; pappus bristles usually brown or tawny, sometimes whitish .... **Hieracium**
        9 Achenes with a long, slender beak from 1/2 as long to longer than the body of the achene 12
          12 Plants strictly scapose, each scape with a solitary head .... **Taraxacum**
          12 Plants caulescent, usually branched above and with several heads ... 13
            13 Beak of the achene with a ring of soft, white hairs at the summit, just below the pappus .... **Pyrrhopappus**
            13 Beak of the achene without a ring of hairs at the summit 14

14 Achenes without scales .... **Crepis**
14 Achenes with a circle of small scales at the base of the beak .... **Chondrilla**
4 Pappus bristles plumose 15
15 Involucral bracts in one series; leafy-stemmed plants with grasslike leaves .... **Tragopogon**
15 Involucral bracts in more than one series; leaves not grasslike 16
16 Leaves mainly cauline; many of the hairs forked at the summit ..... **Picris**
16 Leaves mainly basal; hairs not forked at the summit 17
17 Receptacle chaffy .... **Hypochoeris**
17 Receptacle without chaff .... **Leontodon**

KEY B

1 Receptacle chaffy 2
2 Disk flowers sterile, with undivided style 3
3 Leaves lobed or dissected; achenes only slightly flattened; pappus absent .... **Polymnia**
3 Leaves usually not lobed or dissected; achenes flattened; pappus present or absent 4
4 Rays 5 or 6, 0.7–1.5 cm long; leaves with winged petioles ... **Chrysogonum**
4 Rays more than 6, more than 1.5 cm long; leaves sessile or petioles wingless .... **Silphium**
2 Disk flowers fertile, with divided style 5
5 Involucre of 4 bracts, each 1 cm or more wide; leaves sessile .. **Tetragonotheca**
5 Involucre of more than 4 bracts; leaves sessile or petioled 6
6 Involucral bracts in 2 distinct series, the outer ones more or less spreading 7
7 Achenes wing-margined; pappus awns, if present, not retrorsely barbed .... **Coreopsis**
7 Achenes not wing-margined; pappus awns usually retrorsely barbed .... **Bidens**
6 Involucral bracts all similar, not in 2 distinct series 8
8 Leaves all alternate 9
9 Receptacle strongly conic or columnar; pappus a low crown or none 10
10 Ray flowers subtended by receptacular bracts; leaves pinnatifid .... **Ratibida**
10 Ray flowers not subtended by receptacular bracts; leaves pinnatifid or entire .... **Rudbeckia**
9 Receptacle merely convex; pappus of 2 or 3 awns or scales 11
11 Achenes strongly flattened; stem winged by decurrent petiole bases .... **Verbesina**
11 Achenes not flattened; stem not winged .... **Helianthus**
8 Leaves, at least the lower, opposite 12
12 Rays persistent on the achenes and becoming papery; pappus none .... **Heliopsis**

12 Rays deciduous from the achenes at maturity; pappus present, of 2 awns 13
13 Achenes strongly flattened; stem winged by decurrent petiole bases . . . . . **Verbesina**
13 Achenes not flattened; stem not winged . . . . . **Helianthus**
1 Receptacle without chaff 14
14 Pappus of scales . . . . . **Helenium**
14 Pappus of capillary bristles 15
15 Involucral bracts all of equal length, in one series (sometimes with a few much shorter ones at base) 16
16 Disk flowers fertile, with divided style; stems more or less leafy . . . **Senecio**
16 Disk flowers sterile, with undivided style; stems merely bracteate . . . . . **Tussilago**
15 Involucral bracts imbricate, in more than one series 17
17 Heads large, the disk 3–5 cm wide, the rays 1.5–2.5 cm long . . . . . **Inula**
17 Heads smaller, the disk less than 2 cm wide, the rays less than 1.5 cm long 18
18 Pappus of 2 series of bristles, the outer ones much shorter than the inner 19
19 Ray flowers with an evident pappus like that of the disk flowers . . . . . **Chrysopsis**
19 Ray flowers lacking a pappus . . . . . **Heterotheca**
18 Pappus not differentiated into an inner and an outer series 20
20 Heads more than 1 cm broad from ray apex to ray apex; plant taprooted annual . . . . . **Haplopappus**
20 Heads less than 1 cm broad from ray apex to ray apex; plant fibrous-rooted perennial from a caudex or rhizome 21
21 Leaves more or less glandular-punctate; rays more numerous than the disk flowers and inflorescence a flat-topped cyme . . . . . **Euthamia**
21 Leaves not glandular-punctate (except *S. odora*); plants with either rays less numerous than the disk flowers or inflorescence not a flat-topped cyme . . . . . **Solidago**

KEY C

1 Receptacle without chaff 2
2 Pappus all or in part of capillary bristles 3
3 Involucral bracts in one series, occasionally with a few short basal ones 4
4 Rays less than 2 mm long . . . . . **Conyza**
4 Rays more than 2 mm long . . . . . **Erigeron**
3 Involucral bracts imbricate in several series 5
5 Rays less than 5 mm long, white . . . . . **Solidago**
5 Rays more than 5 mm long, white, blue, or pink 6
6 Disk flowers white; rays 10–25 . . . . . **Solidago**
6 Disk flowers yellow to reddish-purple, usually not white; if white, rays fewer than 10 . . . . . **Aster**

2 Pappus not of capillary bristles, of awns or a short crown or none 7
7 Pappus a short crown or none; heads solitary . . . . **Chrysanthemum**
7 Pappus of 2–4 short awns; heads several in a corymb . . . . **Boltonia**
1 Receptacle chaffy 8
8 Leaves, or many of them, opposite 9
9 Disk flowers sterile, with undivided style; leaves lobed or dissected . . . **Polymnia**
9 Disk flowers fertile, with divided style; leaves not lobed or dissected 10
10 Disk corollas white; pappus a short crown or none; leaves sessile . . . **Eclipta**
10 Disk corollas yellow; pappus of more than 5 well-developed scales, rarely none; leaves petiolate . . . . **Galinsoga**
8 Leaves alternate 11
11 Rays 2–8 cm long . . . . **Echinacea**
11 Rays less than 2 cm long 12
12 Rays more than 8 . . . . **Anthemis**
12 Rays 1–5 13
13 Leaves pinnately dissected . . . . **Achillea**
13 Leaves merely lobed or toothed 14
14 Disk flowers sterile, with undivided style; stem not winged . . . . **Parthenium**
14 Disk flowers fertile, with divided style; stem winged by decurrent petiole bases . . . . **Verbesina**

KEY D

1 Plants thistles, with more or less spiny-margined leaves and usually also with a spiny involucre 2
2 Heads yellow; pappus of awns and bristles . . . . **Cnicus**
2 Heads purple, blue, pink, or white, not yellow; pappus of bristles only 3
3 Pappus of plumose bristles . . . . **Cirsium**
3 Pappus of simple capillary bristles . . . . **Carduus**
1 Plants not thistles, the leaves not spiny-margined 4
4 Pappus partly or wholly of capillary bristles 5
5 Receptacle chaffy or bristly 6
6 Involucral bracts terminating with an inwardly hooked prickle; largest leaves more than 10 cm wide . . . . **Arctium**
6 Involucral bracts not terminating with an inwardly hooked prickle; largest leaves less than 10 cm wide . . . . **Centaurea**
5 Receptacle without chaff 7
7 Heads yellow . . . . **Senecio**
7 Heads white, tan, greenish, or pink to purple, not yellow 8
8 Pappus distinctly double, with an inner series of long bristles and an outer series of very short bristles . . . . **Vernonia**
8 Pappus simple, of similar bristles 9
9 Pappus bristles barbellate or plumose 10
10 Corollas whitish . . . . **Kuhnia**
10 Corollas purplish . . . . **Liatris**
9 Pappus bristles smooth or scabrous, but not plumose 11

11 Involucre of 4 bracts; twining vines .... **Mikania**
11 Involucre of more than 4 bracts; erect or scrambling, but not twining 12
12 Involucre of 1 series of uniform bracts, sometimes with a few much smaller ones at base 13
13 Leaves lanceolate, pinnately veined .... **Erechtites**
13 Leaves reniform or deltoid, palmately veined .... **Cacalia**
12 Involucre of more than 1 series of bracts 14
14 Leaves chiefly or wholly basal, forming a basal mat .... **Antennaria**
14 Leaves chiefly or all cauline, not forming a basal mat 15
15 Stems and lower leaf surfaces white-woolly 16
16 Heads all alike, the outer flowers pistillate and the inner ones perfect .... **Gnaphalium**
16 Heads of 2 sorts, those on some plants wholly staminate, those on other plants largely pistillate .... **Anaphalis**
15 Stems and lower leaf surfaces not white-woolly 17
17 Flowers all perfect and fertile; leaves, at least the lower, usually opposite or whorled .... **Eupatorium**
17 Outer flowers of head pistillate; leaves alternate ... 18
18 Largest leaves more than 1 cm wide ..... **Pluchea**
18 Largest leaves less than 1 cm wide ...... **Conyza**
4 Pappus of a few awns or scales, or absent 19
19 Heads unisexual; pistillate involucre nutlike or burlike 20
20 Leaves cordate to subcordate; pistillate involucre uniformly spiny with numerous hooked spines .... **Xanthium**
20 Leaves cuneate; pistillate involucre tuberculate only at or near apex .... **Ambrosia**
19 Heads bisexual; involucre various 21
21 Receptacle chaffy or bristly 22
22 Involucral bracts in 2 series, the outer ones spreading .... **Bidens**
22 Involucral bracts in 1 to several series, but not biseriate and dimorphic 23
23 Involucral bracts with a modified tip 24
24 Involucral bracts terminating with an inwardly hooked prickle; largest leaves more than 10 cm wide .... **Arctium**
24 Involucral bracts not terminating with an inwardly hooked prickle; largest leaves less than 10 cm wide .... **Centaurea**
23 Involucral bracts without a modified tip 25
25 Heads white, pink, or blue 26
26 Pappus a low crown or none; heads white .... **Eclipta**
26 Pappus of 5 short scales; heads usually pink or blue .... **Marshallia**
25 Heads dull yellow or greenish 27
27 Larger leaves more than 8 cm wide, lobed or dissected .... **Polymnia**

27 Larger leaves less than 8 cm wide, not lobed or dissected . . . . **Iva**

21 Receptacle without chaff 28

28 Leaves pinnatifid or pinnately dissected; corollas yellow or greenish 29

29 Corollas greenish; inflorescence racemose or paniculate; leaves glabrous or white-tomentose beneath . . . . **Artemisia**

29 Corollas yellow; inflorescence corymbiform or heads solitary; leaves glabrous or nearly so 30

30 Receptacle flat or somewhat convex; plant more than 4 dm tall; perennial . . . . **Tanacetum**

30 Receptacle conic; plant less than 4 dm tall; annual . . . **Matricaria**

28 Leaves entire or toothed; corollas white or purple 31

31 Leaves opposite . . . . **Ageratum**

31 Leaves alternate or basal . . . . **Elephantopus**

Achillea L. (Yarrow)

Mulligan, G. A., and I. J. Bassett. 1959. *Achillea millefolium* complex in Canada and portions of the United States. *Canad. J. Bot.* 37:73–79.

A. millefolium L.; fields and waste places; common; Apr–Oct; ALL. Our material is probably referable to subsp. *lanulosa* (Nutt.) Piper, a North American taxon that is distinguished primarily by ploidy level (tetraploid) from the Eurasian hexaploid subsp. *millefolium*.

Ageratum L. (Ageratum)

A. houstonianum Miller*; escaping infrequently to fields and waste places; Jul–Oct; ALL.

Ambrosia L. (Ragweed)

1 Staminate heads sessile . . . . **A. bidentata**

1 Staminate heads pedunculate 2

2 Leaves 3(5)-lobed or unlobed . . . . **A. trifida**

2 Leaves deeply dissected . . . . **A. artemisiifolia**

A. artemisiifolia L., Common R.; fields and waste places; common; Aug–Oct; ALL.

A. bidentata Michx.; fields; infrequent; Aug–Oct; NC, TN.

A. trifida L., Giant R.; fields and waste places; common; Jul–Oct; ALL.

Anaphalis DC. (Pearly Everlasting)

A. margaritacea (L.) Benth. & Hook.; dry open habitats; infrequent; Jul–Sep; TN, VA.

Antennaria Gaertn. (Pussy-toes, Ladies'-tobacco)

Bayer, R. J., and G. L. Stebbins. 1982. A revised classification of *Antennaria* (Asteraceae: Inuleae) of the eastern United States. *Systematic Botany* 7:300–313.

1 Heads solitary . . . . **A. solitaria**

1 Heads 2 or more/inflorescence 2

2 Basal leaves prominently 3–7-nerved 3
3 Pistillate involucres 5–7 mm high; pistillate corollas 3–4 mm high; staminate corollas 2–3.5 mm high; basal leaves tomentose adaxially; young stolons mostly ascending . . . . **A. plantaginifolia**
3 Pistillate involucres 7–10 mm high; pistillate corollas 4–7 mm high; staminate corollas 3.5–5 mm high; basal leaves tomentose or glabrous adaxially; young stolons mostly decumbent . . . . **A. parlinii**
2 Basal leaves prominently 1-nerved 4
4 Stolons 8–12 cm long, procumbent; leaves along stolon smaller than those of basal rosette; basal leaves gradually tapering to the base, nonpetiolate . . . **A. neodioica**
4 Stolons 5–8 cm long, decumbent; leaves along stolon about equal in size to those of basal rosette; basal leaves having a more or less distinct petiole . . . **A. virginica**

A. neodioica Greene; woods and fields; occasional; Mar–May; NC, SC, TN, VA. Represented in our area by subsp. **petaloidea** (Fern.) Bayer & Stebbins.
A. parlinii Fern.; woods and fields; occasional; Mar–May; ALL.
A. plantaginifolia (L.) Richardson; open dry woods; frequent; Mar–May; ALL.
A. solitaria Rydb.; woods; infrequent; Mar–May; ALL.
A. virginica Stebbins; woods and roadsides; infrequent; Mar–May; VA.

Anthemis L. (Dogfennel)

1 Rays pistillate; receptacle chaffy throughout; plant not ill-scented . . . . . . **A. arvensis**
1 Rays sterile; receptacle chaffy only toward middle; plant ill-scented . . . . . . **A. cotula**

A. arvensis L.*, Corn Chamomile; fields and waste places; frequent; Apr–Jul; ALL.
A. cotula L.*, Mayweed; fields and waste places; frequent; May–Jul; ALL.

Arctium L. (Burdock)

A. minus Schkuhr*, Common B.; roadsides and waste places; infrequent; Jul–Oct; NC, SC, TN, VA.

Artemisia L. (Wormwood)

1 Leaves grayish-green beneath; plant perennial . . . . **A. vulgaris**
1 Leaves green and glabrous beneath; plant annual . . . . **A. annua**

A. annua L.*, Sweet W.; fields and waste places; infrequent; Aug–Oct; VA.
A. vulgaris L.*, Mugwort; fields and waste places; infrequent; Aug–Oct; NC, SC, TN, VA.

Aster L. (Aster)

Contributed by F. F. Fusiak

Hill, L. M. 1980. The genus *Aster* (Asteraceae) in Virginia. *Castanea* 45:104–124.
Jones, A. G. 1978. The taxonomy of *Aster* sect. *Multiflori* (Asteraceae). I. Nomenclatural review and formal presentation of taxa. *Rhodora* 80:319–357.
Jones, R. L. 1983. A systematic study of *Aster* section *Patentes* (Asteraceae). *Sida* 10:41–81.
Reveal, J. L., and C. S. Keener. 1981. *Virgulus* Raf. (1837), an earlier name for *Lasallea* Greene (1903) (Asteraceae). *Taxon* 30:648–651.

Semple, J., and L. Brouillet. 1980. A synopsis of North American asters: The subgenera, sections and subsections of *Aster* and *Lasallea*. *Amer. J. Bot.* 67:1010–1026.
Shinners, L. 1945. The genus *Aster* in West Virginia. *Castanea* 10:61–74.

1 Pappus of 2 distinct series, the outer series shorter than the inner series 2
2 Rays violet; leaves thick, linear, 1-nerved, without secondary venation .... **A. linariifolius**
2 Rays white; leaves thin, not linear, distinctly veined 3
3 Achenes pubescent; heads more than 50/plant; plants with creeping rhizomes .... **A. umbellatus**
3 Achenes glabrous; heads less than 40/plant; plants without creeping rhizomes .... **A. infirmus**
1 Pappus of 1 series, not divided into distinct series 4
4 Lower or basal leaves cordate and petiolate 5
5 Inflorescence corymbiform 6
6 Rays white; inflorescence without glands 7
7 Plants with tufts of basal leaves on separate shoots; involucre cylindric; phyllaries ascending, not tightly appressed .... **A. schreberi**
7 Plants without tufts of basal leaves on separate shoots; involucre ovoid; phyllaries tightly appressed 8
8 Rays less than 10/head; involucre 5–8 mm high .... **A. divaricatus** var. **divaricatus**
8 Rays more than 10/head; involucre 7–10 mm high .... **A. divaricatus** var. **chlorolepis**
6 Rays purple or bluish; inflorescence glandular .... **A. macrophyllus**
5 Inflorescence paniculate or racemiform 9
9 Leaves entire 10
10 Heads large; rays more than 20/head; involucre 7–12 mm high; phyllary tips reflexed; cauline leaves linear, greater than 5 cm long, not auriculate clasping .... **A. curtisii**
10 Heads small; rays mostly 10–20/head; involucre less than 7 mm high; phyllary tips appressed; cauline leaves cordate, less than 4 cm long, auriculate clasping .... **A. undulatus**
9 Leaves distinctly serrate or toothed 11
11 Cauline leaves evidently petiolate and cordate to ovate 12
12 Leaves scabrous or hispid above, below, or both 13
13 Upper leaves directly below the inflorescence tapering to the stem, approaching lanceolate and sessile; involucral bracts linear, acute and subulate .... **A. sagittifolius**
13 Upper leaves directly below the inflorescence distinctly petiolate, not tapering to the stem, cordate to ovate; involucral bracts linear, obtuse and not subulate .... **A. cordifolius**
12 Leaves glabrous .... **A. lowrieanus**
11 Cauline leaves sessile and linear to lanceolate .... **A. curtisii**
4 Lower or basal leaves not both cordate and petiolate 14
14 Involucre or peduncles or both glandular 15
15 Cauline leaves distinctly cordate-clasping, almost encircling the stem, approaching perfoliate 16

16 Stem leaves thick, usually less than 8 cm long; disk corollas yellow; anthers yellow . . . . **A. patens**
16 Stem leaves thin, usually over 8 cm long; disk corollas white; anthers purplish . . . . **A. phlogifolius**
15 Cauline leaves merely auriculate-clasping, not encircling the stem 17
17 Inflorescence corymbiform; lower leaves linear to lanceolate, persistent and well developed, lowermost leaves tapering to a well-developed petiole; heads few, less than 20/plant . . . . **A. surculosus**
17 Inflorescence paniculate or heads solitary at tips of leafy elongated branchlets; lower leaves deciduous, those on the branchlets oblong, numerous, and reduced; heads numerous, more than 25/plant 18
18 Rays few, 15–40/head 19
19 Heads 1–2 cm broad; plants distinctly erect . . . **A. grandiflorus**
19 Heads 0.5–1.0 cm broad; plants more or less prostrate . . . . **A. oblongifolius**
18 Rays numerous, more than 50/head . . . . **A. novae-angliae**
14 Involucre and peduncles without glands 20
20 Cauline leaves cordate- or auriculate-clasping 21
21 Cauline leaves distinctly cordate-clasping, almost encircling the stem, approaching perfoliate 22
22 Stem leaves thick, usually less than 8 cm long; disc corollas yellow; anthers yellow . . . . **A. patens**
22 Stem leaves thin, usually more than 8 cm long; disc corollas white; anthers purplish . . . . **A. phlogifolius**
21 Cauline leaves merely auriculate-clasping, not encircling the stem . . 23
23 Basal or lower leaves petiolate; upper leaves tapering to the stem or having winged petioles 24
24 Leaves scabrous above 25
25 Heads numerous, more than 40/plant; leaves ovate to lanceolate; upper leaves with broadly winged petioles and strongly auriculate . . . . **A. prenanthoides**
25 Heads less than 25/plant; leaves linear to lanceolate; upper leaves linear, sessile and merely auriculate . . . **A. surculosus**
24 Leaves glaucous above 26
26 Leaves more than 2.5 cm wide and less than 5× as long as wide . . . . **A. laevis** var. **laevis**
26 Leaves less than 2 cm wide and more than 5× as long as wide . . . . **A. laevis** var. **concinnus**
23 Basal and upper leaves sessile, not petiolate or tapering to the stem 27
27 Stem with spreading hispid hairs visible to the naked eye . . . . **A. puniceus**
27 Stem glabrous . . . . **A. lucidulus**
20 Cauline leaves not cordate- or auriculate-clasping 28
28 Basal leaves elongate, firm and narrow, grasslike, more than 15× as long as wide 29
29 Inflorescence corymbiform . . . . **(A. avitus)**
29 Inflorescence racemiform or spiciform . . . . **A. hemisphaericus**

28 Basal leaves otherwise, not grasslike 30
30 Rays 3–8/head 31
31 Leaves entire 32
32 Leaves glabrous, linear or narrow . . . . **A. solidagineus**
32 Leaves pubescent, broad, spatulate to obovate . . . . **A. tortifolius**
31 Leaves serrate or toothed . . . . **A. paternus**
30 Rays more than 9/head 33
33 Leaves silvery-silky on both sides . . . . **A. concolor**
33 Leaves not silvery-silky 34
34 Achenes glandular . . . . **A. acuminatus**
34 Achenes glabrous or pubescent, not glandular 35
35 Peduncles bracteolate; lower leaves below the inflorescence absent at maturity; leaves within the inflorescence usually less than 2 cm long 36
36 Peduncles glabrous; involucral bracts tightly appressed, not mucronate 37
37 Plants with short or no rhizome; peduncles long (up to 15 cm) and copiously bracteolate; bracts less than 2 cm long . . . . **A. dumosus**
37 Plants with creeping rhizomes; peduncles short (up to 2 cm long) and sparsely bracteolate; bracts leaflike, usually more than 2 cm long . . . **A. vimineus**
36 Peduncles evidently pubescent; involucral bracts loose, usually with mucronate tips 38
38 Involucral bracts pubescent . . . . **A. ericoides**
38 Involucral bracts glabrous . . . . **A. pilosus**
35 Peduncles not bracteolate, if so, then the bracts not numerous; lower leaves below the inflorescence usually persistent, lanceolate and greater than 2 cm long 39
39 Leaves serrate or toothed 40
40 Rays white 41
41 Disc florets deeply lobed, more than 1/2 of the floret 42
42 Leaves pubescent only on the midrib beneath . . . . **A. lateriflorus**
42 Leaves generally pubescent over entire undersurface . . . . **A. ontarionis**
41 Disc florets shallowly lobed, usually less than 2/5 of the floret . . . . **A. lanceolatus**
40 Rays purple or bluish-purple 43
43 Basal leaves 3–10 cm wide and much larger than the stem leaves; achenes pubescent; plants mostly more than 1 m tall . . . . . . **A. tataricus**
43 Basal leaves deciduous at anthesis; achenes glabrous; plants mostly less than 1 m tall . . . . **A. radula**
39 Leaves entire, not evidently serrate 44

44 Rays purple to bluish-purple 45
45 Leaves glaucous . . . **A. laevis** var. **concinnus**
45 Leaves scabrous . . . . **A. praealtus**
44 Rays white . . . . **A. pilosus**

A. acuminatus Michx.; rich woods; common; Jul–Sep; GA, NC, TN, VA.

A. avitus Alexander; granite outcrops; infrequent; Aug–Sep; NC?, SC.

A. concolor L.; dry sandy woods; occasional; Aug–Oct; ALL. *Virgulus concolor* (L.) Reveal & Keener.

A. cordifolius L.; rich woods, woodland edges, and disturbed areas; common; Aug–Oct; GA, NC, TN, VA. This species, *A. lowrieanus* Porter, and *A. sagittifolius* Willd. are extremely variable and form a morphological continuum in need of further study.

A. curtisii T. & G.; dry woods; common; Aug–Oct; GA, NC, SC, TN.

A. divaricatus L. var. **chlorolepis** (Burgess) Ahles; woodlands and open areas at high elevations; frequent; Aug–Oct; GA, NC, SC, TN.

A. divaricatus L. var. **divaricatus**; woodlands and thickets; common; Aug–Oct; ALL.

A. dumosus L.; dry woods, usually with sandy soils; common; Aug–Oct; ALL.

A. ericoides L.; dry open areas; infrequent; Aug–Oct; VA. *Virgulus ericoides* (L.) Reveal & Keener.

A. grandiflorus L.; sandy woods, roadsides, and old fields; occasional; Aug–Oct; NC, TN, VA. *Virgulus grandiflorus* (L.) Reveal & Keener.

A. hemisphaericus Alexander; open woods; infrequent; Aug–Oct; TN. *A. paludosus* Ait. subsp. *hemisphaericus* (Alexander) Cronquist.

A. infirmus Michx.; woodlands; common; Aug–Oct; ALL. *Doellingeria infirma* (Michx.) Greene.

A. laevis L. var. **concinnus** (Willd.) House; dry woods and open areas; occasional; Aug–Oct; GA, NC, TN, VA.

A. laevis L. var. **laevis**; dry woods and open areas; occasional; Aug–Oct; NC, TN, VA.

A. lanceolatus Willd.; woods, roadsides, open disturbed areas; frequent; Aug–Oct; ALL. *A. simplex* Willd.

A. lateriflorus (L.) Britt.; woods, roadsides, and open areas; common; Aug–Oct; ALL.

A. linariifolius L.; dry open woods; common; Aug–Oct; ALL.

A. lowrieanus Porter; woodland edges and open areas; frequent; Aug–Oct; ALL. See taxonomic note under *A. cordifolius*.

A. lucidulus (Gray) Wieg.; moist areas; infrequent; Aug–Oct; NC.

A. macrophyllus L.; rich woods; common; Aug–Oct; ALL.

A. novae-angliae L., New England A.; moist, open wooded areas; frequent; Aug–Oct; GA, NC, TN, VA. *Virgulus novae-angliae* (L.) Reveal & Keener.

A. oblongifolius Nutt.; shale barrens and roadsides; occasional; Aug–Oct; NC, TN, VA. *Virgulus oblongifolius* (L.) Reveal & Keener.

A. ontarionis Wieg.; river bottoms; infrequent; Aug–Oct; TN.

A. patens Ait.; dry woods; common; Aug–Oct; ALL. *Virgulus patens* (Ait.) Reveal & Keener.

A. paternus Cronquist; dry woods; common; Aug–Sep; ALL. *Sericocarpus asteroides* (L.) BSP.

A. phlogifolius Muhl. ex Willd.; moist rich woods; occasional; Aug–Oct; GA, NC, TN, VA. *A. patens* Ait. var. *phlogifolius* (Muhl.) Nees.

A. **pilosus** Willd.; open dry areas; common; Aug–Oct; ALL.
A. **praealtus** Poir.; rich woods; occasional; Sep–Oct; TN.
A. **prenanthoides** Muhl.; moist woods; frequent; Sep–Oct; NC, TN, VA.
A. **puniceus** L.; moist lowlands, roadsides; occasional; Aug–Oct; NC, SC, TN, VA.
A. **radula** Ait.; bogs, streamsides, and swampy woods; infrequent; Jul–Sep; VA.
A. **sagittifolius** Willd.; rich woods and woodland edges; frequent; Aug–Oct; GA, NC, TN, VA. See taxonomic note under *A. cordifolius*.
A. **schreberi** Nees; rich woods and roadsides; occasional; Aug–Oct; TN, VA.
A. **solidagineus** Michx.; dry woods and open areas; frequent; Jul–Sep; ALL. *Sericocarpus linifolius* (L.) BSP.
A. **surculosus** Michx.; sandy woods and open areas; occasional; Aug–Oct; ALL.
A. **tataricus** L. f.*; old fields and roadsides; occasional; Sep–Oct; GA, NC, TN, VA.
A. **tortifolius** Michx.; woodlands and old fields; infrequent; Jun–Sep; GA. *Sericocarpus bifoliatus* (Walt.) Porter.
A. **umbellatus** Mill.; roadsides and woodland edges; common; Aug–Oct; GA, NC, TN, VA. *Doellingeria umbellata* (Mill.) Nees.
A. **undulatus** L.; dry open woods; common; Aug–Oct; ALL.
A. **vimineus** Lam.; moist open areas; infrequent; Aug–Oct; TN, VA.

Bidens L. (Beggar's Ticks)

1 Achenes linear, not widened upward, sometimes narrowed above; rays short, cream-colored .... **B. bipinnata**
1 Achenes otherwise, either broader than linear or cuneate and tapering toward base; rays yellow or none 2
 2 Leaves unlobed, or with 2 small lateral lobes 3
  3 Leaves sessile (except sometimes the lower); heads nodding in age; involucre more than 2 cm wide 4
   4 Rays absent or less than 1.5 cm long .... **B. cernua**
   4 Rays more than 1.5 cm long .... **B. laevis**
  3 Leaves petiolate; heads erect; involucre less than 1.5 cm wide 5
   5 Rays absent or less than 0.5 cm long .... **B. tripartita**
   5 Rays more than 1 cm long .... **B. mitis**
 2 Leaves lobed, dissected, or compound 6
  6 Rays more than 1 cm long 7
   7 Outer involucral bracts coarsely ciliate, appearing almost serrulate .... **B. polylepis**
   7 Outer involucral bracts entire, often finely ciliate .... **B. aristosa**
  6 Rays absent or less than 0.5 cm long 8
   8 Outer involucral bracts 3–5, not ciliate .... **B. discoidea**
   8 Outer involucral bracts 5–16 or more, ciliate at least near base 9
    9 Outer involucral bracts more than 9, coarsely ciliate .... **B. vulgata**
    9 Outer involucral bracts fewer than 9, finely and usually sparsely ciliate .... **B. frondosa**

B. **aristosa** (Michx.) Britton; wet places; infrequent; Sep–Oct; NC, SC, VA.
B. **bipinnata** L., Spanish Needles; moist waste places; common; Jul–Oct; ALL.
B. **cernua** L.; low wet places; infrequent; Jul–Oct; NC, TN, VA.
B. **discoidea** (T. & G.) Britton; wet places; infrequent; Sep–Nov; SC, VA.

B. frondosa L.; wet places and waste places; frequent; Sep–Oct; ALL.
B. laevis (L.) BSP.; wet places; infrequent; Sep–Oct; VA.
B. mitis (Michx.) Sherff; wet places; infrequent; Aug–Oct; NC.
B. polylepis Blake; wet places; occasional; Aug–Oct; NC, SC, TN, VA.
B. tripartita L.; wet places; infrequent; Sep–Oct; NC, SC, TN, VA.
B. vulgata Greene; fields, wet places, and waste places; occasional; Sep–Oct; GA, NC, VA.

Boltonia L'Her.

B. asteroides (L.) L'Her.; moist places; infrequent; Aug–Oct; VA.

Cacalia L. (Indian Plantain)

Pippen, R. W. 1978. *Cacalia. N. Amer. Fl.* II. 10:151–159.

1 Involucral bracts mostly 10–15; flowers 20–40 2
  2 Leaves chiefly cauline, hastate; plants mostly 1–2.5 m tall .... **C. suaveolens**
  2 Leaves mostly basal, not hastate; plants less than 1 m tall .... **C. rugelia**
1 Involucral bracts 5; flowers 5 3
  3 Leaves pale and glaucous beneath; stem smooth or slightly striate .... **C. atriplicifolia**
  3 Leaves green beneath, not glaucous; stem conspicuously grooved .... **C. muhlenbergii**

C. atriplicifolia L., Pale I. P.; woods and pastures; frequent; Jun–Oct; ALL.
C. muhlenbergii (Schultz-Bip.) Fern., Great I. P.; open woods; infrequent; Jun–Oct; GA, NC, TN, VA.
C. rugelia (Shuttleworth ex Chapman) Barkley & Cronquist, Rugel's Ragwort; forest openings at high elevations of the Great Smoky Mts.; infrequent, but locally abundant; Jun–Aug; NC, TN. *Senecio rugelia* (Shuttleworth ex Chapman) Gray.
C. suaveolens L.; wet woods; infrequent; Aug–Oct; NC, VA.

Carduus L. (Thistle)

1 Involucral bracts mostly 2 mm wide or more; heads mostly nodding ..... **C. nutans**
1 Involucral bracts less than 2 mm wide; heads erect .... **C. acanthoides**

C. acanthoides L.*, Plumeless T.; weed of pastures, fields, and roadsides; infrequent; Jun–Oct; NC, VA.
C. nutans L.*, Musk T.; weed of pastures, fields, and roadsides; infrequent; Jun–Oct; TN, VA.

Centaurea L. (Star Thistle, Knapweed)

1 Leaves all pinnatifid, with narrow lobes .... **C. maculosa**
1 Leaves entire or toothed, or some of the larger ones few-lobed 2
  2 Leaves less than 1 cm wide; pappus present; plants annual .... **C. cyanus**
  2 Largest leaves more than 1 cm wide; pappus absent; plants perennial .... **C. dubia**

C. cyanus L.*, Cornflower, Bachelor's Buttons; fields and waste places; infrequent; Apr–Jun; NC, SC, TN, VA.
C. dubia Suter*, Short-fringed K.; fields and roadsides; infrequent; Jul–Sep; VA. *C. nigrescens* Willd.

C. maculosa Lam.*, Spotted K.; fields and waste places; frequent; Jun–Oct; NC, TN, VA.

Chondrilla L. (Skeleton-weed)

C. juncea L.*; fields and waste places; infrequent; Jun–Aug; VA.

Chrysanthemum L. (Ox-eye Daisy)

C. leucanthemum L.; fields and roadsides; common; Apr–Jul; NC, SC, TN, VA.

Chrysogonum L. (Green-and-gold)

1 Plants stoloniferous; stems usually 1 dm tall or less .... **C. virginianum** var. **australe**
1 Plants not stoloniferous; stems to 4 dm tall ...... **C. virginianum** var. **virginianum**

C. virginianum L. var. **australe** (Alexander) Ahles; woods; infrequent; Mar–Jun; NC, VA.

C. virginianum L. var. **virginianum**; woods; infrequent; Mar–Jun; GA, NC, SC, VA.

Chrysopsis (Nutt.) Ell. (Golden Aster)

Semple, J. C. 1981. A revision of the goldenaster genus *Chrysopsis* (Nutt.) Ell. nom. cons. (Compositae-Astereae). *Rhodora* 83:323–384.

Semple, J. C., and F. D. Bowers. 1985. *A revision of the goldenaster genus* Pityopsis *Nutt. (Compositae: Astereae).* Univ. Waterloo Biology Series, no. 29.

1 Achenes slender and elongate; leaves linear, grasslike 2
  2 Basal leaves shorter than stem leaves; mid and upper stem leaves similar .. **C. ruthii**
  2 Basal leaves usually much longer than stem leaves; stem leaves usually reduced .. 3
    3 Peduncles and involucral bracts stipitate-glandular .... **C. graminifolia** var. **aspera**
    3 Peduncles not densely glandular .... **C. graminifolia** var. **latifolia**
1 Achenes relatively short and broad; leaves not linear 4
  4 Leaves and stems rather coarsely hairy, often glandular; plants spreading by slender, creeping rhizomes .... **C. camporum**
  4 Leaves and stems very finely and softly hairy, not glandular; plants fibrous-rooted from a caudex or short, woody rhizome .... **C. mariana**

C. camporum Greene*; fields and roadsides; infrequent; Jul–Oct; TN, VA.

C. graminifolia (Michx.) Ell. var. **aspera** (Shuttlew.) Gray; sandy dry places; infrequent; Aug–Oct; GA, NC, SC. *Pityopsis aspera* (Shuttlew. ex Small) Small var. *adenolepis* (Fern.) Semple & Bowers, *Heterotheca adenolepis* (Fern.) Ahles.

C. graminifolia (Michx.) Ell. var. **latifolia** Fern.; sandy dry places; infrequent; Jul–Oct; ALL. *Heterotheca correllii* (Fern.) Ahles, *Heterotheca nervosa* (Willd.) Shinners, *Pityopsis graminifolia* (Michx.) Ell. var. *latifolia* (Fern.) Semple & Bowers.

C. mariana (L.) Ell.; old fields and woods; frequent; Jun–Oct; ALL. *Heterotheca mariana* (L.) Shinners.

C. ruthii Small; phyllite rocks along the Hiwassee and Ocoee river gorges in Polk Co., TN; rare; Sep–Oct; TN. *Pityopsis ruthii* (Small) Small, *Heterotheca ruthii* (Small) Harms.

Cichorium L. (Chicory)

C. intybus L.*; fields and waste places; common; May–Oct; ALL.

Cirsium Mill. (Thistle)

Moore, R. J., and C. Frankton. 1969. Cytotaxonomy of some *Cirsium* species of the eastern United States, with a key to eastern species. *Canad. J. Bot.* 47:1257–1275.

1 Plants perennial by rhizomes, forming large colonies . . . . **C. arvense**
1 Plants biennial or perennial, not forming large colonies 2
  2 Leaves usually decurrent to the next leaf below . . . . **C. vulgare**
  2 Leaves not decurrent, or only slightly so 3
    3 Involucral bracts not spine-tipped . . . . **C. muticum**
    3 Involucral bracts with an evident spine at least 1 mm long 4
      4 Lower leaf surface villous or hirsute . . . . **C. pumilum**
      4 Lower leaf surface densely white-tomentose 5
        5 Heads relatively large, the involucre 2–3.5 cm high 6
          6 Leaves deeply pinnatifid . . . . **C. discolor**
          6 Leaves merely toothed or shallowly lobed . . . . **C. altissimum**
        5 Heads relatively small, the involucre to 2.2 cm high . . . . **C. carolinianum**

C. altissimum (L.) Spreng.; pastures, woods, and thickets; frequent; Sep–Oct; ALL. *Carduus altissimus* L.

C. arvense (L.) Scop.*, Canada T.; fields and waste places; infrequent; Jul–Oct; NC, TN, VA. *Carduus arvensis* (L.) Robson.

C. carolinianum (Walt.) Fern. & Schubert; open, dry sandy woods; infrequent; May–Jul; GA, NC, SC, TN. *Carduus carolinianus* Walt.

C. discolor (Muhl.) Spreng.; fields and open woods; infrequent; Aug–Oct; NC, SC, TN, VA. *Carduus discolor* (Muhl. ex Willd.) Nutt.

C. muticum Michx., Swamp T.; wet woods and meadows; infrequent; Aug–Oct; NC, SC, TN, VA. *Carduus muticus* (Michx.) Pers.

C. pumilum (Nutt.) Spreng., Pasture T; old fields and open woods; infrequent; Jul–Sep; VA.

C. vulgare (Savi) Tenore*, Bull T.; fields and waste places; frequent; Jul–Oct; NC, SC, TN, VA.

Cnicus L. (Blessed Thistle)

C. benedictus L.*; fields and roadsides; infrequent; Mar–Jun; SC, TN, VA.

Conyza Less. (Horseweed)

1 Stems usually pubescent; involucral bracts all green . . . **C. canadensis** var. **canadensis**
1 Stems glabrous; involucral bracts green tipped with purple . . . . **C. canadensis** var. **pusilla**

C. canadensis (L.) Cronq. var. **canadensis**; waste places; frequent; Jul–Oct; ALL. *Erigeron canadensis* L.

C. canadensis (L.) Cronq. var. **pusilla** (Nutt.) Cronq.; waste places; frequent; Jul–Oct; NC, SC. *Erigeron canadensis* L. var. *pusilla* (Nutt.) Ahles.

Coreopsis L. (Tickseed Sunflower)

Smith, E. B. 1976. A biosystematic survey of *Coreopsis* in eastern United States and Canada. *Sida* 6:123–215.

1 Leaves simple or with a single pair of lateral lobes 2
 2 Leaves coarsely serrate . . . . **C. latifolia**
 2 Leaves entire or with a single pair of lateral lobes 3
  3 Lowermost leaves alternate . . . . **C. linifolia**
  3 All leaves opposite 4
   4 Plant caulescent, with leaves chiefly toward the top of the stem . . . . **C. pubescens**
   4 Plant scapose, with leaves chiefly basal 5
    5 Leaves less than 2× as long as wide; plants stoloniferous . . . **C. auriculata**
    5 Leaves more than 3× as long as wide; plants not stoloniferous . . . . **C. lanceolata**
1 Leaves deeply lobed or dissected 6
 6 Leaves sessile, appearing whorled 7
  7 Leaf segments relatively broad, 5–30 mm wide 8
   8 Stems and leaves pubescent . . . . **C. major** var. **major**
   8 Stems and leaves glabrous . . . . **C. major** var. **rigida**
  7 Leaf segments relatively narrow, less than 5 mm wide . . . . **C. verticillata**
 6 Leaves petioled, not appearing whorled 9
  9 Disk flowers 4-lobed at apex; ray flowers usually with a reddish-brown spot at base; annual . . . . **C. tinctoria**
  9 Disk flowers 5-lobed at apex; ray flowers without a reddish-brown spot at base; perennials 10
   10 Plants less than 1 m tall; leaf segments filiform to narrowly lanceolate; ray flowers toothed at apex . . . . **C. grandiflora**
   10 Plants 1–3 m tall; leaf segments lanceolate to narrowly elliptic; ray flowers entire at apex . . . . **C. tripteris**

C. auriculata L.; woods; infrequent; Apr–Jun; NC, SC, TN, VA.

C. grandiflora Hogg; dry, sandy places; infrequent; May–Jul; NC, SC, VA. Possibly introduced in our area.

C. lanceolata L.; dry sandy places; infrequent; Apr–Jun; GA, NC, SC, VA.

C. latifolia Michx.; rich moist woods; infrequent; Aug–Sep; GA, NC, SC, TN.

C. linifolia Nutt.; wet sandy areas; infrequent; Aug–Oct; GA, NC. *C. gladiata* Walt. var. *linifolia* (Nutt.) Cronq.

C. major Walt. var. **major**; dry open woods and fields; common; Jun–Aug; ALL.

C. major Walt. var. **rigida** (Nutt.) Boynton; dry open woods and fields; occasional; Jun–Aug; NC, SC. *C. major* Walt. var. *stellata* (Nutt.) Robinson.

C. pubescens Ell.; woods and meadows; common; Jul–Sep; ALL.

C. tinctoria Nutt.*, Calliopsis; fields, meadows and roadsides; frequent; NC, SC, TN, VA.

C. tripteris L.; moist woods and meadows; frequent; Jul–Sep; ALL.

C. verticillata L.; dry open woods; May–Jul; NC, SC, VA.

Crepis L. (Hawk's Beard)

Babcock, E. B. 1947. *The genus* Crepis. Univ. Calif. Publ. Bot. Vols. 21 and 22.

1 Involucre glabrous or nearly so; involucral bracts keeled .... **C. pulchra**
1 Involucre pubescent; involucral bracts not keeled .... **C. capillaris**

C. capillaris (L.) Wallroth*; fields and roadsides; infrequent; May–Oct; NC, TN, VA.
C. pulchra L.*; fields and waste places; infrequent; Apr–Jul; NC, SC, VA.

Echinacea Moench (Purple Coneflower)

McGregor, R. L. 1968. The taxonomy of the genus *Echinacea* (Compositae). *Univ. Kansas Sci. Bull.* 48:113–142.

1 Leaves predominantly basal; leaf blades lanceolate, less than 5 cm wide .. **E. pallida**
1 Leaves mostly cauline; leaf blades lanceolate to ovate, more than 5 cm wide 2
  2 Leaves pubescent; chaffy bracts with straight tips .... **E. purpurea**
  2 Leaves glabrous or nearly so; chaffy bracts with incurved tips .... **E. laevigata**

E. laevigata (Boynton & Beadle) Blake; fields and woods; infrequent; May–Jul; SC, VA.
E. pallida (Nutt.) Nutt.; dry open places; infrequent; Jun–Jul; NC, VA.
E. purpurea (L.) Moench; fields and woods; infrequent; Jun–Oct; GA, NC, TN.

Eclipta L.

E. alba (L.) Hassk.*; weed of muddy places; infrequent; Jun–Oct; ALL.

Elephantopus L. (Elephant's Foot)

Clonts, J. A., and S. McDaniel. 1978. *Elephantopus. N. Amer. Fl.* II. 10:196–202.

1 Leaves distinctly cauline .... **E. carolinianus**
1 Leaves mostly basal, cauline leaves greatly reduced .... **E. tomentosus**

E. carolinianus Raeuschel; woods; frequent; Aug–Oct; ALL.
E. tomentosus L.; woods; infrequent; Jul–Sep; ALL.

Erechtites Raf. (Fireweed)

E. hieracifolia (L.) Raf.; old fields, woods, and waste places; frequent; Jul–Oct; ALL.

Erigeron L. (Daisy Fleabane)

Cronquist, A. 1947. Revision of the North American species of *Erigeron*, north of Mexico. *Brittonia* 6:121–302.

1 Stem leaves clasping; pappus of ray and disk flowers alike, of elongate capillary bristles 2
  2 Plant stoloniferous; rays about 50, 1 mm wide or wider .... **E. pulchellus**
  2 Plant not stoloniferous; rays about 100 or more, less than 1 mm wide .... **E. philadelphicus**
1 Stem leaves sessile but not clasping; pappus of the ray and disk flowers unlike, that of the ray flowers lacking capillary bristles 3
  3 Pubescence of the stem mostly long and spreading; leaves usually more than 1 cm wide .... **E. annuus**

3 Pubescence of the stem mostly short and appressed; leaves usually less than 1 cm wide . . . . **E. strigosus**

E. **annuus** (L.) Pers.; roadsides, fields, and waste places; common; May–Oct; ALL.
E. **philadelphicus** L.; fields and waste places; infrequent; Apr–Jun; ALL.
E. **pulchellus** Michx., Robin's Plantain; rich wooded slopes; frequent; Apr–Jun; ALL.
E. **strigosus** Muhl. ex Willd.; roadsides, fields, and waste places; common; Apr–Oct; ALL.

Eupatorium L.

Warnock, M. J. 1987. An index to epithets treated by King and Robinson: Eupatorieae (Asteraceae). *Phytologia* 62:345–431.

1 Leaves all or mostly in whorls of 3–7, generally at least 2 cm wide 2
2 Leaves relatively short, less than 12 cm long, strongly triplinerved, and abruptly contracted to the petiole . . . . **E. dubium**
2 Leaves longer, the largest more than 15 cm long, not triplinerved, and gradually narrowed to the petiole 3
3 Flowers mostly 9–22/head; inflorescence flat-topped . . . . **E. maculatum**
3 Flowers mostly 4–7/head; inflorescence convex 4
4 Stem solid, purplish only at the nodes . . . . **E. purpureum**
4 Stem hollow, purplish throughout . . . . **E. fistulosum**
1 Leaves mostly opposite or alternate (if whorled, then less than 2 cm wide) 5
5 Flowers 9 or more/head 6
6 Receptacle conic; flowers more than 35/head, commonly blue . . **E. coelestinum**
6 Receptacle flat or merely convex; flowers fewer than 35/head, normally white or pinkish 7
7 Involucral bracts not strongly imbricate, the principal ones subequal in length, with a few shorter outer ones sometimes also present 8
8 Petioles of the principal leaves usually more than 2 cm long; leaves dull above . . . . **E. rugosum**
8 Petioles of the principal leaves less than 2 cm long; leaves shiny above . . . . **E. aromaticum**
7 Involucral bracts imbricate in 3 or more series, the outer ones less than ½ as long as the inner 9
9 Flowers pink or purplish . . . . **E. incarnatum**
9 Flowers white 10
10 Leaves petiolate . . . . **E. serotinum**
10 Leaves sessile, usually connate-perfoliate . . . . **E. perfoliatum**
5 Flowers fewer than 9/head, mostly 5/head 11
11 Principal leaves pinnatifid to ternate or deeply dissected 12
12 Widest leaf segments less than 0.5 mm wide . . . . **E. capillifolium**
12 Widest leaf segments more than 1 mm wide . . . . **E. compositifolium**
11 Leaves entire to deeply toothed, but not pinnatifid 13
13 Involucral bracts acuminate to attenuate 14
14 Larger leaves mostly 1.5–3 cm wide; involucre 8–11 mm high 15
15 Leaves usually pubescent, mostly coarsely serrate, tending to be obtuse or rounded at tip . . . . **E. album** var. **album**

15 Leaves sparsely pubescent or subglabrous, serrate to subentire, often acute . . . . **E. album** var. **vaseyi**

14 Larger leaves mostly less than 1.5 cm wide; involucre 5–7 mm high . . . . **E. leucolepis**

13 Involucral bracts acute to rounded, not acuminate 16

16 Leaves rounded to truncate at base 17

17 Plant glabrous below the inflorescence . . . . **E. sessilifolium**

17 Plant pubescent below the inflorescence 18

18 Leaves narrow, mostly lanceolate or lance-ovate; upper leaves and main branches of the inflorescence tending to be alternate; chiefly in wet soil . . . . **E. rotundifolium** var. **saundersii**

18 Leaves broad, ovate to subrotund; upper leaves and main branches of the inflorescence mostly opposite 19

19 Principal pair of lateral veins of the leaf diverging from the base of the midrib; leaves blunt, tending to be broadly cuneate or almost truncate at the base . . . . **E. rotundifolium** var. **rotundifolium**

19 Principal pair of lateral veins of the leaf diverging distinctly above the base of the midrib; leaves mostly acute . . . . **E. rotundifolium** var. **ovatum**

16 Leaves tapering to a narrow base 20

20 Plant with conspicuously tuberous-thickened, short rhizomes . . . . **E. mohrii**

20 Plant without tuberous-thickened rhizomes 21

21 Leaves mostly 2.5–7× as long as wide, the larger ones more than 1 cm wide, all opposite or the upper ones alternate . . . . **E. altissimum**

21 Leaves mostly 6–40× as long as wide, seldom more than 1 cm wide, mostly in whorls of 3 or 4, but sometimes merely opposite or even alternate above 22

22 Plants usually 1 m tall or less; leaves usually in whorls of 4, but sometimes in whorls of 3 or opposite or even alternate . . . . **E. hyssopifolium** var. **hyssopifolium**

22 Plants up to 1.5 m tall; leaves usually in whorls of 3, opposite or alternate, sometimes in whorls of 4 . . . . **E. hyssopifolium** var. **laciniatum**

E. album L. var. **album**; dry open woods and fields; frequent; Jun–Sep; ALL.

E. album L. var. **vaseyi** (T. C. Porter) Cronq.; dry open woods and fields; occasional; Jun–Sep; GA, TN, VA.

E. altissimum L.; woods and fields; infrequent; Aug–Oct; SC, TN, VA. *E. saltuense* Fern. may represent a hybrid between *E. altissimum* and other closely related taxa.

E. aromaticum L.; dry woods and fields; frequent; Aug–Oct; ALL. *Ageratina aromatica* (L.) Spach.

E. capillifolium (Lam.) Small, Dogfennel; fields and pastures; infrequent; Sep–Oct; GA, NC, SC, TN.

E. coelestinum L., Mistflower; low woods and meadows; infrequent; Jul–Oct; NC, SC, TN, VA. *Conoclinium coelestinum* (L.) DC.

E. compositifolium Walt., Dogfennel; fields and open woods; infrequent; Sep–Oct; SC.

E. **dubium** Willd., Joe-Pye-weed; meadows and open woods; infrequent; Jul–Oct; GA, NC. *Eupatoriadelphus dubius* (Willd.) K. & R.

E. **fistulosum** Barratt, Joe-Pye-weed; bottomlands and moist woods; common; Jul–Oct; ALL. *Eupatoriadelphus fistulosus* (Barratt) K. & R.

E. **hyssopifolium** L. var. **hyssopifolium**; dry fields and open places; infrequent; Jul–Oct; ALL.

E. **hyssopifolium** L. var. **laciniatum** A. Gray; dry fields and open places; infrequent; Jul–Oct; GA, VA.

E. **incarnatum** Walt.; woods and wet places; infrequent; Jul–Oct; TN, VA. *Fleischmannia incarnata* (Walt.) K. & R.

E. **leucolepis** (DC.) T. & G.; wet places; infrequent; Jul–Oct; VA.

E. **maculatum** L., Joe-Pye-weed; moist places; infrequent; Jul–Oct; NC, SC, TN, VA. *Eupatoriadelphus maculatus* (L.) K. & R.

E. **mohrii** Greene; moist places; infrequent; Jul–Oct; VA.

E. **perfoliatum** L., Boneset; moist or wet places; frequent; Aug–Oct; ALL.

E. **purpureum** L., Joe-Pye-weed; wooded slopes; common; Jul–Oct; ALL. *Eupatoriadelphus purpureus* (L.) K. & R.

E. **rotundifolium** L. var. **ovatum** (Bigel.) Torr.; woods; infrequent; Aug–Oct; ALL. *E. pubescens* Muhl.

E. **rotundifolium** L. var. **rotundifolium**; woods; infrequent; Aug–Oct; ALL.

E. **rotundifolium** L. var. **saundersii** (T. C. Porter) Cronq.; woods; infrequent; Aug–Oct; GA, NC, VA. *E. pilosum* Walt.

E. **rugosum** Houttuyn, White Snakeroot; rich woods; common; Jul–Oct; ALL. Plants of upper elevations with larger, more numerously flowered heads may be segregated as var. *roanense* (Small) Fern.

E. **serotinum** Michx.; old fields and waste places; infrequent; Aug–Oct; ALL.

E. **sessilifolium** L.; woods; occasional; Jul–Oct; GA, NC, TN, VA.

Euthamia Nutt. (Flat-topped Goldenrod)

Sieren, D. J. 1981. The taxonomy of the genus *Euthamia*. *Rhodora* 83:551–579.

1 Leaves with 3(–5) prominent nerves; rays (11–)17–22(–35) .... **E. graminifolia**

1 Leaves with 1 prominent and sometimes a pair of inconspicuous lateral veins; rays 7–15 .... **E. tenuifolia**

E. **graminifolia** (L.) Nutt.; open moist ground; infrequent; Aug–Sep; NC, VA. *Solidago graminifolia* (L.) Salisb.

E. **tenuifolia** (Pursh) Greene; open moist ground; infrequent; Sep–Oct; VA. *Solidago tenuifolia* Pursh.

Galinsoga Ruiz & Pavon (Peruvian Daisy)

Canne, J. M. 1977. A revision of the genus *Galinsoga* (Compositae: Heliantheae). *Rhodora* 79:319–389.

1 Ray corolla to 2.5 mm long; outer involucral bracts 1 or 2, with herbaceous margins; inner chaffy bracts usually entire .... **G. quadriradiata**

1 Ray corolla absent or up to 1.5 mm long; outer involucral bracts 2–4, with scarious margins; inner chaffy bracts deeply 3-lobed .... **G. parviflora**

G. **parviflora** Cav.*; waste places; infrequent; May–Oct; NC, SC, TN, VA.

G. quadriradiata Ruiz & Pavon*; waste places; frequent; May–Oct; ALL. *G. ciliata* (Raf.) Blake.

Gnaphalium L. (Cudweed)

1 Leaves lanceolate; pappus bristles distinct and falling separately 2
  2 Stems white-woolly, glands hidden or none . . . . **G. obtusifolium**
  2 Stems green, conspicuously glandular . . . . **G. helleri**
1 Leaves spatulate or oblanceolate; pappus deciduous as a unit . . . . **G. purpureum**

G. helleri Britton; sandy open places in woods; infrequent; Sep–Oct; GA, SC, TN, VA.
G. obtusifolium L., Rabbit-tobacco; sandy waste places; frequent; Aug–Oct; ALL.
G. purpureum L.; fields, roadsides and waste places; frequent; Mar–Jul; ALL.

Haplopappus Cass.

H. divaricatus (Nutt.) Gray; sandy fields; infrequent; Aug–Oct; GA, SC.

Helenium L. (Sneezeweed)

Bierner, M. W. 1972. Taxonomy of *Helenium* sect. *Tetrodus* and a conspectus of North American *Helenium* (Compositae). *Brittonia* 24:331–355.
______. 1974. The taxa of *Helenium* in Tennessee. *Castanea* 39:346–349.

1 Leaves 0.5–3 mm wide; disk corollas yellow; plant annual . . . . **H. amarum**
1 Leaves more than 3 mm wide; disk corollas yellow or reddish; plant perennial 2
  2 Rays pistillate, with well-developed style; disk corollas yellow . . . . **H. autumnale**
  2 Rays neutral and sterile, lacking style; disk corollas usually reddish 3
    3 Pappus scales not awned; heads usually 1–5/plant . . . . **H. brevifolium**
    3 Pappus scales with a terminal awn; heads usually more than 5/plant . . . . **H. flexuosum**

H. amarum (Raf.) H. Rock; fields and waste places; frequent; Aug–Oct; ALL.
H. autumnale L.; wet fields; infrequent; Aug–Oct; ALL. *H. virginicum* Blake.
H. brevifolium (Nutt.) A. Wood; wet areas; infrequent; May–Jun; NC, SC, VA.
H. flexuosum Raf.; moist places; occasional; May–Oct; NC, SC, TN, VA.

Helianthus L. (Sunflower)

Beatley, J. C. 1963. The sunflowers (genus *Helianthus*) in Tennessee. *J. Tennessee Acad. Sci.* 38:135–154.
Heiser, C. B., Jr., D. M. Smith, S. B. Clevenger, and W. C. Martin, Jr. 1969. The North American sunflowers (*Helianthus*). *Mem. Torrey Bot. Club* 22: (3) 1–218.

1 Plant a tap-rooted annual with mostly alternate, ovate, cordate, long-petiolate leaves and usually red-purple disk flowers . . . . **H. annuus**
1 Plant a fibrous-rooted to tuberous-rooted perennial from a rhizome, tuber, or crown bud; other features not as above 2
  2 Disk flowers red or purple 3
    3 Cauline leaves numerous and not conspicuously reduced upward, sessile, linear to lance-ovate, mostly alternate; involucral bracts narrow, 1–3 mm wide . . . . **H. angustifolius**
    3 Cauline leaves reduced upward, mostly opposite; involucral bracts wider, 2.5–5.5 mm wide . . . . **H. atrorubens**

2 Disk flowers yellow 4
4 Leaves mostly basal, the upper ones reduced . . . . . **H. occidentalis**
4 Leaves well distributed along stem, only gradually reduced upward 5
5 Stems more or less glabrous below the inflorescence 6
6 Heads 0.5–1.5 cm wide; rays 5–8; rhizomes short or none 7
7 Lower leaf surface pubescent and with yellowish, subsessile glands . . . . . **H. microcephalus**
7 Lower leaf surface glabrous, glaucous, and lacking glands 8
8 Leaves distinctly petiolate, the petiole 1–4 cm long . . . . . **H. glaucophyllus**
8 Leaves sessile or subsessile, the petiole less than 1 cm long . . . . . **H. laevigatus**
6 Heads 1.5–3.5 cm wide; rays 10–25; rhizomes present 9
9 Leaves sessile or subsessile, the petiole less than 1 cm long 10
10 Leaves widest near the truncate or rounded base . . . . . **H. divaricatus**
10 Leaves tapering to a narrow base . . . . . **H. giganteus**
9 Leaves distinctly petiolate, the petiole 1–4 cm long 11
11 Leaves relatively narrow, usually at least 3× as long as wide 12
12 Lower leaf surface pale, glabrous, and glaucous . . **H. strumosus**
12 Lower leaf surface green, pubescent, not glaucous . . . . . **H. grosseserratus**
11 Leaves relatively broad, less than 3× as long as wide, often more than 4 cm wide 13
13 Involucral bracts extending beyond disk; leaves generally strongly serrate . . . . . **H. decapetalus**
13 Involucral bracts not extending beyond disk; leaves inconspicuously serrulate . . . . . **H. strumosus**
5 Stem pubescent below the inflorescence 14
14 Leaves with wide, clasping bases, densely soft-pubescent . . . . . **H. mollis**
14 Leaf bases not clasping, leaves not densely soft-pubescent 15
15 Involucral bracts conspicuously reflexed . . . . . **H. resinosus**
15 Involucral bracts not conspicuously reflexed 16
16 Leaves relatively large and long-petiolate; rhizomes tuber-bearing . . . . . **H. tuberosus**
16 Leaves smaller or short-petiolate; rhizomes not tuber-bearing 17
17 Plants with short rhizomes; leaves mostly alternate 18
18 Stems spreading-hirsute; leaves flat . . . . . **H. giganteus**
18 Stems rather loosely white-strigose; leaves often folded together lengthwise . . . . . **H. maximilianii**
17 Plants with elongate rhizomes; leaves mostly opposite . . . . . **H. hirsutus**

H. angustifolius L.; moist or open places; infrequent; Jul–Oct; ALL.

H. annuus L.*, Common S.; fields, roadsides, and waste places; occasional; Jun–Oct; NC, VA.

H. atrorubens L.; open woods and fields; frequent; Jul–Oct; ALL.

H. decapetalus L.; woods and along streams; occasional; Jul–Oct; ALL.

H. divaricatus L.; woods and roadsides; occasional; Jun–Aug; ALL.

H. giganteus L.; wet places; frequent; Jul–Oct; ALL.

H. glaucophyllus D. M. Smith; moist woods and roadsides; infrequent; Sep–Oct; NC, TN.

H. grosseserratus Martens*; roadsides and fields; infrequent; Jul–Sep; TN, VA.

H. hirsutus Raf.; woods and roadsides; occasional; Jul–Oct; NC, SC, TN, VA.

H. laevigatus T. & G.; shale barrens; infrequent; Sep–Oct; VA.

H. maximilianii Schrader*; roadsides and waste places; infrequent; Aug–Oct; TN, VA.

H. microcephalus T. & G.; woods, road banks, and fields; common; Aug–Oct; ALL.

H. mollis Lam.*; roadsides and fields; infrequent; Jul–Sep; NC, TN, VA.

H. occidentalis Riddell; dry, sandy open places; infrequent; Aug–Oct; NC.

H. resinosus Small; woods and meadows; infrequent; Jun–Oct; GA, NC, SC. *H. tomentosus* Michx., misapplied.

H. strumosus L.; woods and open places; frequent; Jul–Sep; ALL.

H. tuberosus L., Jerusalem Artichoke; woodland borders and waste places; occasional; Jul–Oct; ALL. *H.* × *laetiflorus* Pers., hybrid of *H. tuberosus* × *H. pauciflorus*, an occasional escape from cultivation.

Heliopsis Pers. (Ox-eye)

H. helianthoides (L.) Sweet; woods and fields; common; May–Oct; ALL.

Heterotheca Cassini

H. subaxillaris (Lam.) Britton & Rusby; sandy fields and roadsides; infrequent; Jul–Oct; SC, VA. Perhaps not native in our area but likely to become more common.

Hieracium L. (Hawkweed)

1 Corollas red-orange . . . . **H. aurantiacum**
1 Corollas yellow 2
  2 Heads solitary or rarely 2/scape . . . . **H. pilosella**
  2 Heads several to numerous, 5–30 or more 3
    3 Basal leaves few or absent, stem leafy to inflorescence . . . . **H. paniculatum**
    3 Leaves chiefly basal or on lower half of stem 4
      4 Basal leaves narrowly oblanceolate, seldom more than 3 cm wide; heads in compact flat-topped corymbs 5
        5 Leaves glaucous, sparsely hairy to glabrous . . . . **H. florentinum**
        5 Leaves not glaucous, pubescent . . . . **H. caespitosum**
      4 Basal leaves widely oblanceolate to lance-ovate, often more than 3 cm wide; heads in loose concave or convex corymbs or in panicles 6
        6 Lower part of stem glabrous; basal leaves purple-veined . . . . . **H. venosum**
        6 Lower part of stem pubescent; leaves not purple-veined 7
          7 Leaves entirely basal, with at most 1 cauline leaf . . . . (**H. traillii**)
          7 Leaves distributed on lower half of stem 8
            8 Flowers 20–40/head; achenes narrowed toward apex . . . . . **H. gronovii**
            8 Flowers 40–100/head; achenes widest toward apex . . . . . . **H. scabrum**

H. aurantiacum L.*, King Devil, Devil's Paintbrush; fields and roadsides; infrequent; May–Jul; NC, VA.

H. caespitosum Dumort*, King Devil; fields and roadsides; occasional; May–Jul; ALL. *H. pratense* Tausch.

H. florentinum All.*, King Devil; fields and roadsides; infrequent; May–Sep; GA, NC, SC, VA. *H. piloselloides* Vill.

H. gronovii L.; dry open woods and fields; frequent; Jul–Oct; ALL.
H. paniculatum L.; woods; frequent; Jul–Oct; ALL.
H. pilosella L.*, Mouse-ear; fields and open woods; occasional; May–Jul; GA, NC, TN, VA.
H. scabrum Michx.; dry woods; occasional; Jul–Oct; ALL.
H. traillii Greene; shale barrens; infrequent; Apr–Jul; VA.
H. venosum L., Rattlesnake Weed; dry open woods; frequent; Apr–Jul; ALL.

Hypochoeris L. (Cat's Ear)

H. radicata L.*; fields and waste places; infrequent; Apr–Jul; NC, SC, TN, VA.

Inula L. (Elecampane)

I. helenium L.*; fields and roadsides; infrequent; May–Jul; NC, TN, VA.

Iva L. (Marsh Elder)

I. annua L.; waste places; infrequent; Sep–Nov; SC, TN. Perhaps not native to our area.

Krigia Schreber (Dwarf Dandelion)

Shinners, L. H. 1947. Revision of the genus *Krigia* Schreber. *Wrightia* 1:187–206.

1 Plants perennial; pappus bristles 15–40; involucre 7–14 mm high 2
  2 Plant with cauline leaves; inflorescence usually branched with several heads 3
    3 Larger leaves more than 1 cm wide . . . . **K. biflora**
    3 Larger leaves less than 1 cm wide . . . . **K. montana**
  2 Leaves basal; heads solitary and scapose . . . . **K. dandelion**
1 Plants annual; pappus bristles 5–8 or none; involucre 3–7 mm high 4
  4 Pappus clearly present; plant scapose or leafy only near base . . . . **K. virginica**
  4 Pappus absent or vestigial; leaves cauline as well as basal . . . . **K. oppositifolia**

K. biflora (Walt.) Blake; woods and fields; occasional; May–Jul; GA, NC, TN, VA.
K. dandelion (L.) Nutt.; rocky woods and fields; infrequent; Apr–May; TN, VA.
K. montana (Michx.) Nutt.; moist cliff crevices and stream banks; infrequent; May–Sep; GA, NC, SC.
K. oppositifolia Raf.; low moist places; infrequent; Mar–Jun; SC, TN.
K. virginica (L.) Willd.; woods and fields; frequent; May–Jul; ALL.

Kuhnia L. (False Boneset)

K. eupatorioides L.; woods and thickets; frequent; Jun–Oct; NC, SC, TN, VA.

Lactuca L. (Lettuce)

1 Achenes with several prominent nerves on each side; beak of the achene various . . . 2
  2 Achenes with a short stout beak less than 1/2 as long as the body, or beakless; flowers usually blue 3
    3 Pappus light brown; leaves auriculate-clasping . . . . **L. biennis**
    3 Pappus white; leaves without clasping bases . . . . **L. floridana**
  2 Achenes with a filiform beak at least as long as the body; flowers yellow (but commonly blue when dried) 4
    4 Leaves prickly-margined . . . . **L. serriola**
    4 Leaves not prickly-margined . . . . **L. saligna**

1 Achenes with only a single prominent nerve on each side; beak of the achene slender, at least 1/2 as long as the body 5
5 Involucre less than 1.5 cm high in fruit; achenes 4.5–6.5 mm long (including beak) . . . . **L. canadensis**
5 Involucre more than 1.5 cm high in fruit; achenes 7–10 mm long (including beak) . . . 6
6 Stem leaves much reduced; flowers usually blue . . . . **L. graminifolia**
6 Stem leaves not greatly reduced; flowers usually yellow . . . . **L. hirsuta**

L. biennis (Moench) Fern., Tall Blue L.; pastures, thickets, and woodland borders; infrequent; Aug–Oct; NC, VA.
L. canadensis L., Wild L.; fields, woods, and waste places; common; Jun–Oct; NC, SC, TN, VA.
L. floridana (L.) Gaertn., Woodland L.; thickets and woods; occasional; Aug–Oct; ALL.
L. graminifolia Michx.; fields and roadsides; Apr–Jul; infrequent; NC.
L. hirsuta Muhl., Wild L.; woods and clearings; infrequent; May–Oct; NC.
L. saligna L.*, Willow-leaved L.; fields and waste places; infrequent; Aug–Oct; VA.
L. serriola L.*, Prickly L.; fields and waste places; common; Jun–Oct; NC, SC, TN, VA. Represented in our area by var. **integrata** Gren. & Godr. *L. scariola* L.

Lapsana L. (Nipplewort)

L. communis L.*; fields and woods; infrequent; Jun–Aug; NC, VA.

Leontodon L.

L. taraxacoides (Vill.) Merat*; lawns and waste places; infrequent; Jul–Oct; NC, TN.

Liatris Schreber (Blazing Star, Gay Feather)

1 Pappus plumose, the barbels mostly more than 0.5 mm long; involucre 2 cm or more long . . . . **L. squarrosa**
1 Pappus more or less barbellate, the barbels mostly less than 0.5 mm long; involucre less than 2 cm long 2
2 Heads usually with 14 or more flowers; larger leaves more than 1 cm wide 3
3 Involucral bracts glabrous, with wide, scarious margins . . . . **L. aspera**
3 Involucral bracts pubescent, or with narrow, ciliate margins 4
4 Flowers mostly 25–80/head, heads seldom more than 20 . . . . **L. scariosa**
4 Flowers mostly 14–24/head, heads often more than 20 5
5 Middle and outer involucral bracts appressed . . . . **L. turgida**
5 Middle and outer involucral bracts loose or squarrose above middle . . . . **L. squarrulosa**
2 Heads usually with 3–14 flowers; leaves often all less than 1 cm wide 6
6 Inside of corolla tube glabrous 7
7 Heads sessile; flowers 5–14/head . . . . **L. spicata**
7 Heads pedunculate; flowers 3–6/head . . . . **L. microcephala**
6 Inside of corolla tube pubescent 8
8 Pappus 1/2 the length of the corolla tube or less . . . . **L. helleri**
8 Pappus more than 1/2 the length of the corolla tube 9
9 Involucral bracts keeled, firm in texture, and acute, often with a terminal mucro . . . . **L. regimontis**
9 Involucral bracts not keeled, thinner in texture, without a terminal mucro . . 10
10 Larger leaves usually more than 10 mm wide . . . . **L. turgida**
10 Larger leaves usually less than 7 mm wide . . . . **L. graminifolia**

L. aspera Michx.; sandy or rocky open woods; infrequent; Jul–Aug; NC, SC, TN, VA.
L. graminifolia Willd.; dry open woods and fields; infrequent; Sep–Oct; GA, NC, SC, VA.
L. helleri (Porter) Porter; open rocky outcrops; infrequent; Jul–Aug; NC.
L. microcephala (Small) K. Schumann; exposed rocky places and meadows; infrequent; Jul–Sep; NC, TN.
L. regimontis (Small) K. Schumann; sandy fields and woods; infrequent; Aug–Oct; SC.
L. scariosa (L.) Willd.; dry open woods; infrequent; Aug–Sep; ALL.
L. spicata (L.) Willd.; wet meadows; infrequent; Jul–Sep; ALL.
L. squarrosa (L.) Michx.; dry open places; infrequent; Jun–Aug; ALL.
L. squarrulosa Michx.; dry woods and open places; infrequent; Jul–Aug; SC, TN.
L. turgida Gaiser; dry woods; infrequent; Jul–Aug; GA, VA.

Marshallia Schreber (Barbara's Buttons)

Channell, R. B. 1957. A revisional study of the genus *Marshallia*. *Contr. Gray Herb.* 181:41–132.

1 Involucral and receptacular bracts obtuse .... **M. obovata**
1 Involucral bracts acute; receptacular bracts linear 2
  2 Leaves thin, the middle and upper not distinctly reduced; internodes 10–25 .... **M. trinervia**
  2 Leaves thick, the middle and upper decreasing in size; internodes 4–12 .... **M. grandiflora**

M. grandiflora Beadle & Boynton; stream banks, bogs, and flood plains; infrequent; Jun–Jul; NC.
M. obovata (Walter) Beadle & Boynton; meadows, woodlands, and stream banks; infrequent; Apr–Jun; GA, NC, SC, TN.
M. trinervia (Walt.) Trelease; stream banks and damp woods; infrequent; May–Jul; NC.

Matricaria L.

M. matricarioides (Lessing) Porter*, Pineapple-weed; roadsides and waste places; infrequent; Jun–Oct; NC, TN, VA.

Mikania Willd. (Climbing Hempweed)

M. scandens (L.) Willd.; wet woods and marshes; infrequent; Jul–Oct; SC, VA.

Parthenium L. (Wild Quinine)

1 Heads 4–7 mm wide; leaves and stems glabrous or with short, appressed pubescence .... **P. integrifolium**
1 Heads 7–10 mm wide; leaves and stems with longer, spreading pubescence .... **P. hispidum**

P. hispidum Raf.; woods and fields; infrequent; May–Aug; VA.
P. integrifolium L.; woods and fields; infrequent; May–Aug; ALL.

Picris L.

P. hieracioides L.*, Bitterweed; fields and waste places; infrequent; May–Oct; NC, TN.

Pluchea Cass. (Marsh Fleabane)

P. camphorata (L.) DC.; wet woods and fields; infrequent; Aug–Oct; SC, TN, VA.

Polymnia L. (Leaf Cup)

Wells, J. R. 1965. A taxonomic study of *Polymnia* (Compositae). *Brittonia* 17:144–159.

1 Rays yellow; achenes striate with many nerves . . . . **P. uvedalia**
1 Rays white or none; achenes prominently 3-ribbed . . . . **P. canadensis**

P. canadensis L.; moist woods in calcareous soils; infrequent; Jul–Oct; NC, TN, VA.
P. uvedalia (L.) L., Bearsfoot; woods and meadows; common; Jul–Oct; ALL.

Prenanthes L. (Rattlesnake Root)

Fusiak, F., and E. E. Schilling. 1984. Systematics of the *Prenanthes roanensis* complex (Asteraceae: Lactuceae). *Bull. Torrey Bot. Club* 111:338–348.

1 Flowers 4–8/head; inner involucral bracts mostly 4–8/head 2
  2 Inner involucral bracts glabrous; tips of involucral bracts green; inflorescence paniculate . . . . **P. altissima**
  2 Inner involucral bracts pubescent; tips of involucral bracts black; inflorescence thyrsoid . . . . **P. roanensis**
1 Flowers more than 8/head; inner involucral bracts more than 7/head 3
  3 Pappus cinnamon-brown; rays white to pale pink . . . . **P. alba**
  3 Pappus pale tan or cream-colored; rays not white 4
    4 Involucre glabrous; outer involucral bracts triangular-ovate . . . . . . **P. trifoliolata**
    4 Involucre sparsely hairy; outer involucral bracts lanceolate . . . . . **P. serpentaria**

P. alba L., White Lettuce; wooded slopes and road banks; infrequent; Aug–Oct; NC, VA.
P. altissima L.; woods; common; Aug–Oct; ALL.
P. roanensis (Chickering) Chickering; rich woods and open slopes at high elevations; infrequent; Aug–Oct; NC, TN, VA.
P. serpentaria Pursh, Lion's-foot; woods, thickets, and roadsides; frequent; Aug–Oct; ALL.
P. trifoliolata (Cass.) Fern., Gall-of-the-earth; woods; occasional; Aug–Oct; ALL.

Pyrrhopappus DC. (False Dandelion)

P. carolinianus (Walter) DC.; fields and waste places; occasional; Mar–Jun; ALL.

Ratibida Raf. (Coneflower)

R. pinnata (Vent.) Barnhart; fields and dry woods; infrequent; Jun–Aug; TN.

Rudbeckia L.

Perdue, R. E., Jr. 1957. Synopsis of *Rudbeckia* subgenus *Rudbeckia*. *Rhodora* 59:293–299.

1 Disk flowers greenish-yellow; stems glabrous or nearly so . . . . **R. laciniata**
1 Disk flowers purplish-black; stems pubescent or glabrous 2
  2 Chaffy bracts conspicuously awn-pointed; lower leaves deeply lobed or dissected . . . . **R. triloba**
  2 Chaffy bracts not awn-pointed; lower leaves not lobed or dissected 3
    3 Chaffy bracts acute; stems coarsely hairy; style tips elongate and subulate . . . . **R. hirta**
    3 Chaffy bracts obtuse; stems appressed-pubescent or glabrous; style tips short and blunt 4

4 Surface of upper ½ of chaffy bracts glabrous, the margin pubescent . . . . **R. fulgida**

4 Surface and margin of upper ½ of chaffy bracts densely pubescent . . . . **R. heliopsidis**

R. fulgida Ait.; woods and fields; infrequent; Aug–Oct; NC, SC, TN, VA.

R. heliopsidis T. & G., Black-eyed Susan; woods and meadows; infrequent; Jul–Sep; SC.

R. hirta L., Black-eyed Susan; fields and meadows; common; May–Jul; ALL. A highly variable species, the infraspecific classification not yet sorted out satisfactorily.

R. laciniata L., Coneflower; moist woods and meadows; frequent; Jul–Oct; ALL.

R. triloba L.; woods and fields; frequent; Jul–Oct; NC, SC, TN, VA. *R. beadlei* Small.

Senecio L. (Groundsel, Ragwort)

Barkley, T. M. 1978. *Senecio*. *N. Amer. Fl.* II. 10:50–139.

1 Rays absent; plant annual . . . . **S. vulgaris**

1 Rays present; plant perennial or biennial 2

2 Leaves 2–3× pinnatifid or more finely dissected . . . . **S. millefolium**

2 Leaves at most once-pinnatifid, the divisions wider 3

3 Stem and lower leaf surfaces tomentose 4

4 Basal leaf blades up to 3.5 cm long . . . . (**S. antennariifolius**)

4 Basal leaf blades usually more than 3.5 cm long 5

5 Cauline leaves pinnatifid . . . . **S. plattensis**

5 Cauline leaves entire or toothed, but not pinnatifid . . . . **S. tomentosus**

3 Stem (except at base) and lower leaf surfaces not tomentose 6

6 Basal leaves widest at base, cordate or truncate 7

7 Basal leaf blades 1.5–3.5× as long as wide, subcordate to truncate . . . . **S. schweinitzianus**

7 Basal leaf blades less than 1.5× as long as wide, deeply cordate . . **S. aureus**

6 Basal leaves not widest at base, the blade narrowed to a cuneate base 8

8 Basal leaves obovate; plant stoloniferous . . . . **S. obovatus**

8 Basal leaves lanceolate; plant not stoloniferous 9

9 Heads fewer than 20; stem often glabrous at base . . . . . . **S. pauperculus**

9 Heads more than 20; stem densely woolly at base . . . . **S. anonymus**

S. anonymus A. Wood, Squaw-weed; open woods and fields; common; May–Jun; ALL. *S. smallii* Britton.

S. antennariifolius Britton; shale barrens; infrequent; Apr–Jun; VA.

S. aureus L., Golden G.; moist woods and fields; infrequent; Mar–Jun; ALL.

S. millefolium T. & G.; rock outcrops; infrequent; Apr–Jun; GA, NC, SC.

S. obovatus Muhl. ex Willd.; wooded slopes, especially calcareous ones; infrequent; Apr–Jun; ALL.

S. pauperculus Michx.; wet meadows; infrequent; Apr–May; GA, VA. *S. crawfordii* Britt.

S. plattensis Nutt.; dry open places; infrequent; May–Jun; VA.

S. schweinitzianus Nutt.; high balds; infrequent; May–Jul; NC, TN. Disjunct from extreme northeastern U.S. and adjacent Canada. *S. robbinsii* Oakes.

S. tomentosus Michx.; dry open places; infrequent; Apr–Jun; SC, VA.

S. vulgaris L.*; waste places; infrequent; Mar–Jun; GA, NC, SC, VA.

Silphium L. (Rosin Weed)

1 Leaves connate-perfoliate; stem square in cross section .... **S. perfoliatum**
1 Leaves not connate-perfoliate; stem more or less round in cross section 2
2 Leaves mostly basal; stem leaves reduced .... **S. compositum**
2 Leaves mostly or all cauline, not much reduced 3
3 Stem pubescent .... **S. asteriscus**
3 Stem essentially glabrous, often glaucous 4
4 Chaffy bracts with gland-tipped hairs .... **S. dentatum**
4 Chaffy bracts without conspicuous gland-tipped hairs .... **S. trifoliatum**

S. asteriscus L.; open woods and clearings; infrequent; Jun–Sep; GA, NC, TN.
S. compositum Michx.; dry sandy soils; occasional; May–Sep; ALL.
S. dentatum Ell.; dry woods; infrequent; May–Aug; NC, SC.
S. perfoliatum L.; moist woods; infrequent; Jun–Aug; NC, VA.
S. trifoliatum L.; woods and fields; infrequent; Jun–Sep; GA, NC, TN, VA.

Solidago L. (Goldenrod)

1 Inflorescence corymbiform, flat or round-topped 2
2 Rays white .... **S. ptarmicoides**
2 Rays yellow 3
3 Involucral bracts striate-nerved, plants mostly 5–20 dm tall, leaves scabrous .... **S. rigida**
3 Involucral bracts not striate-nerved, plants mostly 1–4 dm tall, leaves smooth ... . **S. spithamea**
1 Inflorescence thyrsoid, racemose, paniculiform, or of axillary clusters; not corymbiform 4
4 Inflorescence racemose, either of small, axillary clusters, or a terminal thyrse, neither nodding at the summit nor with recurved-secund branches 5
5 Leaves mostly basal, cauline leaves reduced or absent 6
6 Outer involucral bracts with conspicuous squarrose tips .... **S. squarrosa**
6 Outer involucral bracts not squarrose 7
7 Involucral bracts more than 1.5 mm wide at midlength ...... **S. glomerata**
7 Involucral bracts less than 1.5 mm wide at midlength 8
8 Involucral bracts less than 0.75 mm wide at midlength 9
9 Leaves finely pubescent; rays mostly 9–16/head .... **S. puberula**
9 Leaves glabrous; rays mostly 6–9/head .... **S. roanensis**
8 Involucral bracts more than 0.75 mm wide at midlength 10
10 Achenes short-hairy .... **S. gracillima**
10 Achenes essentially glabrous at maturity 11
11 Leaves and stem pubescent 12
12 Rays yellow .... **S. hispida**
12 Rays white .... **S. bicolor**
11 Leaves and usually stem essentially glabrous 13
13 Lowest leaves less than 7× as long as wide; plants of upland habitats 14
14 Inflorescence very narrow, spiciform-thyrsoid; middle cauline leaves mostly 0.5–2 cm wide .... **S. erecta**

14 Inflorescence denser and broader; middle cauline leaves usually more than 2 cm wide .... **S. speciosa**

13 Lowest leaves mostly 7–15× as long as wide; plants of bog habitats .... **S. uliginosa**

5 Leaves chiefly cauline 15

15 Achenes glabrous at maturity; leaves entire or few-toothed .... **S. petiolaris**

15 Achenes short-hairy; leaves generally toothed 16

16 Leaves relatively broad, 1–2.5× as long as wide .... **S. flexicaulis**

16 Leaves relatively narrow, 2.5–10× as long as wide 17

17 Leaves mostly 2.5–3× as long as wide .... **S. flaccidifolia**

17 Leaves mostly 3–10× as long as wide 18

18 Stems terete, glaucous .... **S. caesia**

18 Stems striate-angled, grooved, not glaucous 19

19 Heads relatively small, the involucre 3–5 mm high .... **S. curtisii**

19 Heads relatively large, the involucre 4.5–7 mm high .... **S. lancifolia**

4 Inflorescence terminal and paniculiform, the branches often recurved-secund ... 20

20 Pappus bristles very short and firm, much shorter than the achene; basal and lowermost cauline leaves distinctly cordate and slender-petiolate .. **S. sphacelata**

20 Pappus of well-developed bristles, usually as long as or longer than the achenes; leaves not at once cordate and slender-petiolate 21

21 Leaves mostly basal, the cauline leaves reduced or absent 22

22 Stem (and often also the leaves) pubescent .... **S. nemoralis**

22 Stem (and often also the leaves) glabrous or nearly so below the inflorescence 23

23 Leaves strongly scabrous above; stem more or less strongly angled, at least below .... **S. patula**

23 Leaves glabrous or at most faintly scabrous above; stem terete or striate, not angled 24

24 Basal and lowermost cauline leaves gradually tapering to the petiole; heads with or without chaff 25

25 Rays 1–8; achenes glabrous or sometimes slightly hairy at apex 26

26 Plant puberulent in the inflorescence; inflorescence much longer than broad .... **S. uliginosa**

26 Plant glabrous throughout; inflorescence about as long as broad .... **S. pinetorum**

25 Rays mostly 7–13; achenes short-hairy .... **S. juncea**

24 Basal and lowermost cauline leaves rather abruptly contracted to the petiole; heads without chaff 27

27 Leaves glabrous or somewhat strigose; disk flowers mostly 8–14 .... **S. arguta**

27 Leaves loosely hirsute at least on the midrib and main veins beneath; disk flowers mostly 4–7 .... **S. ulmifolia**

21 Leaves chiefly cauline 28

28 Leaves not triple-nerved 29

29 Leaves entire, ordinarily anise-scented when bruised ..... **S. odora**
29 Leaves toothed, not anise-scented 30
30 Stems glabrous below the inflorescence; rays mostly 3–5 .... **S. ulmifolia**
30 Stems hirsute; rays mostly 6–11 .... **S. rugosa**
28 Leaves more or less triple-nerved 31
31 Rays mostly 4–8 .... **S. radula**
31 Rays mostly 8–17 32
32 Stem puberulent; leaves puberulent beneath and scabrous above .... **S. canadensis**
32 Stem glabrous and glaucous; leaves glabrous or with hairs restricted to the main veins .... **S. gigantea**

S. arguta Aiton; open woods and dry meadows; Jul–Oct; ALL. Plants with pubescent achenes may be segregated from the typical variety as var. *caroliniana* A. Gray (leaves glabrous) or var. *boottii* (Hook.) Palmer & Steyerm. (leaves strigose); shale barren ecotypes may be referred to *S. harrisii* Steele.

S. bicolor L., Silverrod; dry woods and open rocky places; frequent; Sep–Oct; NC, TN, VA.

S. caesia L., Blue-stem G.; moist woods; infrequent; Sep–Oct; ALL.

S. canadensis L.; open fields; common; Sep–Oct; ALL. *S. altissima* L.

S. curtisii T. & G.; rich woods at middle elevations; occasional; Aug–Oct; ALL.

S. erecta Pursh; dry woods; occasional; Aug–Oct; ALL.

S. flaccidifolia Small; moist woods; occasional; Sep–Oct; NC, TN.

S. flexicaulis L.; rich woods; occasional; Aug–Oct; GA, NC, TN, VA.

S. gigantea Aiton; old fields; common; Jul–Oct; ALL.

S. glomerata Michx., Skunk G.; balds and wet open places; Aug–Oct; NC, TN.

S. gracillima T. & G.; pine barrens; infrequent; Sep–Oct; VA.

S. hispida Muhl.; dry woods and open places; infrequent; Sep–Oct; VA.

S. juncea Aiton; open woods and fields; infrequent; Jul–Oct; GA, NC, VA.

S. lancifolia (T. & G.) Chapman; rich woods; infrequent; Aug–Oct; NC.

S. nemoralis Aiton; dry woods and open places; frequent; Sep–Oct; ALL.

S. odora Aiton; dry open woods; occasional; Jul–Oct; ALL.

S. patula Muhl.; swamps and wet meadows; occasional; Sep–Oct; ALL.

S. petiolaris Aiton; woods and open places; infrequent; Sep–Oct; GA, NC, SC.

S. pinetorum Small; dry woods; infrequent; Jul–Oct; NC, VA.

S. ptarmicoides (Nees) Boivin; dry open places; infrequent; Aug–Oct; GA.

S. puberula Nutt.; wet open places; infrequent; Sep–Oct; ALL.

S. radula Nutt.; dry woods; infrequent; Aug–Oct; GA.

S. rigida L.; open places; infrequent; Aug–Oct; GA, NC, VA.

S. roanensis T. C. Porter; woods and openings; occasional; Aug–Oct; ALL.

S. rugosa Miller; woods and meadows; frequent; Sep–Oct; ALL. Plants of our area have been segregated as subsp. *rugosa* (leaves relatively thin, rays 8–11) or subsp. *aspera* (Ait.) Cronq. (leaves relatively thick and firm, rays 6–8).

S. speciosa Nutt.; open fields and woodland borders; infrequent; Sep–Oct; ALL.

S. sphacelata Raf.; open woods and rocky places; infrequent; Aug–Sep; GA, NC, TN, VA.

S. spithamea M. A. Curtis, Skunk G.; balds; infrequent; Jul–Aug; NC, TN.

S. squarrosa Muhl.; rocky woods; infrequent; Aug–Sep; NC, VA.

S. uliginosa Nutt.; wet woods; infrequent; Sep–Oct; GA, NC, TN, VA.
S. ulmifolia Muhl.; woods; infrequent; Sep–Oct; NC, TN, VA.

Sonchus L. (Sow Thistle)

1 Plant perennial by rhizomes; heads more than 3 cm wide .... **S. arvensis**
1 Plant annual; heads less than 3 cm wide 2
 2 Leaf bases rounded; achenes several-nerved but not rugose .... **S. asper**
 2 Leaf bases sagittate; achenes transversely tuberculate-rugose .... **S. oleraceus**

S. arvensis L.*, Perennial S. T.; waste places; infrequent; Jun–Oct; NC, VA.
S. asper (L.) Hill*, Prickly S. T.; fields and waste places; frequent; Mar–Oct; NC, SC, TN, VA.
S. oleraceus L.*, Common S. T.; fields and waste places; infrequent; Apr–Oct; SC, VA.

Tanacetum L. (Tansy)

T. vulgare L.*; roadsides and waste places; infrequent; Aug–Oct; NC, VA.

Taraxacum Wiggers (Dandelion)

1 Achenes reddish-brown; leaves tending to be deeply incised .... **T. laevigatum**
1 Achenes olive-green or greenish-brown; leaves usually less deeply incised .... **T. officinale**

T. laevigatum (Willd.) DC.*, Red-seeded D.; pastures, lawns, and waste places; infrequent; Mar–Oct; NC, SC, TN, VA. *T. erythrospermum* Andrz.
T. officinale Weber*, Common D.; lawns and waste places; common; Feb–Oct; ALL.

Tetragonotheca L. (Squarehead)

T. helianthoides L.; dry open woods; infrequent; Apr–Jul; GA, NC, SC.

Tragopogon L. (Goat's Beard)

Ownbey, M. 1950. Natural hybridization and amphiploidy in the genus *Tragopogon*. *Amer. J. Bot.* 37:487–499.

1 Flowers purple .... **T. porrifolius**
1 Flowers yellow 2
 2 Peduncles enlarged and fistulose above in flower and fruit; involucral bracts extending beyond the rays .... **T. dubius**
 2 Peduncles not enlarged in flower and scarcely so in fruit; involucral bracts not extending beyond the rays .... **T. pratensis**

T. dubius Scop.*; fields and waste places; infrequent; Apr–Jul; NC, VA.
T. porrifolius L.*, Salsify, Vegetable Oyster; fields and waste places; infrequent; Apr–Jul; NC, VA.
T. pratensis L.*; fields and waste places; infrequent; Apr–Jun; VA.

Tussilago L. (Coltsfoot)

T. farfara L.*; waste places; infrequent; Mar–May; NC, TN, VA. The flowers appear before the basal leaves develop.

Verbesina L. (Flatseed Sunflower)

1 Rays white . . . . **V. virginica**
1 Rays yellow 2
  2 Involucral bracts well developed; leaves opposite . . . . **V. occidentalis**
  2 Involucral bracts few, inconspicuous; leaves alternate . . . . **V. alternifolia**

V. alternifolia (L.) Britton; woods and thickets; occasional; Aug–Sep; ALL.

V. occidentalis (L.) Walt., Yellow Crown-beard; thickets, woods, and waste places; common; Aug–Oct; NC, SC, TN, VA.

V. virginica L.; thickets, woods, and waste places; occasional; Jul–Oct; GA, NC, SC, TN.

Vernonia Schreber (Ironweed)

Jones, S. B., Jr., and W. Z. Faust. 1978. *Vernonia. N. Amer. Fl.* II. 10:180–195.

1 Stems glabrous and glaucous; pappus pale straw-colored . . . . **V. flaccidifolia**
1 Stems sparsely to densely pubescent, especially on inflorescence branches; pappus brown to purple 2
  2 Heads with 9–20 flowers; involucral bracts short-acute . . . . **V. gigantea**
  2 Heads with 30–65 flowers; involucral bracts long-acuminate 3
    3 Pappus pale tan to whitish; lower surface of leaves much paler than upper surface . . . . **V. glauca**
    3 Pappus brown to purple; lower surface of leaves not much paler than upper surface . . . . **V. noveboracensis**

V. flaccidifolia Small; upland woods and open disturbed sites; infrequent; Jun–Sep; GA, TN.

V. gigantea (Walt.) Trelease ex Branner & Coville; pastures, roadsides, wet woods; infrequent; Aug–Oct; TN, VA. *V. altissima* Nutt.

V. glauca (L.) Willd.; woods; infrequent; Jun–Sep; GA, SC, VA.

V. noveboracensis (L.) Michx.; low woods and wet fields; frequent; Aug–Oct; ALL.

Xanthium L. (Cocklebur)

1 Leaves broad, cordate or deltoid at base, spineless . . . . **X. strumarium**
1 Leaves lanceolate, tapering to base, bearing a 3-forked axillary spine . . **X. spinosum**

X. spinosum L.*; waste places; infrequent; Jul–Oct; VA.

X. strumarium L.*, Common C.; fields and waste places; common; Jul–Oct; ALL. Some authors recognize two varieties in our area: var. *glabratum* (DC.) Cronq., with mature pistillate involucre glabrous and 1.5–2 cm long, and var. *canadense* (Mill.) T. & G., with mature pistillate involucre pubescent and 2.3–3 cm long.

BALSAMINACEAE (Touch-me-not Family)

Impatiens L. (Touch-me-not)

1 Flowers orange, rarely orange-yellow or white . . . . **I. capensis**
1 Flowers yellow, rarely cream or white . . . . **I. pallida**

I. capensis Meerb., Spotted T.; moist shaded sites and along streams; common; Jun–Sep; ALL.

I. pallida Nutt., Pale T.; moist shaded sites and along streams, especially at mid to high elevations; common; Jun–Sep; NC, TN, VA.

BERBERIDACEAE (Barberry Family)

1 Shrubs .... **Berberis**
1 Herbs 2
  2 Flowers solitary 3
    3 Flower arising on stem between 2 deeply lobed, peltate leaves ... **Podophyllum**
    3 Flower scapose from underground stem; leaves 2-parted .... **Jeffersonia**
  2 Flowers in cymes or racemes 4
    4 Inflorescence cymose, flowers white .... **Diphylleia**
    4 Inflorescence racemose, flowers green, often maroon tinged .... **Caulophyllum**

Berberis L. (Barberry)

1 Thorns 3-pronged; leaves toothed .... **B. canadensis**
1 Thorns simple; leaves entire .... **B. thunbergii**

B. canadensis Mill., American B.; open sites along streams and woodlands; occasional; Apr–May; ALL.

B. thunbergii DC.*, Japanese B.; escaping to woodlands; infrequent; Mar–Apr; NC, SC, VA.

Caulophyllum Michx. (Blue Cohosh)

Loconte, H., and W. H. Blackwell, Jr. 1984. Berberidaceae in Ohio. *Castanea* 49:39–43.

1 Pistils 3.5–5 mm long, the stylar portion 0.8–1.5 mm long; sepals 6–9 mm long; main inflorescence with 4–18 flowers; lst leaf bi- or triternate; leaflets 5–10 cm long .... **C. giganteum**
1 Pistils 1.3–3 mm long, the stylar portion 0.3–1 mm long; sepals 3–6.5 mm long; main inflorescence with 5–70 flowers; lst leaf tri- or quadternate; leaflets 3–8 cm long .... **C. thalictroides**

C. giganteum (Farw.) Loconte & Blackwell; rich woods; infrequent; Apr–May; NC, TN, VA. *C. thalictroides* (L.) Michx. var. *giganteum* Farw. More common north of our area.

C. thalictroides (L.) Michx.; rich woods; common; Apr–May; ALL.

Diphylleia Michx. (Umbrella Leaf)

D. cymosa Michx.; seeps in rich woods; occasional; Apr–May; ALL.

Jeffersonia Bart. (Twinleaf)

J. diphylla (L.) Pers.; rich woods on basic soils; occasional; Mar–Apr; NC, TN, VA.

Podophyllum L. (May Apple)

P. peltatum L.; rich woods and meadows; common; Apr–May; ALL.

BETULACEAE (Birch Family)

Hardin, J. W. 1971. Studies of the southeastern United States flora. I. Betulaceae. *J. Elisha Mitchell Sci. Soc.* 87:39–41.

1 Shrubs, generally less than 5 m tall 2
  2 Pistillate bracts woody; nut less than 3 mm broad .... **Alnus**
  2 Pistillate bracts not woody; nut 1–1.5 cm broad .... **Corylus**

1 Medium-sized trees, generally more than 5 m tall 3
  3 Fruits winged; fruiting structure conelike . . . . . **Betula**
  3 Fruits not winged; fruiting structure racemose or spicate 4
    4 Fruiting bract 2- or 3-lobed; bark smooth . . . . . **Carpinus**
    4 Fruiting bract not lobed; bark shredding . . . . . **Ostrya**

Alnus Ehrhart (Alder)

1 Fruit broadly winged; buds sessile . . . . . **A. crispa**
1 Fruit narrowly winged; buds stalked 2
  2 Fruiting catkins erect; leaves broadest at or above the middle . . . . . **A. serrulata**
  2 Fruiting catkins drooping; leaves broadest at or below the middle . . . . . **A. rugosa**

A. crispa (Ait.) Pursh, Mountain A.; high-elevation balds and meadows; infrequent; May–Jun; Roan Mt., NC, TN. Disjunct from northern populations by about 800 miles.

A. rugosa (DuRoi) Spreng., Speckled A.; wet areas; infrequent; May–Jun; VA.

A. serrulata (Ait.) Willd., Smooth A.; streams and marshes; common; Feb–Mar; ALL.

Betula L. (Birch)

1 Bark and twigs with odor of wintergreen when crushed 2
  2 Leaves 2–6 cm long, with 4–6 pairs of veins . . . . . **B. uber**
  2 Leaves 7–15 cm long, with 8–12 pairs of veins 3
    3 Bark yellowish, scaly on older trees; fruiting bracts ciliate; buds acute . . . . . **B. alleghaniensis**
    3 Bark reddish-brown and tight; fruiting bracts glabrous; buds sharp-pointed . . . . . **B. lenta**
1 Bark and twigs not aromatic 4
  4 Leaves with 8 or more pairs of lateral veins; bark reddish-brown, exfoliating . . . . . **B. nigra**
  4 Leaves with 5–8 pairs of lateral veins; bark whitish to gray 5
    5 Bark not exfoliating; leaves glabrous beneath . . . . . **B. populifolia**
    5 Bark exfoliating; leaves pubescent on veins beneath 6
      6 Twigs densely pubescent; leaf base cuneate to rounded . . . . . **B. papyrifera** var. **papyrifera**
      6 Twigs glabrous to slightly pubescent; leaf base cordate . . . . . **B. papyrifera** var. **cordifolia**

B. alleghaniensis Britt., Yellow B.; rich woods especially at higher elevations; frequent; Apr–May; ALL. *B. lutea* Michx. f.

B. lenta L., Sweet B.; rich woods generally at lower elevations; frequent; Mar–Apr; ALL.

B. nigra L., River B.; low woods and along streams; occasional; Mar–Apr; ALL.

B. papyrifera Marsh. var. **cordifolia** (Regel) Fern.; fir forests; infrequent; May–Jun; NC, TN. Closely related to the following taxon and also suspected of being a fertile hybrid between *B. papyrifera* and *B. alleghaniensis*.

B. papyrifera Marsh. var. **papyrifera**, Paper B.; rich woods; infrequent; Apr–May; VA.

B. populifolia Marsh., Gray B.; low woods; infrequent and perhaps not native to our area; May–Jun; NC, SC, VA.

B. uber (Ashe) Fern., Ashe's B.; stream banks; infrequent; May–Jun; VA. Extremely rare and considered to be extinct until rediscovered in 1975. The first U.S. tree to be classified as endangered.

Carpinus L. (Blue Beech, Ironwood)

C. caroliniana Walt.; along streams and low woods; common; Mar–Apr; ALL. Represented primarily by subsp. **virginiana** (Marshall) Furlow. In our area, this subsp. overlaps and frequently intergrades with the more southern subsp. *caroliniana*. For more details see J. J. Furlow, The *Carpinus caroliniana* complex in North America. II. Systematics. *Systematic Botany* 12 (1987): 416–434.

Corylus L. (Hazel Nut)

1 Fruiting bracts forming a beak up to 5 cm long; petioles not glandular ... **C. cornuta**
1 Fruiting bracts 0.5–1.5 cm long; petioles glandular .... **C. americana**

C. americana Walt.; woodland margins; frequent; Feb–Mar; ALL.
C. cornuta Marsh., Beaked H. N.; thickets and woodlands; occasional; Feb–Mar; ALL.

Ostrya Scopoli (Hop Hornbeam)

O. virginiana (Mill.) K. Koch; woodlands; common; Apr–May; ALL. The fruits resemble hops.

BIGNONIACEAE (Bignonia Family)

1 Trees; leaves simple .... **Catalpa**
1 Vines; leaves compound 2
  2 Leaves with 2 leaflets and terminal, branched tendrils .... **Bignonia**
  2 Leaves with 7–15 leaflets, tendrils absent .... **Campsis**

Bignonia L. (Cross Vine)

B. capreolata L.; open woods, roadsides, thickets; common; Apr–May; ALL. *Anisostichus capreolata* (L.) Bureau.

Campsis Lour. (Trumpet Creeper)

C. radicans (L.) Seemans; woodland margins, fencerows, thickets; common; May–Jul; ALL.

Catalpa Scop.

1 Corolla 4 cm or less broad, the lower lobe not notched; capsule about 1 cm thick .... **C. bignonioides**
1 Corolla 5 cm or more broad, the lower lobe notched; capsule about 1.5 cm thick .... **C. speciosa**

C. bignonioides Walt.*, Catalpa; river banks and low woods; occasional; May–Jun; ALL. Naturalized in our area.
C. speciosa Warder ex Engelm.*, Cigar Tree; lowlands and waste places; frequent; May–Jun; NC, SC, TN, VA. Naturalized in our area.

BORAGINACEAE (Borage Family)

1 Flowers zygomorphic 2
  2 Corolla blue; stamens exserted .... **Echium**
  2 Corolla white; stamens included within the corolla tube .... **Myosotis**

1 Flowers actinomorphic 3
3 Nutlets bristly or prickly 4
4 Corolla 6–12 mm broad; nutlets 5–8 mm long . . . . . **Cynoglossum**
4 Corolla 2–3 mm broad; nutlets 3–4 mm long . . . . . **Hackelia**
3 Nutlets without prickles, either smooth or pitted 5
5 Plants glabrous . . . . . **Mertensia**
5 Plants pubescent 6
6 Racemes bractless or bracteate at the base only . . . . . **Myosotis**
6 Racemes leafy-bracteate 7
7 Style shorter than the corolla tube; corolla lobes rounded . . . . . **Lithospermum**
7 Style longer than the corolla tube; corolla lobes acute . . . . . **Onosmodium**

Cynoglossum L.

1 Flowers pale blue to white; leaves primarily on the lower portion of the stem . . . . . **C. virginianum**
1 Flowers reddish-purple; leaves produced throughout the stem length . . . **C. officinale**

C. officinale L.*, Hound's-tongue; fields and roadsides; occasional; May–Jul; NC, VA.
C. virginianum L., Wild Comfrey; woodlands and clearings; frequent; Apr–Jun; ALL.

Echium L. (Blue Weed)

E. vulgare L.*; fields and waste places; frequent; Jun–Sep; GA, NC, TN, VA.

Hackelia Opiz (Stickseed)

H. virginiana (L.) Johnston; woodlands; occasional; Jun–Sep; GA, NC, TN, VA.

Lithospermum L.

1 Corolla yellow-orange . . . . . **L. canescens**
1 Corolla white or greenish-white 2
2 Leaves linear-lanceolate, less than 1 cm wide, lateral veins not evident . . . . . **L. arvense**
2 Leaves lanceolate to elliptic, more than 1 cm wide, lateral veins evident . . . . . **L. latifolium**

L. arvense L.*; waste places; occasional; Mar–Jun; NC, SC, TN, VA.
L. canescens (Michx.) Lehmann, Hoary Puccoon; dry woodlands, often on basic soils; infrequent; Apr–May; VA.
L. latifolium Michx., Gromwell; dry woods and thickets; infrequent; May–Jun; VA.

Mertensia Roth (Bluebells)

M. virginica (L.) Pers.; moist alluvial woods, often on basic soils; occasional; Apr–May; TN, VA.

Myosotis L.

1 Calyx with a few appressed hairs; plants of wet or muddy sites 2
2 Corolla 5–10 mm broad . . . . . **M. scorpioides**
2 Corolla 2–5 mm broad . . . . . **M. laxa**
1 Calyx sparsely or densely pubescent with uncinate hairs; plants of dry or mesic sites 3

3 Corolla blue; calyx actinomorphic . . . . **M. arvensis**
3 Corolla white; calyx slightly zygomorphic 4
4 Fruiting pedicels more or less erect, the lowest, 2 cm or less apart . . . **M. verna**
4 Fruiting pedicels spreading, the lowest, 2 cm or more apart . . . . **M. macrosperma**

M. arvensis (L.) Hill*; waste places; infrequent; May–Aug; NC, VA.

M. laxa Lehmann; shallow water and wet grounds; occasional; May–Sep; NC, TN, VA.

M. macrosperma Engelm., Scorpion Grass; open woods and fields; infrequent; Apr–May; NC, SC, TN, VA.

M. scorpioides L.*, Forget-me-not; shallow water and wet grounds; occasional; May–Aug; NC, TN, VA. Similar to but slightly larger than *M. laxa*.

M. verna Nutt.; open woods and fields; infrequent; Apr–Jun; NC, TN, VA. Scarcely differing from *M. macrosperma*.

Onosmodium Michx. (Marbleseed)

1 Corolla lobes yellow to orange, longer than wide . . . . **O. virginianum**
1 Corolla lobes white to greenish-white, about as long as wide . . . . . **O. hispidissimum**

O. hispidissimum Mackenzie; dry woodlands; infrequent; Jun–Jul; TN, VA.

O. virginianum (L.) A. DC.; dry woodlands; occasional; Jun–Jul; GA, NC, SC, VA.

BRASSICACEAE (Mustard Family)

1 Petals yellow or absent 2
2 Fruit less than 4× as long as wide 3
3 Leaves entire or slightly toothed 4
4 Plants pubescent, trichomes stellate or forked; fruits 2–several-seeded 5
5 Fruit orbicular, flattened . . . . **Alyssum**
5 Fruit obovoid, inflated . . . . **Camelina**
4 Plants glabrous or with a few simple, unbranched hairs; fruit 1-seeded . . . **Isatis**
3 Leaves toothed, or at least the lower pinnatifid 6
6 Stem pubescence forked; leaf divisions less than 4 mm wide . . . . **Descurainia**
6 Stem glabrous or with simple hairs; leaf divisions more than 5 mm wide . . . . **Rorippa**
2 Fruit more than 4× as long as wide 7
7 Plants pubescent with stellate or forked trichomes 8
8 Leaves pinnately dissected . . . . **Descurainia**
8 Leaves simple, the margin entire or slightly toothed . . . . **Erysimum**
7 Plants glabrous or with simple unbranched hairs 9
9 Flowers less than 5 mm wide 10
10 Fruit appressed to the stem . . . . **Sisymbrium**
10 Fruit divergent . . . . **Rorippa**
9 Flowers 7 mm or more wide 11
11 Fruit indehiscent, sutures not evident, the beak (seedless portion) 1–3 cm long . . . . **Raphanus**
11 Fruit dehiscent, sutures evident 12
12 Stem leaves entire or shallowly lobed, sometimes clasping . . . **Brassica**
12 Stem leaves prominently lobed to pinnatifid 13

13 Stem leaves auriculate clasping, the terminal lobe wider than the laterals; mature fruits usually less than 7 cm long ...... **Barbarea**
13 Stem leaves not auriculate clasping, the terminal lobe not wider than the laterals; mature fruits up to 10 cm long .... **Sisymbrium**
1 Petals white, pink, or purple 14
14 Fruit short and broad, not more than 3× as long as wide 15
15 Fruit triangular, broadest at the apex .... **Capsella**
15 Fruit broadest at or near the middle 16
16 Fruit slightly longer than wide, not emarginate at the apex; plants with stellate or forked trichomes .... **Draba**
16 Fruit nearly round, emarginate at the apex; plants glabrous or pubescent, but rarely with stellate or forked trichomes 17
17 Seeds 1 or 2/fruit .... **Lepidium**
17 Seeds 3 or more/fruit .... **Thlaspi**
14 Fruit at least 4× as long as wide 18
18 Leaves simple, rarely lobed, the margin entire or toothed, but not deeply divided 19
19 Petals purple .... **Hesperis**
19 Petals white to pink 20
20 Plants decumbent and stoloniferous at base .... **Cardamine**
20 Plants upright, not stoloniferous 21
21 Leaves chiefly basal; petals less than 4 mm long .... **Arabidopsis**
21 Leaves primarily cauline, or if leaves chiefly basal, the petals 6 mm or more long 22
22 All leaves petiolate, the blade rhombic, about as broad as long .... **Alliaria**
22 All leaves, or at least the upper, sessile, blades ovate to lanceolate 23
23 Plant bulbous at the base; fruit terete .... **Cardamine**
23 Plant not bulbous at the base; fruit flat or subterete .. **Arabis**
18 Leaves either ternate, palmately divided, or pinnatifid 24
24 Plants aquatic, rooting at the nodes .... **Nasturtium**
24 Plants terrestrial, not rooting at the nodes 25
25 Stems 4–8 dm tall; fruit indehiscent .... **Raphanus**
25 Stems 1–3 dm tall; fruit dehiscent 26
26 Leaves palmate or ternately divided .... **Dentaria**
26 Leaves, at least the basal, pinnately divided or lobed 27
27 Stems glabrous below .... **Cardamine**
27 Stems pubescent below 28
28 Leaflets 3–5; seeds not winged .... **Cardamine**
28 Leaflets numerous; seeds narrowly winged .... **Sibara**

Alliaria Scopoli (Garlic Mustard)

A. petiolata (Bieb.) Cavara & Grande*; waste places; occasional; Apr–May; TN, VA. *A. officinalis* Andrz.

Alyssum L. (Yellow Alyssum)

A. alyssoides L.*; fields and disturbed sites; occasional; Jun–Sep; VA.

Arabidopsis Heynh. (Mouse-ear Cress)

A. thaliana (L.) Heynh.*; disturbed sites; frequent; Mar–May; NC, SC, TN, VA.

Arabis L.

1 Fruits erect, appressed to the stem at maturity .... **A. glabra**
1 Fruits not appressed to the stem at maturity 2
  2 Fruits less than 4 cm long at maturity; stems freely branched at base; stem leaves 5 mm or less wide .... **A. lyrata**
  2 Fruits 4 cm or more long at maturity; stems not branched at base; stem leaves more than 5 mm wide 3
    3 Leaves less than 3× as long as wide, densely pubescent .... **A. patens**
    3 Leaves more than 3× as long as wide, glabrous or slightly pubescent 4
      4 Stem leaves pubescent; fruits 2–3 mm wide .... **A. canadensis**
      4 Stem leaves glabrous; fruits 1–2 mm wide 5
        5 Leaves auriculate at the base, lanceolate to oblong-lanceolate; fruit 1-nerved for about 1/3 of its length .... **A. laevigata** var. **laevigata**
        5 Leaves essentially sessile, linear to linear-lanceolate; fruit 1-nerved to the middle or beyond .... **A. laevigata** var. **burkii**

A. canadensis L., Sicklepod; woodlands, usually on basic soils; frequent; May–Jun; ALL.

A. glabra (L.) Bernh.; woodland margins; infrequent; May–Jun; NC.

A. laevigata (Muhl.) Poir. var. **burkii** Porter; rocky woodlands; infrequent; May–Jun; NC. Closely related to the shale barren endemic, *A. serotina* Steele; see T. F. Wieboldt, The shale barren endemic, *Arabis serotina* (Brassicaceae). *Sida* 12 (1987): 381–389.

A. laevigata (Muhl.) Poir. var. **laevigata**; woodlands and rock outcrops; frequent; May–Jun; NC, TN, VA.

A. lyrata L., Rockcress; basic soils on rock outcrops; occasional; Mar–Jun; NC, SC, TN, VA.

A. patens Sulliv.; moist woodland slopes; infrequent; May–Jun; NC, VA.

Barbarea R. Br. (Winter Cress)

1 Basal leaves with 1–4 pairs of lateral lobes; fruits up to 3 cm long ...... **B. vulgaris**
1 Basal leaves much dissected, with up to 10 or more pairs of lateral lobes; fruits 4–8 cm long .... **B. verna**

B. verna (Mill.) Aschers.*; fields and waste places; frequent; Mar–May; NC, SC, TN, VA.

B. vulgaris R. Br.*; fields and waste places; frequent; Mar–May; ALL. *B. arcuata* (Opiz) Reichemb.

Brassica L.

1 Stem leaves, at least the upper, auriculate-clasping .... **B. napus**
1 Stem leaves not auriculate-clasping .... **B. juncea**

B. juncea (L.) Cosson*, Indian Mustard; fields and waste places; occasional; Apr–Jun; NC, TN, VA.

B. napus L.*, Turnip, Rape; fields and roadsides; occasional; Mar–May; NC, SC, TN, VA. *B. campestris* L., *B. rapa* L.

Camelina Crantz (False Flax)

C. microcarpa Andrz.*; waste places; occasional; Apr–May; NC, TN, VA.

Capsella Medicus (Shepherd's Purse)

C. bursa-pastoris (L.) Medicus*; lawns and waste places; common; Mar–Jun; ALL.

Cardamine L. (Bitter Cress)

1 Cauline leaves simple or with 1 or 2 lateral lobes 2
  2 Plant erect, from a bulbous base . . . . **C. bulbosa**
  2 Plant decumbent and rooting at the nodes; roots fibrous . . . . **C. rotundifolia**
1 Cauline leaves with 3–many segments 3
  3 Petals 6–10 mm long; leaflets up to 5; perennials 4
    4 Leaf base auriculate; stem glabrous at the base . . . . **C. clematitis**
    4 Leaf base not auriculate; stem pubescent at the base . . . . **C. flagellifera**
  3 Petals 1–3 mm long; leaflets numerous; annuals or biennials 5
    5 Petioles ciliate at the base . . . . **C. hirsuta**
    5 Petioles not ciliate at the base 6
      6 Terminal leaflet broader than the lateral leaflets; leaflets decurrent along the rachis . . . . **C. pensylvanica**
      6 Terminal leaflet similar to the lateral leaflets in size and shape; leaflets not decurrent along the rachis . . . . **C. parviflora**

C. bulbosa (Schreber ex Muhl.) BSP.; damp woods and openings; occasional; Mar–May; NC, TN, VA.
C. clematitis Shuttlew.; seeps at high elevations; occasional; Apr–May; ALL.
C. flagellifera O. E. Schulz; rich woodlands; occasional; Apr–May; GA, NC, TN. *C. hugeri* Small.
C. hirsuta L.*; lawns and waste areas; common; Jan–May; ALL.
C. parviflora L.; dry or wet sandy areas; occasional; Mar–May; NC, SC, TN, VA.
C. pensylvanica Muhl. ex Willd.; margin of swamps and wet woods; frequent; Mar–May; NC, TN, VA.
C. rotundifolia Michx.; seeps and stream banks; infrequent; Apr–May; NC, TN, VA.

Dentaria L. (Toothwort)

1 Stem above the cauline leaves pubescent . . . . **D. laciniata**
1 Stem above the cauline leaves glabrous 2
  2 Cauline leaves biternate, the segments narrow, 2–5 mm wide . . . . **D. multifida**
  2 Cauline leaves ternate, the segments 1 cm or more wide 3
    3 Rhizome continuous, not jointed; basal and cauline leaves similar in shape . . . . **D. diphylla**
    3 Rhizome distinctly segmented; basal leaves ovate, stem leaves linear to lanceolate . . . . **D. heterophylla**

D. diphylla Michx.; rich woods; frequent; Apr–May; ALL. *Cardamine diphylla* (Michx.) Wood.
D. heterophylla Nutt.; rich alluvial woods; occasional; Apr–May; NC, TN, VA. *Cardamine angustata* O. E. Schulz.
D. laciniata Muhl. ex Willd.; rich alluvial woods; common; Apr–May; ALL. *Cardamine concatenata* (Michx.) Ahles, *C. laciniata* (Muhl. ex Willd.) Wood.

D. multifida Muhl. ex Ell.; rich woods, usually on basic soils; infrequent; Apr–May; TN. *Cardamine angustata* O. E. Schulz var. *multifida* (Muhl.) Ahles, *C. multifida* Muhl.

Descurainia Webb & Berth. (Tansy Mustard)

1 Fruit shorter than the subtending pedicel . . . . **D. pinnata**
1 Fruit 2–several times longer than the subtending pedicel . . . . **D. sophia**

D. pinnata (Walt.) Britt.; disturbed sites; infrequent; Feb–May; VA.
D. sophia (L.) Webb*; fields and waste places; infrequent; Apr–Jun; NC, VA.

Draba L.

1 Stem leafless; petals bifid . . . . **D. verna**
1 Stem leafy; petals entire 2
  2 Fruit twisted, style persistent; leaves toothed; mat-forming perennials . . . . **D. ramosissima**
  2 Fruit straight, style deciduous; leaves entire; annuals 3
    3 Fruits glabrous or nearly so . . . . **D. brachycarpa**
    3 Fruits densely stellate-pubescent . . . . (**D. aprica**)

D. aprica Beadle, Open Ground Whitlow Grass; shallow sandy soils near red cedars; infrequent; Mar–Apr; GA, SC (Piedmont).
D. brachycarpa Nutt. ex T. & G.; lawns and waste places; infrequent; Feb–Apr; NC, SC, TN, VA.
D. ramosissima Desv., Rocktwist; dry, usually calcareous bluffs; occasional; Apr–May; NC, TN, VA.
D. verna L.*, Whitlow Grass; lawns and waste places; frequent; Feb–Apr; NC, SC, TN, VA.

Erysimum L. (Wormseed Mustard)

E. cheiranthoides L.*; roadsides and waste places; infrequent; Jun–Jul; NC, VA.

Hesperis L. (Dame's Rocket)

H. matronalis L.*; roadsides and open woods; occasional; Apr–Jun; TN, VA.

Isatis L.

I. tinctoria L.*; disturbed sites; infrequent; May–Jun; VA. Used as a source of blue dye.

Lepidium L.

1 Stem leaves clasping, auriculate at the base . . . . **L. campestre**
1 Stem leaves tapering at the base . . . . **L. virginicum**

L. campestre (L.) R. Br.*, Cow Cress; waste places; frequent; Mar–Jun; ALL.
L. virginicum L.*, Peppergrass; waste places; common; Apr–Jun; ALL.

Nasturtium R. Br. (Water Cress)

N. officinale R. Br.*; streams and springs; frequent; Apr–Jun; NC, TN, VA. Often used as a green salad.

Raphanus L. (Wild Radish)

R. raphanistrum L.*; waste places; infrequent; Mar–Jun; NC, SC, TN, VA.

Rorippa Scopoli (Yellow Cress)

Stuckey, R. L. 1972. Taxonomy and distribution of the genus *Rorippa* (Cruciferae) in North America. *Sida* 4:279–430.

1 Fruits sessile or pedicels less than 1 mm long; petals absent .... **R. sessiliflora**
1 Fruits pedicellate, the pedicels 3 mm or more long; petals present 2
  2 Petals more than 3 mm long, longer than the sepals; plants rhizomatous and clonal .... **R. sylvestris**
  2 Petals less than 3 mm long, equal to or shorter than the sepals; plants solitary, not rhizomatous .... **R. palustris**

R. palustris (L.) Besser; marshes and flood plains; occasional; May–Oct; NC, SC, TN, VA. *R. islandica* (L.) Borbas.

R. sessiliflora (Nutt. in T. & G.) Hitchcock; mesic sites; infrequent; Apr–Jul; VA.

R. sylvestris (L.) Besser*; mesic sites; infrequent; May–Sep; VA.

Sibara Greene

S. virginica (L.) Rollins; lawns and waste places; occasional; Mar–May; NC, SC, TN. *Arabis virginica* (L.) Poir.

Sisymbrium L.

1 Fruits widest at the base, appressed to the rachis, 1–1.5 cm long .... **S. officinale**
1 Fruits linear, spreading, 5–10 cm long .... **S. altissimum**

S. altissimum L.*, Tumble Mustard; waste places; infrequent; May–Jun; NC, VA.

S. officinale (L.) Scopoli*, Hedge Mustard; fields and waste places; frequent; May–Jul; NC, SC, TN, VA. The var. **leiocarpum** DC., with glabrous fruits and rachises, is the most frequently encountered variant in our area.

Thlaspi L. (Penny Cress)

1 Mature fruits 10–14 mm long; pedicels ascending; stem leaves 8 or more .. **T. arvense**
1 Mature fruits 4–6 mm long; pedicels horizontal; stem leaves 5 or less .. **T. perfoliatum**

T. arvense L.*; fields and waste places; frequent; Mar–May; GA, NC, TN, VA.

T. perfoliatum L.*; fields and waste places; occasional; Mar–May; NC, TN, VA.

BUXACEAE (Box Family)

Pachysandra Michx. (Allegheny Spurge)

P. procumbens Michx.; calcareous woodlands; infrequent; Mar–Apr; GA, NC, SC, TN.

CABOMBACEAE (Water Shield Family)

Brasenia Schreber (Water Shield)

B. schreberi Gmel.; ponds and slow streams; infrequent; May–Sep; NC, SC. Disjunct from the Coastal Plain.

CACTACEAE (Cactus Family)

Opuntia Miller (Prickly Pear)

O. humifusa (Raf.) Raf.; sandy or rocky dry sites; frequent; Jul–Aug; NC, SC, TN, VA. *O. compressa* (Salisbury) Macbride.

CALLITRICHACEAE (Water Starwort Family)

Callitriche L.

1 Plants essentially terrestrial .... **C. terrestris**
1 Plants submerged or on wet mud 2
  2 Margin of each carpel rounded; fruit about as long as wide ...... **C. heterophylla**
  2 Margin of each carpel slightly winged; fruit slightly longer than wide ... **C. verna**

C. heterophylla Pursh; pools or running water; frequent; Mar–Nov; NC, SC, TN, VA.
C. terrestris Raf. emend. Torr.; damp shaded soil; infrequent; Mar–Jun; NC, VA. *C. deflexa* A. Br.
C. verna L.; pools or running water; infrequent; Mar–Nov; VA. *C. palustris* L.

CALYCANTHACEAE (Calycanthus Family)

Calycanthus L. (Sweetshrub)

1 Lower leaf surface, twigs, and petioles pubescent .... **C. floridus** var. **floridus**
1 Lower leaf surface, twigs, and petioles glabrous or with a few scattered hairs .... **C. floridus** var. **glaucus**

C. floridus L. var. **floridus**; slopes and coves of deciduous woods and along streams; infrequent; Apr–May; GA, SC, TN.
C. floridus L. var. **glaucus** (Willd.) T. & G.; slopes and coves of deciduous woods and along streams; frequent; Apr–Jun; ALL. *C. floridus* L. var. *laevigatus* (Willd.) T. & G., *C. floridus* L.var. *oblongifolius* (Nutt.) Boufford & Spongberg.

Populations in north central Georgia bearing yellowish-green flowers have been described as *C. brockiana* Ferry & Ferry f.; for details, see R. J. Ferry, Sr., and R. J. Ferry, Jr., *Calycanthus brockiana* (Calycanthaceae), a new spicebush from north central Georgia. *Sida* 12 (1987): 339–341.

CAMPANULACEAE (Harebell Family)

1 Flowers zygomorphic .... **Lobelia**
1 Flowers actinomorphic 2
  2 Leaves sessile, about as broad as long; corolla widely spreading .... **Triodanis**
  2 Leaves petiolate or, if sessile, linear; corolla rotate or campanulate ... **Campanula**

Campanula L. (Bellflower)

1 Flowers in spikes or racemes 2
  2 Corolla rotate; style curved .... **C. americana**
  2 Corolla campanulate; style straight .... **C. rapunculoides**

1 Flowers in loose panicles 3
3 Style exserted; stems erect or arching . . . . **C. divaricata**
3 Style equal to or shorter than the corolla; stems reclining or trailing . . . . **C. aparinoides**

C. americana L.; moist woods; common; Jun–Jul; ALL.
C. aparinoides Pursh; bogs and wet meadows; occasional; Jun–Aug; NC, TN, VA.
C. divaricata Michx.; rocky slopes; common; Jul–Oct; ALL.
C. rapunculoides L.*; waste grounds; occasional; Jun–Aug; NC, VA.

Lobelia L.

1 Flowers scarlet . . . . **L. cardinalis**
1 Flowers blue or white 2
2 Flowers 18–40 mm long, including calyx 3
3 Calyx margins densely hirsute; filament tube 12–15 mm long . . . . **L. siphilitica**
3 Calyx margins entire or glandular-toothed; filament tube 6–11 mm long 4
4 Calyx margins glandular-dentate; flowers few, 3–20 . . . . **L. glandulosa**
4 Calyx margins entire or with a few teeth; flowers many, 20–100 5
5 Stems essentially glabrous, especially throughout the inflorescence . . . . **L. amoena**
5 Stems short-pubescent throughout . . . . **L. puberula**
2 Flowers 7–18 mm long, including calyx 6
6 Median stem leaves less than 5 mm wide . . . . **L. nuttallii**
6 Median stem leaves more than 8 mm wide 7
7 Stems villous or hirsute, become branched with age; capsule swollen . . . . **L. inflata**
7 Stems glabrous or nearly so, not branching; capsule not swollen . . . **L. spicata**

L. amoena Michx.; roadsides and stream banks; occasional; Sep–Oct; GA, NC, SC, TN.
L. cardinalis L., Cardinal Flower; marshes, wet ditches, and low woods; common; Jul–Sep; ALL.
L. glandulosa Walt.; moist sites; infrequent; Sep–Oct; NC.
L. inflata L., Indian Tobacco; open woodlands and fields; common; Jul–Oct; ALL.
L. nuttallii R. & S.; moist sites; infrequent; May–Oct; GA, NC, SC.
L. puberula Michx.; open woodlands and meadows; frequent; Aug–Oct; ALL. Scarcely distinct from *L. amoena*.
L. siphilitica L., Great Blue Lobelia; moist open woods and roadsides; frequent; Jul–Oct; ALL.
L. spicata Lam.; woodlands and fields; frequent; Jul–Oct; ALL.

Triodanis Raf. (Venus' Looking Glass)

1 Capsule pores at or near middle . . . . **T. perfoliata** var. **perfoliata**
1 Capsule pores near apex . . . . **T. perfoliata** var. **biflora**

T. perfoliata (L.) Nieuwland var. **biflora** (R. & P.) Bradley; disturbed sites; infrequent; May–Jun; NC, SC, TN. *Specularia biflora* (R. & P.) F. & M.
T. perfoliata (L.) Nieuwland var. **perfoliata**; disturbed sites; common; May–Jun; ALL. *Specularia perfoliata* (L.) DC.

CANNABACEAE (Hemp Family)

Humulus L. (Hops)

1 Principal leaves 3-lobed; base of fruit with dense, yellow glands .... **H. lupulus**
1 Principal leaves 5–7-lobed; fruit inconspicuously glandular at base .... **H. japonicus**

H. japonicus Sieb. & Zucc.*, Japanese H.; escaping to waste places; occasional; Jun–Sep; TN, VA.
H. lupulus L.*; waste places and moist thickets; occasional; Jul–Aug; GA, VA.

CAPPARACEAE (Caper Family)

Polanisia Raf. (Clammyweed)

P. dodecandra (L.) DC.; possibly adventive on disturbed, often gravelly sites; infrequent; Jun–Sep; VA.

CAPRIFOLIACEAE (Honeysuckle Family)

1 Herbs (or sometimes scarcely woody at base), trailing or erect 2
  2 Flowers sessile; leaves deciduous; stems erect .... **Triosteum**
  2 Flowers pedicellate; leaves evergreen; stems trailing .... **Linnaea**
1 Shrubs or woody vines 3
  3 Leaves pinnately compound .... **Sambucus**
  3 Leaves simple 4
    4 Corolla rotate or subrotate; styles nearly absent .... **Viburnum**
    4 Corolla campanulate, bilabiate, tubular, or funnelform; styles elongate 5
      5 Leaves serrate; fruit a woody capsule .... **Diervilla**
      5 Leaves entire (occasionally deeply lobed but not serrate); fruit fleshy 6
        6 Corolla tubular or funnelform, often zygomorphic; berries several-seeded .... **Lonicera**
        6 Corolla campanulate, actinomorphic; drupes 2-seeded ... **Symphoricarpos**

Diervilla Miller (Bush Honeysuckle)

1 Leaves with petioles 5 mm or more long .... **D. lonicera**
1 Leaves sessile or with petioles less than 5 mm long 2
  2 Branchlets and lower leaf surfaces glabrous .... **D. sessilifolia** var. **sessilifolia**
  2 Branchlets and lower leaf surfaces hairy .... **D. sessilifolia** var. **rivularis**

D. lonicera Mill.; woodlands and bluffs; occasional; Jun–Jul; GA, NC, TN, VA.
D. sessilifolia Buckley var. **rivularis** (Gattinger) Ahles; rocky woodlands and dry river bluffs; infrequent; Jun–Jul; NC, TN. Similar to the following taxon but usually restricted to more xeric sites.
D. sessilifolia Buckley var. **sessilifolia**; stream banks and mesic woodlands; occasional; Jun–Jul; GA, NC, SC, TN.

Linnaea L. (Twinflower)

L. borealis L.; moist woodlands; infrequent; Sep; TN. Not seen in our area since 1892, presumably from the Great Smoky Mountains National Park.

Lonicera L. (Honeysuckle)

1 Inflorescence axillary, flowers paired with fused ovaries; leaves not connate 2
  2 Plants climbing or trailing; fruit black .... **L. japonica**
  2 Plants erect, shrubby; fruit red 3
    3 Peduncles shorter or about as long as the flower; leaves pubescent beneath; cultivar .... **L. morrowii**
    3 Peduncles longer than the flowers; leaves essentially glabrous beneath; native ... . **L. canadensis**
1 Inflorescence a terminal spike; ovaries not fused; some leaves connate 4
  4 Corolla 3–5.5 cm long, crimson, the lobes nearly equal .... **L. sempervirens**
  4 Corolla less than 3 cm long, orange, yellow, or tinged with red 5
    5 Corolla lobes tinged with purple or red, distinctly gibbous at the base; leaves glaucous beneath .... **L. dioica**
    5 Corolla lobes yellow, scarcely or not gibbous at the base; leaves green to slightly glaucous beneath .... **L. flava**

L. canadensis Bartram, Fly H.; moist woods and bogs at high elevations; occasional; May–Jun; GA, NC, TN, VA.

L. dioica L.; woodlands and thickets; occasional; Jun–Aug; GA, NC, TN, VA.

L. flava Sims, Yellow H.; woodlands and thickets; infrequent; Apr–May; NC, SC.

L. japonica Thunb.*, Japanese H.; widespread and one of our most noxious weeds; common; Apr–Jun; ALL.

L. morrowii Gray*; occasionally escaping to woodlands; May–Jun; TN, VA.

L. sempervirens L.*, Coral H.; persistent and frequently escaping; Apr–Jul; ALL. Native to the eastern U.S. but probably introduced in our area.

Sambucus L. (Elderberry)

1 Cymes flat-topped; fruit blue-black .... **S. canadensis**
1 Cymes paniculate; fruit red .... **S. pubens**

S. canadensis L., Common E.; open, usually moist sites; common; May–Jul; ALL.

S. pubens Michx., Red E.; open woodlands, usually at higher elevations than the preceding species; Apr–Jun; GA, NC, TN, VA. *S. racemosa* L. var. *pubens* (Michx.) Koehne.

Symphoricarpos Duhamel

1 Corolla 3–4 mm long; fruits pink to purple .... **S. orbiculatus**
1 Corolla 5–9 mm long; fruits white .... **S. albus**

S. albus (L.) Blake, Garden Snowberry; doubtfully native; infrequent; May–Jul; VA.

S. orbiculatus Moench., Coral Berry; open woodlands, usually on basic or neutral soils; frequent; Jul–Sep; ALL.

Triosteum L. (Horse Gentian)

1 Stems soft pubescent (canescent) .... **T. perfoliatum**
1 Stems hispid or hirsute 2
  2 Leaves lanceolate, broadest near the middle, less than 5 cm wide; flowers usually yellow .... **T. angustifolium**
  2 Leaves ovate, broadest near the base, usually more than 5 cm wide; flowers purplish .... **T. aurantiacum**

T. angustifolium L.; open woodlands, usually on basic soils; occasional; Apr–May; TN, VA.

T. aurantiacum Bicknell; open woodlands, usually on basic soils; occasional; May–Jun; GA, NC, TN, VA.

T. perfoliatum L.; open woodlands; usually on basic soils; frequent; May–Jun; ALL.

These taxa are variable and in need of critical study.

Viburnum L.

1 Outer flowers large, showy, and sterile; drupes red . . . . **V. lantanoides**
1 Flowers all similar in size; drupes black or blue 2
  2 Leaves 3-lobed, palmately veined . . . . **V. acerifolium**
  2 Leaves not lobed, pinnately veined 3
    3 Leaves coarsely dentate; lateral veins simple or forked and extending to the leaf margin 4
      4 Upper leaves with petioles more than 5 mm long; stipules present or absent . . . . **V. dentatum**
      4 Upper leaves sessile or subsessile; stipules present . . . . . . **V. rafinesquianum**
    3 Leaves entire, serrate, or denticulate; lateral veins joining before reaching the leaf margin 5
      5 Cymes on peduncles 5–50 mm long 6
        6 Leaves entire to crenulate, oblong to lanceolate . . . . **V. nudum**
        6 Leaves mostly denticulate or rarely entire, ovate to lanceolate . . . . **V. cassinoides**
      5 Cymes sessile or peduncles less than 5 mm long 7
        7 Leaves prominently acuminate . . . . **V. lentago**
        7 Leaves acute to slightly acuminate 8
          8 Leaves glabrous or essentially so, membranaceous, dull . . **V. prunifolium**
          8 Leaves, especially when young, rusty-brown pubescent beneath, coriaceous, lustrous . . . . **V. rufidulum**

V. acerifolium L., Maple-leaf Arrowwood; deciduous woodlands; frequent; Apr–Jun; ALL.

V. cassinoides L., Wild Raisin; woodlands and clearings; frequent; May–Jun; ALL.

V. dentatum L.; moist woodlands and stream banks; occasional; Mar–May; ALL. *V. dentatum* L. var. *lucidum* Ait.

V. lantanoides Michx., Witch Hobble; coves and mesic woods at mid to high elevations; occasional; Apr–Jun; GA, NC, TN, VA. *V. alnifolium* Marsh.

V. lentago L.; woods and stream banks; infrequent; May–Jun; GA, VA.

V. nudum L.; moist woods and streams; occasional; Apr–May; NC, SC, VA. Very closely related to *V. cassinoides*.

V. prunifolium L.; moist woods and stream banks; frequent; Mar–Apr; ALL.

V. rafinesquianum Schultes; rocky woodlands, often on calcareous soils; occasional; Apr–May; TN, VA. Closely related to *V. dentatum*.

V. rufidulum Raf., Blue Haw; woodlands and thickets; infrequent; Mar–Apr; NC, SC, TN. Related to and occasionally hybridizing with *V. prunifolium*.

A taxonomically difficult genus with several sections in need of further study.

CARYOPHYLLACEAE (Pink Family)

1 Fruit a single-seeded utricle, indehiscent; petals absent 2
 2 Stipules present . . . . **Paronychia**
 2 Stipules absent . . . . **Scleranthus**
1 Fruit a few- to many-seeded capsule; petals usually present 3
 3 Sepals separate 4
  4 Petals deeply cleft or bifid 5
   5 Capsule ovoid, opening to the base by twice as many valves as styles; styles 3, rarely 5 . . . . **Stellaria**
   5 Capsule cylindric, somewhat asymmetrical, open at the apex by a ring of small teeth; styles 5 . . . . **Cerastium**
  4 Petals entire or merely erose, reduced or occasionally absent 6
   6 Inflorescence umbellate . . . . **Holosteum**
   6 Inflorescence cymose or racemose 7
    7 Styles 5; petals reduced . . . . **Sagina**
    7 Styles 3; petals conspicuous . . . . **Arenaria**
 3 Sepals fused 8
  8 Styles 2 9
   9 Leaves more than 1 cm wide . . . . **Saponaria**
   9 Leaves less than 1 cm wide 10
    10 Calyx 5–15-nerved; bracts broadly ovate . . . . **Petrorhagia**
    10 Calyx with 30 or more nerves; bracts linear . . . . **Dianthus**
  8 Styles 3–5 or flowers unisexual 11
   11 Styles 3, rarely more . . . . **Silene**
   11 Styles 5 or flowers unisexual 12
    12 Calyx lobes longer than the tube . . . . **Agrostemma**
    12 Calyx lobes shorter than the tube . . . . **Lychnis**

Agrostemma L. (Corn Cockle)

A. githago L.*; roadsides and fields; frequent; May–Jul; ALL.

Arenaria L. (Sandwort)

1 Capsule 6-valved; leaves ovate . . . . **A. serpyllifolia**
1 Capsule 3-valved; leaves mostly oblong, linear, or subulate 2
 2 Leaves rigid, clustered into distinct fascicles . . . . **A. stricta**
 2 Leaves strictly opposite 3
  3 Pedicels and sepals glandular-pubescent; stems lax . . . . **A. godfreyi**
  3 Pedicels and sepals glabrous; stems erect 4
   4 Petals 3 mm or less long . . . . **(A. alabamensis)**
   4 Petals more than 3 mm long 5
    5 Upper stem leaves oblong-ovate, less than 7 mm long; calyx rounded at base . . . . **A. uniflora**
    5 Upper stem leaves linear, more than 7 mm long; calyx truncate at base . . 6
     6 Mat-forming perennials with sterile, basal shoots present at anthesis . . . . **A. groenlandica**
     6 Winter annuals; basal leaves absent at anthesis . . . . **A. glabra**

A. alabamensis (McCormick, Bozeman & Spongberg) Wyatt; wet areas on granite outcrops; infrequent; Apr–May; NC (Piedmont).

A. glabra (Michx.) Fern.; rock outcrops; infrequent; Apr–May; NC, SC.

A. godfreyi Shinners; wet ditches, bogs, and meadows; infrequent and disjunct from the Coastal Plain; May–Jun; TN.

A. groenlandica (Retz.) Sprengel; rock outcrops; occasional; Apr–Jun; NC, SC, TN, VA.

A. serpyllifolia L.*; disturbed sites; common; Mar–Jun; ALL.

A. stricta Michx.; rocky woodlands; occasional; Jun–Jul; VA.

A. uniflora (Walt.) Muhl.; granite outcrops; rare and local; Apr–May; NC, SC. More commonly found on granite outcrops on the Piedmont of GA, NC, SC.

Cerastium (Mouse-ear Chickweed)

1 Perennials, often with basal branches 2
  2 Petals longer than the sepals .... **C. arvense**
  2 Petals shorter than the sepals .... **C. fontanum**
1 Annuals or winter annuals, without basal branches 3
  3 Petals 1.5–2.5× as long as the sepals; sepals obtuse 4
    4 Pedicel about as long as the sepals at anthesis ..... **C. nutans** var. **brachypodum**
    4 Pedicel longer than the sepals at anthesis .... **C. nutans** var. **nutans**
  3 Petals about as long as the sepals; sepals acute to acuminate 5
    5 Flowers glomerate; pedicels equal to or shorter than the capsule ....
    .... **C. glomeratum**
    5 Flowers in loose cymes; pedicels longer than the capsule 6
      6 Margins of floral bracts scarious .... **C. semidecandrum**
      6 Margins of floral bracts not scarious .... **C. brachypetalum**

C. arvense L.; open rocky sites; infrequent; Apr–Jul; VA.

C. brachypetalum Pers.*; disturbed sites; infrequent; Apr–Jun; NC, VA.

C. fontanum Baumg.*; waste places; frequent; Mar–Jun; NC, SC, TN, VA. Represented in our area by subsp. **triviale** (Link) Jalas. *C. vulgatum* L., *C. holosteoides* Fries var. *vulgare* (Fries) Wahlenb.

C. glomeratum Thuillier*; waste places and often a troublesome lawn weed; common; Mar–Jun; ALL. *C. viscosum* L.

C. nutans Raf. var. **brachypodum** Engelm.; woodlands and disturbed sites; infrequent; Apr–May; NC, VA.

C. nutans Raf. var. **nutans**; rich woodlands; frequent; Apr–May; ALL.

C. semidecandrum L.*; disturbed sites; occasional; Apr–Jun; NC, TN, VA.

Dianthus L. (Deptford Pink)

D. armeria L.*; fields and waste places; common; May–Jul; ALL.

Holosteum L. (Jagged Chickweed)

H. umbellatum L.*; waste places and lawns; occasional; Mar–May; NC, SC, TN, VA.

Lychnis L. (Evening Campion)

L. alba Miller*; disturbed sites; frequent; May–Aug; ALL.

Paronychia Miller

1 Sepals with a slender awn at the tip 2
  2 Plants silky-pubescent; calyx pubescent . . . . **P. argyrocoma**
  2 Plants not silky-pubescent; calyx glabrous . . . . **P. virginica**
1 Sepals with only a cusp or mucro at the tip 3
  3 Stems puberulent . . . . **P. fastigiata**
  3 Stems glabrous . . . . **P. canadensis**

P. argyrocoma (Michx.) Nutt., Silverling; open rocky slopes; occasional; Apr–Jul; GA, NC, TN, VA. *P. argyrocoma* (Michx.) Nutt. var. *albimontana* (Fern.) Maguire.

P. canadensis (L.) Wood; woodlands and disturbed sites; frequent; Jun–Oct; NC, SC, TN, VA.

P. fastigiata (Raf.) Fern.; woodlands and disturbed sites; frequent; Jun–Oct; ALL. *P. montana* (Small) Pax & Hoffm.

P. virginica Sprengel; rocky slopes, often on basic soils; infrequent; Jun–Aug; VA.

Petrorhagia (Ser.) Link

P. prolifer (L.) Ball & Heywood*; disturbed sites; occasional; Jun–Sep; VA. *Dianthus prolifer* L.

Sagina L. (Pearlwort)

S. decumbens (Ell.) T. & G.; disturbed sites; infrequent; Mar–Jun; ALL.

Saponaria L. (Bouncing Bet)

S. officinalis L.*; disturbed sites; common; May–Jul; NC, SC, TN, VA.

Scleranthus L.

S. annuus L.*; disturbed sites; infrequent; Apr–Oct; NC, SC, TN, VA.

Silene L.

1 Stem leaves whorled . . . . **S. stellata**
1 Stem leaves opposite 2
  2 Calyx glabrous or essentially so 3
    3 Calyx inflated 4
      4 Flowers solitary in the upper axils . . . . **S. nivea**
      4 Flowers in cymose panicles . . . . **S. cucubalus**
    3 Calyx not inflated 5
      5 Cymes open, few-flowered; stems with sticky internodes . . . . . . **S. antirrhina**
      5 Cymes congested; stems not sticky . . . . **S. armeria**
  2 Calyx pilose 6
    6 Petals deep red, crimson, or scarlet . . . . **S. virginica**
    6 Petals white or pink 7
      7 Petals entire or slightly erose; stems less than 2.5 dm tall . . . . . **S. caroliniana**
      7 Petals bifid or deeply cleft; stems more than 2.5 dm tall 8
        8 Petals deeply cleft into (usually) 8 linear segments; perennials . . . . **S. ovata**
        8 Petals bifid; annuals or biennials 9
          9 Calyx in fruit less than 1.5 cm long; hirsute biennial; flowers open during day . . . . **S. dichotoma**

9 Calyx in fruit 2.5 cm or more long; viscid-pubescent annual; flowering at dusk . . . . **S. noctiflora**

S. antirrhina L., Sleepy Catchfly; fields, railways, and disturbed sites; common; Apr–Jul; NC, SC, TN, VA.
S. armeria L.*; occasionally escaping to waste places; Jun–Oct; NC, VA.
S. caroliniana Walt., Wild Pink; dry rocky woodlands; occasional; Apr–Jun; TN, VA. Represented in our area by var. **pensylvanica** (Michx.) Fern.
S. cucubalus Wibel*; waste places; occasional; May–Aug; NC, SC, TN, VA.
S. dichotoma Ehrh.*; disturbed sites; occasional; May–Aug; NC, TN, VA.
S. nivea (Nutt.) Otth., Snowy Campion; moist woods; infrequent; Jun–Jul; VA.
S. noctiflora L.*; waste areas; infrequent; Jun–Aug; NC, VA.
S. ovata Pursh; rich woodlands; infrequent; Aug–Sep; GA, NC, TN.
S. stellata (L.) Ait. f., Starry Campion; open woodlands; common; Jun–Sep; ALL.
S. virginica L., Fire Pink; open woodlands; common; Apr–Jun; ALL.

Stellaria L. (Chickweed)

1 Styles 5 . . . . **S. aquatica**
1 Styles 3 2
2 Leaves ovate, broadly elliptic, or oblanceolate; stems puberulent 3
3 Petals less than 3.5 mm long; leaves ovate, the lower distinctly petiolate . . . . **S. media**
3 Petals more than 3.5 mm long; leaves elliptic, sessile or occasionally petiolate 4
4 Sepals acuminate, equal to or longer than the petals . . . . **S. corei**
4 Sepals obtuse to acute, shorter than the petals . . . . **S. pubera**
2 Leaves narrow, linear, or lanceolate; stems glabrous 5
5 Petals shorter than the sepals; stems decumbent, often rooting at the nodes . . . . **S. alsine**
5 Petals longer than the sepals; stems erect or arching, not rooting at the nodes . . . 6
6 Sepals 2–4 mm long; seeds smooth; inflorescence few-flowered . . . . **S. longifolia**
6 Sepals 4–7 mm long; seeds ridged; inflorescence many-flowered . . . . **S. graminea**

S. alsine Grimm; moist open areas; infrequent; Jun–Sep; NC, TN.
S. aquatica (L.) Scopoli*; moist sites; infrequent; Jun–Oct; NC, TN, VA. *Myosoton aquaticum* (L.) Moench.
S. corei Shinners, Tennessee C.; rich woods and seepages; occasional; Apr–Jun; NC, TN. *S. tennesseensis* (Mohr) Strausbaugh & Core.
S. graminea L.*; fields and waste places; frequent; May–Aug; NC, SC, TN, VA.
S. longifolia Willd.; moist open areas; infrequent; May–Jul; TN, VA.
S. media (L.) Villars*; disturbed sites; common; Jan–Dec, mostly Mar–May; ALL.
S. pubera Michx., Giant C.; rich woods; occasional; Apr–Jun; GA, NC, SC, TN.

CELASTRACEAE (Stafftree Family)

1 Leaves alternate . . . . **Celastrus**
1 Leaves opposite 2

2 Leaves lanceolate to obovate, more than 3 cm long, deciduous or semievergreen; seed surrounded by a scarlet or orange aril .... **Euonymus**
2 Leaves linear, less than 3 cm long, evergreen; seed with white, basal aril .... **Paxistima**

Celastrus L. (Bittersweet)

1 Leaves ovate-oblong; flowers 4–many/inflorescence .... **C. scandens**
1 Leaves suborbicular to broadly obovate; flowers usually less than 4/inflorescence .... **C. orbiculatus**

C. orbiculatus Thunb.*; thickets and woodland margins; infrequent; May–Jun; GA, NC, SC, VA.

C. scandens L.; thickets, river banks, and woodland margins; occasional; May–Jun; GA, NC, TN, VA.

Euonymus L.

1 Flowers 4-parted; fruit not tuberculate .... **E. atropurpureus**
1 Flowers 5-parted; fruit tuberculate 2
2 Low trailing shrub; leaves widest above the middle, petiole more than 2 mm long .. .. **E. obovatus**
2 Erect or arching shrub; leaves widest about the middle, petiole less than 2 mm long .... **E. americanus**

E. americanus L., Strawberry Bush; open rich woodlands; common; May–Jun; ALL.

E. atropurpureus Jacq., Burning Bush; rich woods and field margins; infrequent; May–Jun; GA, NC, TN, VA.

E. obovatus Nutt., Trailing Strawberry Bush; rich mountain woods; occasional; May–Jun; GA, NC, TN.

Paxistima Raf. (Canby's Mountain Lover)

P. canbyi Gray; calcareous rocky slopes; infrequent; Apr–May; NC, TN, VA.

CERATOPHYLLACEAE (Hornwort Family)

Ceratophyllum L. (Hornwort)

C. demersum L.; quiet waters; infrequent; May–Sep; VA.

CHENOPODIACEAE (Goosefoot Family)

1 Leaves represented by small, opposite scales; plant succulent .... **Salicornia**
1 Leaves present, mostly alternate; plants not succulent 2
2 Leaves mostly linear, 4 mm or less wide; stems villous above .... **Kochia**
2 Leaves mostly linear to wider, 4 mm or more wide; stems glabrous, glandular-pubescent, or farinose above 3
3 Flowers bisexual; fruit surrounded by 5 entire sepals .... **Chenopodium**
3 Flowers unisexual; fruit surrounded by 2 bracteoles .... **Atriplex**

Atriplex L. (Spear Scale)

A. patula L.; margins of saline marshes and ponds; infrequent; Jul–Oct; VA (Saltville).

Chenopodium L.

1 Leaves with resinous glands, aromatic .... **C. ambrosioides**
1 Leaves glabrous or farinose, but not glandular 2
  2 Lower leaf surface farinose 3
    3 Leaves entire, linear to lanceolate .... **C. desiccatum**
    3 Leaves usually lobed or toothed, lanceolate to deltoid .... **C. album**
  2 Lower leaf surface glabrous or essentially so 4
    4 Seeds 1.5–2.5 mm broad .... **C. gigantospermum**
    4 Seeds less than 1.5 mm broad 5
      5 Pericarp easily removed from the seed; leaves thin .... **C. standleyanum**
      5 Pericarp adherent to the seed; leaves thick .... **C. murale**

C. album L.*, Lamb's Quarters; waste places; common; Jun–Nov; NC, SC, TN, VA.
C. ambrosioides L.*, Mexican Tea; waste places; common; Jul–Nov; NC, SC, TN, VA.
C. desiccatum A. Nels., Narrow-leaved Goosefoot; waste places; infrequent; May–Nov; VA. Represented in our area by var. **leptophylloides** (Murr.) Wahl.
C. gigantospermum Aellen, Maple-leaved Goosefoot; rocky woods and sheltered bluffs; infrequent; May–Oct; NC, VA.
C. murale L.*, Nettle-leaved Goosefoot; waste places; occasional; May–Oct; TN, VA.
C. standleyanum Aellen; woodlands and roadsides; occasional; Jul–Oct; VA.

Kochia Roth

K. scoparia (L.) Schrader*; disturbed sites; infrequent; Jun–Sep; VA.

Salicornia L. (Glasswort)

S. europaea L.; margins of saline marshes and ponds; infrequent; Jul–Oct; VA (Saltville).

CISTACEAE (Rockrose Family)

1 Shrubs .... **Hudsonia**
1 Herbs 2
  2 Petals yellow and showy or absent .... **Helianthemum**
  2 Petals red to maroon .... **Lechea**

Helianthemum Miller (Frostweed)

Daoud, H. S., and R. L. Wilbur. 1965. A revision of the North American species of *Helianthemum* (Cistaceae). *Rhodora* 67:63–82, 201–216, 255–312.

1 External sepals of cleistogamous flowers 0.6 mm or more long; capsule sharply triangular in cross section .... **H. bicknellii**
1 External sepals of cleistogamous flowers less than 0.6 mm long; capsule weakly angled or rounded in cross section 2
  2 Seeds 1–3/cleistogamous capsule; stems arising from a horizontal rootstock .... **H. propinquum**
  2 Seeds 5 or more/cleistogamous capsule; stems arising from an erect caudex .... **H. canadense**

H. bicknellii Fern.; barrens, woodlands, and fields; infrequent; Jun–Jul; GA, NC, TN, VA.

H. canadense (L.) Michx.; sandy woods and slopes; infrequent; Apr–Jul; GA, NC, TN, VA.
H. propinquum Bicknell; sandy fields and woods; infrequent; May–Jul; NC, TN, VA.

Hudsonia L.

H. montana Nutt.; shrub balds; Jun–Jul; a very rare NC endemic.

Lechea L. (Pinweed)

Wilbur, R. L., and H. S. Daoud. 1961. The genus *Lechea* (Cistaceae) in the southeastern United States. *Rhodora* 63:103–118.

1 Stems spreading-pubescent .... **L. mucronata**
1 Stems glabrous or appressed-pubescent 2
  2 Outer 2 sepals equal to or longer than the inner 3 3
    3 Cauline leaves 1 mm or less wide; capsule completely enclosed by the sepals; stigma not exposed after anthesis .... **L. tenuifolia**
    3 Cauline leaves more than 1 mm wide; capsule slightly to conspicuously exserted; stigma exposed after anthesis .... **L. minor**
  2 Outer 2 sepals shorter than the inner 3 4
    4 Capsule 1.5× as long as broad; base of fruiting calyx yellowish and differing in texture and color from the calyx lobes .... **L. racemulosa**
    4 Capsule mostly less than 1.5× as long as broad; base of fruiting calyx not differing in color or texture from the calyx lobes 5
      5 Fruiting calyx obovoid; seeds dark brown, smooth .... **L. pulchella**
      5 Fruiting calyx subglobose; seeds pale brown, covered by a gray membrane ... . **L. intermedia**

L. intermedia Leggett; sandy clearings; infrequent; Jul–Oct; VA.
L. minor L.; sandy woods and fields; infrequent; Jul–Aug; GA, NC.
L. mucronata Raf.; open woods and fields; infrequent; Jun–Aug; NC. *L. villosa* Ell.
L. pulchella Raf.; sandy woods; infrequent; Jun–Aug; VA. *L. leggettii* Britt. & Holl.
L. racemulosa Michx.; dry woods, fields, and disturbed sites; common; Jul–Aug; ALL.
L. tenuifolia Michx.; dry woods; infrequent; Jun–Aug; GA, SC.

CLETHRACEAE (White Alder Family)

Clethra L. (Sweet Pepperbush)

C. acuminata Michx.; rich moist woods and stream banks; common; Jul–Aug; ALL.

CLUSIACEAE (St. John's-wort Family)
(Includes Hypericaceae)

Contributed by David H. Webb

Adams, P. 1973. Clusiaceae of the southeastern United States. *J. Elisha Mitchell Sci. Soc.* 89:62–71.

1 Petals yellow; stamens few–numerous, if stamens fascicled, then intervening glands absent .... **Hypericum**

1 Petals flesh-colored, mauve-purple, or pinkish; stamens 9, in 3 fascicles, each fascicle with 3 stamens alternating with 3 orange glands .... **Triadenum**

Hypericum L.

1 Petals and sepals 4; sepals in unequal pairs 2
 2 Styles and carpels 3; upper leaves clasping or cordate, more than 1 cm wide .... **H. crux-andreae**
 2 Styles and carpels 2; leaves narrowed at the base, less than 1 cm wide 3
  3 Plants erect, branching above ground level, often 1–1.5 m tall; leaves widest near the middle .... **H. hypericoides**
  3 Plants forming low compact mats with decumbent stems, rarely more than 0.3 m tall; leaves widest above the middle .... **H. stragulum**
1 Petals and sepals 5; calyx nearly equal 4
 4 Woody shrubs or subshrubs 5
  5 Leaves and sepals with a narrow notch or groove at the base 6
   6 Flowers large, showy, 2.5–4 cm wide, terminal and solitary or in 3-flowered dichasia; capsules 6–9 mm wide .... **H. frondosum**
   6 Flowers smaller, less than 2 cm wide, in 3–many-flowered terminal or subterminal dichasia or both; capsules 1.5–6 mm wide 7
    7 Mature capsules slender-conic, not exceeding 6 mm long and 3 mm wide; seeds reddish-brown .... **H. densiflorum**
    7 Mature capsules lance-ovoid, more than 7 mm long and 3.5 mm wide; seeds dark brown or black .... **H. prolificum**
  5 Leaves and sepals without a narrow notch or groove at base 8
   8 Plants decumbent, mat-forming, with numerous erect branches; flowers solitary or in 3-flowered dichasia .... **H. buckleyi**
   8 Plants erect; inflorescence a many-flowered dichasium .... **H. nudiflorum**
 4 Annual or perennial herbs 9
  9 Leaves and sepals with black and translucent glands 10
   10 Stems much branched with stout lateral branches; leaves mostly less than 2.5 cm long; seeds rough .... **H. perforatum**
   10 Stems sparsely branched with small lateral branches; leaves usually more than 2.5 cm long; seeds smooth or nearly so 11
    11 Capsules less than 6 mm long; widespread at lower elevations .... **H. punctatum**
    11 Capsules more than 6 mm long; plants mostly above 3,500 ft. 12
     12 Styles less than 4 mm long; petals usually 7–9 mm long .... **H. mitchellianum**
     12 Styles more than 4 mm long; petals usually 11–16 mm long .... **H. graveolens**
  9 Leaves and sepals with translucent glands only, black glands absent 13
   13 Styles appressed; stigmas minute; perennials with prominent rhizomes .... **H. ellipticum**
   13 Styles divergent; stigmas capitate; annual, or if perennial, lacking a prominent rhizome 14
    14 Stamens fewer than 25; styles 0.4–2 mm long 15
     15 Leaves scalelike to linear-subulate; flowers axillary and appearing racemose 16

16 Leaves linear-subulate, 0.5–2.2 cm long; capsule ovoid . . . . **H. drummondii**

16 Leaves scalelike, 1–4 mm long; capsule conical . . . . **H. gentianoides**

15 Leaves linear to elliptic to ovate; inflorescence basically a compound dichasium 17

17 Leaves linear, 1–3-nerved; capsule conical . . . . **H. canadense**

17 Leaves elliptic to ovate, mostly 5-nerved; capsule ellipsoid . . . . **H. mutilum**

14 Stamens numerous (50–80); styles 2–4 mm long . . . . . **H. denticulatum**

H. buckleyi M. A. Curtis; seepage slopes, moist crevices, ditches at high elevations; occasional; Jun–Aug; GA, NC, SC.

H. canadense L.; bogs, wet meadows, ditches; infrequent; Jul–Sep; GA, NC, TN, VA.

H. crux-andreae (L.) Crantz, St. Peter's-wort; moist to dry woods and ditches; occasional; Jul–Oct; GA, NC, SC. *Ascyrum stans* Michx., *H. stans* (Michx.) P. Adams & Robson.

H. densiflorum Pursh; wet meadows, stream banks, dry woods, and balds; infrequent; Jun–Aug; ALL.

H. denticulatum Walt.; fields, roadsides, and outcrops; occasional; Jun–Aug; ALL. Represented in our area by var. **acutifolium** (Ell.) Blake.

H. drummondii (Grev. & Hook.) T. & G.; dry fields and roadsides; occasional; Jul–Sep; NC, SC, TN, VA.

H. ellipticum Hook.; bogs and wet meadows; infrequent; Jun–Aug; TN. Disjunct from West Virginia.

H. frondosum Michx.; bluffs and calcareous outcroppings; occasional; Jun–Jul; TN.

H. gentianoides (L.) BSP.; dry fields, roadsides, and outcrops; frequent; Jul–Sep; ALL.

H. graveolens Buckley; moist slopes and dry roadsides at high elevations; occasional; Jun–Aug; NC, TN. Apparently hybridizes with *H. mitchellianum*.

H. hypericoides (L.) Crantz, St. Andrew's Cross; dry woods; occasional; Jun–Aug; ALL. *Ascyrum hypericoides* L.

H. mitchellianum Rydb.; moist slopes and dry roadsides at high elevations; occasional; Jun–Aug; NC, TN, VA. Apparently hybridizes with *H. graveolens*.

H. mutilum L.; marshes, ditches, and wet meadows; common; Jun–Oct; ALL.

H. nudiflorum Michx.; stream banks and woods openings; occasional; Jun–Jul; GA, NC.

H. perforatum L.*; fields and roadsides; frequent; Jun–Aug; NC, SC, TN, VA.

H. prolificum L.; rocky slopes, fields, and roadsides; frequent; Jun–Aug; ALL.

H. punctatum Lam.; fields, roadsides, ditches, and open woods; common; Jun–Sep; ALL.

H. stragulum P. Adams & Robson; dry woods, rocky slopes and road embankments; occasional; Jun–Aug; ALL. *Ascyrum hypericoides* L. var. *multicaule* (Michx.) Fern., *H. hypericoides* (L.) Crantz subsp. *multicaule* (Michx. ex Willd.) Robson.

Triadenum Raf.

1 Leaves sessile, cordate-clasping; filaments united basally . . . . **T. virginicum**

1 Leaves petioled, narrowed at the base; filaments united to above the middle . . . . **T. walteri**

T. virginicum (L.) Raf.; bogs and wet meadows; occasional; Jul–Sep; NC, TN, VA.

Hypericum virginicum L. A complex often divided into var. *virginicum* and var. *fraseri* (Spach) Fern.

T. walteri (J. G. Gmelin) Gleason; marshes and low woods; occasional; Aug–Sep; NC, SC. *Hypericum walteri* J. G. Gmelin.

CONVOLVULACEAE (Morning Glory Family)

1 Plants yellow-orange, parasitic; leaves reduced to scales or absent .... **Cuscuta**
1 Plants green; leaves variously developed and not reduced to scales 2
  2 Styles 2, united near the base .... **Bonamia**
  2 Style 1 3
    3 Stigma 1, capitate or somewhat lobed .... **Ipomoea**
    3 Stigmas 2, not capitate 4
      4 Calyx concealed by 2 large bracts; fruit 1-locular .... **Calystegia**
      4 Calyx not concealed by bracts; fruit 2-locular .... **Convolvulus**

Bonamia Thouars

B. humistrata (Walt.) Gray; waste places; infrequent; Jun–Aug; SC.

Calystegia R. Br. (Bindweed)

1 Stems erect or slightly arching .... **C. spithamaea**
1 Stems trailing or twining 2
  2 Stems and leaves densely soft-pubescent .... **C. sericata**
  2 Stems and leaves glabrous or nearly so .... **C. sepium**

C. sepium (L.) R. Br., Hedge B.; fields and disturbed sites; frequent; May–Aug; NC, SC, TN, VA.

C. sericata (House) Bell; slopes and clearings; infrequent; Jun–Jul; GA, NC, SC.

C. spithamaea Pursh, Low B.; fields and disturbed sites; occasional; May–Jun; NC, SC, TN, VA.

Convolvulus L. (Bindweed)

C. arvensis L.; disturbed sites; occasional; May–Oct; NC, SC, TN, VA.

Cuscuta L. (Dodder, Love Vine)

1 Flowers subtended by 1–5 bracts; sepals separate .... **C. compacta**
1 Flowers bractless; sepals united at base 2
  2 Calyx lobes prominently imbricate and forming prominent angles at the sinuses ... . **C. pentagona**
  2 Calyx lobes not obviously angled 3
    3 Flowers all or mostly 4-merous 4
      4 Tips of corolla lobes inflexed; flowers with a papillose appearance .. **C. corylii**
      4 Tips of corolla lobes erect; flowers not papillose in appearance .... **C. polygonorum**
    3 Flowers all or mostly 5-merous 5
      5 Corolla lobes acute, tips sometimes inflexed .... **C. campestris**
      5 Corolla lobes obtuse, tips usually not inflexed 6
        6 Flowers 2–4 mm long; capsule beakless .... **C. gronovii**
        6 Flowers 4–6 mm long; capsule beak 1–1.5 mm long .... **C. rostrata**

C. campestris Yuncker; frequent; Jun–Nov; NC, SC, TN, VA.
C. compacta Juss.; occasional; Aug–Nov; ALL.
C. corylii Engelm.; infrequent; Jul–Nov; NC, VA.
C. gronovii Willd.; frequent; Aug–Nov; ALL.
C. pentagona Engelm.; occasional; Jun–Nov; GA, NC, VA.
C. polygonorum Engelm.; infrequent; Jun–Oct; VA.
C. rostrata Shuttlew. ex Engelm.; occasional; Aug–Sep; GA, NC, SC, VA.

These taxonomically difficult species parasitize a wide range of herbaceous and woody hosts and occur in numerous habitats. Identification should be attempted with fresh material.

Ipomoea L. (Morning Glory)

1 Stamens exserted; corolla salverform . . . . **I. coccinea**
1 Stamens included; corolla funnelform 2
  2 Sepals densely hirsute below 3
    3 Leaves usually 3-lobed; calyx lobes linear, 15–25 mm long . . . . . . **I. hederacea**
    3 Leaves entire; calyx lobes acute to acuminate, 10–15 mm long . . . . **I. purpurea**
  2 Sepals glabrous or sparsely pubescent throughout 4
    4 Perennial; sepals glabrous; corolla 5–8 cm long . . . . **I. pandurata**
    4 Annual; sepals ciliate; corolla less than 5 cm long . . . . **I. lacunosa**

I. coccinea L., Red M. G.*; waste places; frequent; Aug–Nov; NC, SC, TN, VA.
I. hederacea (L.) Jacq.*; waste places; occasional; Jul–Nov; ALL.
I. lacunosa L.; moist fields and waste places; occasional; Sep–Nov; ALL.
I. pandurata (L.) Meyer, Man Root; roadsides and other waste places; frequent; May–Jul; ALL. Tubers often reach up to 6 dm long.
I. purpurea (L.) Roth, Common M. G.*; waste places; frequent; Jul–Sep; ALL.

CORNACEAE (Dogwood Family)

Cornus L. (Dogwood)

Wilson, J. S. 1965. Variation of three taxonomic complexes of the genus *Cornus* in eastern United States. *Trans. Kansas Acad. Sci.* 67:747–817.

1 Leaves alternate (often crowded toward the tips of the branches) . . . . . **C. alternifolia**
1 Leaves opposite 2
  2 Flowers surrounded by 4 large white or pinkish bracts; fruits red 3
    3 Shrub or tree . . . . **C. florida**
    3 Herb, from woody rhizomes . . . . **C. canadensis**
  2 Flowers not subtended by large showy bracts; fruits blue or white 4
    4 Sepal length usually more than 1 mm; pith of second-year stems brown . . . . **C. amomum**
    4 Sepal length usually less than 1 mm; pith of second-year stems tan or white . . . 5
      5 Young leaves downy-pubescent beneath . . . . **C. rugosa**
      5 Young leaves with appressed pubescence (rarely glabrous) 6
        6 Leaves with 5–7 lateral veins on each side of the midrib . . . **C. stolonifera**
        6 Leaves with 3–4 lateral veins on each side of the midrib 7

7 Young branches maroon; fruit light blue .... **C. foemina** subsp. **foemina**
7 Young branches tan; fruit white .... **C. foemina** subsp. **racemosa**

C. alternifolia L. f., Alternate-leaved D.; rich woods; common; May–Jun; ALL.

C. amomum Mill., Silky Cornel; marshes and stream banks; common; May–Jun; ALL.

C. canadensis L., Bunchberry; moist woods; infrequent; Jun–Jul; VA.

C. florida L., Flowering D.; woodlands; common; Apr–May; ALL. The cultivar with pink bracts is referred to as forma *rubra* (Weston) Palmer & Stey.

C. foemina Mill. subsp. **foemina**; swamps and river banks; infrequent; May–Jun; SC, VA. *C. stricta* Lam.

C. foemina Mill. subsp. **racemosa** (Lam.) J. S. Wilson; moist woods and meadows; occasional; May–Jun; SC, VA. *C. racemosa* Lam.

C. rugosa Lam.; moist woods and stream banks; infrequent; May–Jun; VA.

C. stolonifera Michx., Red Osier; moist woods and stream banks; infrequent; May–Jun; VA.

CRASSULACEAE (Orpine Family)

Sedum L. (Stonecrop)

Clausen, R. T. 1975. Sedum *of North America north of the Mexican Plateau.* Ithaca, N.Y.: Cornell Univ. Press.

1 Stems and leaves red or maroon; carpels fused at base .... **S. smallii**
1 Stems and leaves green, occasionally tinged with red; carpels separate 2
2 Flowers unisexual; plants usually dioecious .... **S. rosea**
2 Flowers bisexual 3
3 Flowers yellow 4
4 Leaves alternate .... **S. acre**
4 Leaves whorled .... **S. sarmentosum**
3 Flowers white to pink 5
5 Leaves of flowering stems 3–12 cm long; ovaries stipitate ... **S. telephioides**
5 Leaves of flowering stems less than 3 cm long; ovaries not stipitate 6
6 Stem leaves, at least the lower, in whorls of 3 .... **S. ternatum**
6 Stem leaves alternate 7
7 Leaves of flowering stems auriculate at the base .... **S. pulchellum**
7 Leaves of flowering stems not auriculate at the base 8
8 Ratio of width to thickness of leaves of flowering stem less than 1.7; leaves usually green .... **S. nevii**
8 Ratio of width to thickness of leaves of flowering stem more than 2; leaves sometimes glaucous .... **S. glaucophyllum**

S. acre L.*; rarely persistent on dry rocky sites; May–Jun; NC, TN.

S. glaucophyllum Clausen; damp, primarily basic, rocky sites; occasional; May–Jun; NC, VA.

S. nevii Gray; open rocky sites; Apr–May; rare and restricted to the Ocoee River gorge, TN.

S. pulchellum Michx.; in limestone gravel along roadsides; infrequent; Apr–May; TN. Possibly introduced to our area.

S. rosea (L.) Scopoli; rocky open cliffs at high elevations; infrequent; Jul–Aug; NC, TN(?). This northern disjunct is nearly extirpated from our area.

S. sarmentosum Bunge*; persistent on rocky or shallow soils; occasional; Apr–May; NC, SC, TN, VA.

S. smallii (Britt.) Ahles; around vernal pools on granite outcrops and more common on the Piedmont; infrequent; NC, SC. *Diamorpha cymosa* (Nutt.) Britt., *D. smallii* Britt. Evidence favoring the placement of this taxon in the segregate genus *Diamorpha* is found in R. L. Wilbur, What do we know about *Diamorpha smallii* (Crassulaceae), "one of the better-known taxa in the southeastern flora"? *Sida* 13 (1988): 1–16.

S. telephioides Michx., Live-for-ever; cliffs and rocky woods; frequent; Aug–Oct; NC, SC, TN, VA.

S. ternatum Michx.; moist rocky woods and flood plains; common; Apr–May; NC, SC, TN, VA. Our most commonly encountered species.

CUCURBITACEAE (Gourd Family)

1 Fruit not spiny . . . . **Melothria**
1 Fruit spiny 2
  2 Fruits clustered, 1-seeded; petals 5 . . . . **Sicyos**
  2 Fruits single, 4-seeded; petals 6 . . . . **Echinocystis**

Echinocystis T. & G. (Wild Cucumber)

E. lobata (Michx.) T. & G.; stream bank thickets; infrequent; Jul–Oct; NC.

Melothria L. (Creeping Cucumber)

M. pendula L.; damp wood margins and thickets; occasional; Jul–Sep; NC, SC, TN, VA.

Sicyos L. (Bur Cucumber)

S. angulatus L.; stream banks and waste places; frequent; Jul–Sep; NC, SC, TN, VA.

DIAPENSIACEAE (Diapensia Family)

1 Flowers in racemes; leaves orbicular, marginal teeth not mucronate . . . . **Galax**
1 Flowers solitary; leaves widely elliptic, marginal teeth distinctly mucronate . . . . **Shortia**

Galax Raf.

G. urceolata (Poir.) Brummit; open woods; common; May–Jul; ALL. *G. aphylla* L.

Shortia T. & G. (Oconee Bells)

S. galacifolia T. & G.; rich woods along stream banks; infrequent; Mar–Apr; GA, NC, SC, VA(?).

DIPSACACEAE (Teasel Family)

Dipsacus L. (Wild Teasel)

D. fullonum L.*; roadsides, old fields, and waste places; occasional; Jul–Sep; NC, TN, VA. *D. sylvestris* Hudson.

DROSERACEAE (Sundew Family)

Drosera L. (Sundew)

1 Leaf blade at maturity suborbicular, wider than long; seeds about 6× as long as broad .... **D. rotundifolia**
1 Leaf blade at maturity spatulate to obovate, longer than broad; seeds about 2× as long as broad 2
  2 Petiole glabrous or with inconspicuous, sessile glands; petals white .... **D. intermedia**
  2 Petiole with few–many long, non-gland-tipped hairs; petals pink .... **D. capillaris**

D. capillaris Poir.; wet sandy ditches and bogs; infrequent; May–Aug; NC.

D. intermedia Hayne; wet sandy ditches and bogs; infrequent; Jul–Aug; NC, SC. This and the preceding species are primarily Coastal Plain in distribution.

D. rotundifolia L., Round-leaved S.; peaty bogs and permanently wet slopes; occasional; Jul–Sep; ALL.

EBENACEAE (Ebony Family)

Diospyros L. (Persimmon)

D. virginiana L.; dry woodlands and old fields; common; May–Jun; ALL.

ELAEAGNACEAE (Oleaster Family)

Elaeagnus L. (Oleaster)

E. umbellata Thunb.*; persistent after cultivation; occasional; Apr–May; ALL.

ELATINACEAE (Waterwort Family)

Elatine L. (Waterwort)

E. triandra Schkuhr; shallow water and pond margins; infrequent; Jul–Sep; NC. Perhaps introduced in our area.

ERICACEAE (Heath Family)
(Includes Monotropaceae and Pyrolaceae)

1 Plants parasitic, lacking chlorophyll; leaves reduced to scales 2
  2 Petals united; fruit a berry .... **Monotropsis**
  2 Petals separate; fruit a capsule .... **Monotropa**
1 Plants with chlorophyll; leaves not reduced to scales 3
  3 Petals separate 4
    4 Low shrubs .... **Leiophyllum**
    4 Perennial herbs, with subterranean rhizomes 5
      5 Inflorescence corymbose; style short, straight; cauline leaves present .... **Chimaphila**
      5 Inflorescence racemose; style elongate, straight or curved downward; leaves all basal .... **Pyrola**

3 Petals fused or, if appearing separate, then the ovary inferior 6
6 Ovary superior 7
7 Flowers 4-merous; fruit 4-celled . . . . **Menziesia**
7 Flowers 5-merous; fruit 5-celled 8
8 Corolla salverform, campanulate or rotate, widest at the apex 9
9 Stems creeping; fruit berrylike . . . . **Epigaea**
9 Stems erect; fruit dry 10
10 Corolla actinomorphic; the anthers (in bud) held in pockets of the corolla . . . . **Kalmia**
10 Corolla slightly zygomorphic; anthers exserted, not held in corolla pockets . . . . **Rhododendron**
8 Corolla tubular, ovoid, globose, or urceolate, usually constricted at the apex 11
11 Flowers solitary in leaf axils 12
12 Lower leaf surface lepidote; leaves not aromatic; fruit a capsule . . . . **Chamaedaphne**
12 Lower leaf surface essentially glabrous; leaves with the odor of wintergreen; fruit a berry . . . . **Gaultheria**
11 Flowers in terminal or axillary panicles, racemes or corymbs 13
13 Leaves 1–3 cm long; stems prostrate . . . . **Arctostaphylos**
13 Leaves more than 3 cm long; stems erect 14
14 Flowers globose, 2–4 mm long; leaves deciduous . . . . . . **Lyonia**
14 Flowers urceolate, 4–8 mm long; leaves deciduous or evergreen 15
15 Flowers in axillary or terminal racemes or both . . **Leucothoe**
15 Flowers in terminal panicles 16
16 Leaves deciduous; trees . . . . **Oxydendrum**
16 Leaves evergreen; shrubs . . . . **Pieris**
6 Ovary inferior 17
17 Low, trailing shrub, leaves with the odor of wintergreen . . . . **Gaultheria**
17 Erect shrubs, or if trailing, the leaves without the odor of wintergreen . . . 18
18 Leaves gland-dotted beneath . . . . **Gaylussacia**
18 Leaves not gland-dotted beneath 19
19 Leaves deciduous . . . . **Vaccinium**
19 Leaves evergreen 20
20 Leaves entire . . . . **Vaccinium**
20 Leaves finely serrulate . . . . **Gaylussacia**

Arctostaphylos Adans. (Bearberry)

A. uva-ursi (L.) Spreng.; granitic slopes; a rare northern disjunct; May–Jun; VA.

Chamaedaphne Moench (Leatherleaf)

C. calyculata (L.) Moench; bogs; infrequent; Mar–Apr; NC. *Cassandra calyculata* (L.) D. Don. Disjunct from the Coastal Plain.

Chimaphila Pursh (Pipsissewa)

1 Leaves lanceolate, mottled with white or gray along the midrib . . . . **C. maculata**
1 Leaves oblanceolate, green throughout . . . . **C. umbellata**

C. maculata (L.) Pursh, Spotted Wintergreen, Pipsissewa; dry upland woods; common; May–Jun; ALL.
C. umbellata (L.) Barton; upland woods; occasional; May–Jun; VA.

Epigaea L. (Trailing Arbutus)

E. repens L.; dry, often acid woodlands; common; Feb–May; ALL.

Gaultheria L. (Wintergreen)

G. procumbens L.; dry to moist acid woods; frequent; Jun–Aug; ALL.

Gaylussacia HBK. (Huckleberry)

1 Leaves evergreen, nonglandular . . . . **G. brachycera**
1 Leaves deciduous, with yellow glands below or on both surfaces 2
  2 Bracts equal to or longer than the pedicels, persistent; glands stalked . . . **G. dumosa**
  2 Bracts shorter than the pedicels, early deciduous; glands sessile 3
    3 Leaves glandular on both surfaces . . . . **G. baccata**
    3 Leaves glandular on the lower surface only 4
      4 Leaves thin-membranaceous, acuminate at the apex . . . . **G. ursina**
      4 Leaves subcoriaceous, acute, rounded or notched at the apex . . . **G. frondosa**

G. baccata (Wang.) K. Koch, Black H.; xeric woods; frequent; Apr–Jun; ALL.
G. brachycera (Michx.) Gray, Box H.; xeric woods; infrequent and more common north and west of our area; May–Jun; VA. Each patch represents a single plant.
G. dumosa (Andrz.) T. & G.; sandy woodlands; occasional; Mar–Jun; NC, SC, VA.
G. frondosa (L.) T. & G.; xeric woodlands; infrequent; Mar–May; SC, VA.
G. ursina (M. A. Curtis) T. & G. ex Gray; woodlands at mid to upper elevations; occasional; May–Jun; GA, NC, SC, TN.

Kalmia L.

Southall, R. M., and J. W. Hardin. 1974. A taxonomic revision of *Kalmia* (Ericaceae). *J. Elisha Mitchell Sci. Soc.* 90:1–23.

1 Flowers in lateral clusters; leaves mostly whorled . . . . **K. carolina**
1 Flowers in terminal clusters; leaves alternate . . . . **K. latifolia**

K. carolina Small, Sheepkill; bogs and low woods; occasional; Apr–Jun; GA, NC, TN, VA. *K. angustifolia* L. var. *caroliniana* (Small) Fern.
K. latifolia L., Mountain Laurel, Ivy; primarily dry rocky woods; common; Apr–Jun; ALL.

Leiophyllum (Pers.) Hedwig f. (Sand Myrtle)

L. buxifolium (Bergius) Ell.; rocky woods and exposed bluffs; occasional; Apr–May; GA, NC, SC, TN. A highly variable taxon sometimes divided into two poorly defined varieties.

Leucothoe D. Don

1 Leaves evergreen . . . . **L. fontanesiana**
1 Leaves deciduous 2
  2 Anthers with 2 awns; capsule angled on the sutures . . . . **L. recurva**

2 Anthers with 4 awns; capsule round on the sutures .... **L. racemosa**

L. fontanesiana (Steud.) Sleumer, Dog Hobble; acid soils along streams; frequent; Apr–May; GA, NC, SC, TN. *L. editorum* Fern. & Schub., *L. axillaris* (Lam.) D. Don var. *editorum* Fern. & Schub.

L. racemosa (L.) Gray, Fetter Bush; stream banks and low woods; infrequent; May–Jun; GA, NC, VA. More common on the Piedmont and Coastal Plain.

L. recurva (Buckley) Gray, Fetter Bush; rocky woods especially at high elevations; frequent; Apr–Jun; ALL.

Lyonia Nutt. (Maleberry)

Judd, W. S. 1981. A monograph of *Lyonia* (Ericaceae). *J. Arnold Arbor.* 62:63–209, 315–436.

1 Inflorescence naked or with only a few leafy bracts .... **L. ligustrina** var. **ligustrina**
1 Inflorescence with conspicuous bracts or at least the lower inflorescences bracteate ... . **L. ligustrina** var. **foliosiflora**

L. ligustrina (L.) DC. var. **foliosiflora** (Michx.) Fern.; moist woodlands; infrequent and more common on the Coastal Plain; Apr–Jun; GA, SC.

L. ligustrina (L.) DC. var. **ligustrina**; stream banks and woodlands; frequent; Apr–Jun; ALL.

Menziesia Smith (Minnie Bush)

M. pilosa (Michx.) Juss.; woodlands and heath communities at high elevations; frequent; May–Jul; GA, NC, TN, VA.

Monotropa L.

1 Flowers solitary; stems usually white .... **M. uniflora**
1 Flowers in racemes; stems yellowish to nearly red .... **M. hypopithys**

M. hypopithys L., Pinesap; woodlands; frequent; May–Oct; ALL. *M. lanuginosa* Michx.

M. uniflora L., Indian Pipe; woodlands; frequent; Jun–Oct; ALL.

Monotropsis Schweinitz ex Ell. (Sweet Pinesap)

M. odorata Schweinitz; woodlands, often hidden under leaf litter; occasional; Feb–Apr, sometimes fall flowering; ALL. *M. lehmaniae* Burnham.

Oxydendrum DC. (Sourwood)

O. arboreum (L.) DC.; dry to mesic woodlands; common; Jun–Jul; ALL.

Pieris D. Don (Fetter Bush)

P. floribunda (Pursh) B. & H.; high-elevation balds and thickets; occasional; May–Jun; ALL.

Pyrola L. (Shinleaf)

1 Style and filaments straight; anthers surrounding the style .... **P. secunda**
1 Style curved downward; anthers not surrounding the style 2
2 Sepals longer than broad .... **P. americana**

2 Sepals as broad as or broader than long 3
3 Leaves 1–3 cm long, the blade less than 2.5 cm wide . . . . **P. chlorantha**
3 Leaves 3–9 cm long, the blade more than 2.5 cm wide . . . . **P. elliptica**

P. americana Sweet; dry woods or moist bogs; frequent; Jun–Aug; NC, TN, VA. *P. rotundifolia* L. var. *americana* (Sweet) Fern.

P. chlorantha Swartz; dry woods; infrequent; Jun–Aug; VA. *P. virens* Schweig.

P. elliptica Nutt.; dry woods; infrequent; Jun–Aug; NC?, VA.

P. secunda L.; moist woods and bogs; infrequent; Jun–Jul; VA. *Orthilia secunda* (L.) House.

Rhododendron L.

Contributed by W. S. Judd and K. A. Kron

1 Leaves evergreen, entire 2
2 Leaves with conspicuous brown scales beneath . . . . **R. minus**
2 Leaves without brown scales beneath 3
3 Calyx lobes 2–6 mm long; leaves broadest above the middle, acute at the apex . . . . **R. maximum**
3 Calyx lobes less than 2 mm long; leaves widest near the middle, rounded or obtuse at the apex . . . . **R. catawbiense**
1 Leaves deciduous, serrulate or ciliate 4
4 Corolla tube shorter than the lobes; stamens 5–7 . . . . **R. vaseyi**
4 Corolla tube equal to or longer than the lobes, stamens 5 5
5 Corolla yellow, orange, or red 6
6 Flowers appearing before or with the leaves . . . . **R. calendulaceum**
6 Flowers appearing after the leaves have expanded . . . . . . **R. cumberlandense**
5 Corolla white to deep pink 7
7 Flowers appearing after the leaves 8
8 Twigs glabrous . . . . **R. arborescens**
8 Twigs strigose . . . . **R. viscosum**
7 Flowers appearing with or before the leaves 9
9 Corolla hirsute, or if stipitate-glandular, the bud scales glabrous abaxially and the flowers pinkish . . . . **R. periclymenoides**
9 Corolla stipitate-glandular and plants without the above combination of characters 10
10 Corolla lobes about as long as to only slightly shorter than the tube, the tube gradually expanded into the limb; margin of leaves conspicuously ciliate . . . . **R. prinophyllum**
10 Corolla lobes conspicuously shorter than the tube, the tube usually abruptly expanded into the limb; margin of leaves obscurely crisped-ciliate . . . . **R. canescens**

R. arborescens (Pursh) Torr., Sweet/Smooth Azalea; bogs, stream banks, dry woods, and balds; occasional; May–Jul; ALL.

R. calendulaceum (Michx.) Torr., Flame Azalea; woodlands; frequent; May–Jul; ALL.

R. canescens (Michx.) Sweet, Piedmont Azalea; moist or dry woods; infrequent; Mar–May; GA, NC, TN.

R. catawbiense Michx., Mountain Rosebay; rocky exposed ridges and balds at mid to upper elevations; frequent; Apr–Jun; ALL.

R. cumberlandense E. L. Braun; mixed hardwoods; occasional; May–Jul; GA, NC, TN.

R. maximum L., Great Laurel; stream banks and rich woods; common; Jun–Aug; ALL.

R. minus Michx.; exposed ridges, cliffs, and balds; occasional; Apr–Jun; GA, NC, SC, TN. *R. carolinianum* Rehder.

R. periclymenoides (Michx.) Shinners, Wild Azalea, Pinxter Flower; dry or moist woods; common; Mar–May; ALL. *R. nudiflorum* (L.) Torr.

R. prinophyllum (Small) Millais, Roseshell/Early Azalea, Election Pink; dry to moist woods; infrequent; May–Jun; NC, VA. *R. roseum* (Loisel.) Rehder.

R. vaseyi Gray, Pinkshell Azalea; bogs and high-elevation spruce forests; infrequent; May–Jun; NC.

R. viscosum (L.) Torr., Swamp/Clammy Azalea; dry or moist woods, bogs, and stream banks; occasional; May–Jul; ALL. *R. viscosum* (L.) Torr. var. *serrulatum* (Small) Ahles.

Some intermediates and local variants occur within the deciduous-leaved taxa, especially in disturbed areas, high-elevation meadows, and balds.

Vaccinium L. (Blueberry)

Vander Kloet, S. P. 1978. The taxonomic status of *Vaccinium pallidum*, the hillside blueberries including *Vaccinium vacillans*. *Canad. J. Bot.* 56:1559–1574.

______. 1980. The taxonomy of the highbush blueberry, *Vaccinium corymbosum*. *Canad. J. Bot.* 58:1187–1201.

1 Corolla lobes 4, the lobes longer than the tube and usually reflexed 2
  2 Leaves deciduous, more than 2 cm long; erect shrubs .... **V. erythrocarpum**
  2 Leaves evergreen, less than 2 cm long; trailing shrubs .... **V. macrocarpon**
1 Corolla lobes 5, the lobes shorter than the tube and not reflexed 3
  3 Corolla open-campanulate, broadest at the apex; anthers awned 4
    4 Stamens exserted; leaves deciduous, dull green or glaucous ..... **V. stamineum**
    4 Stamens included; leaves thick, persistent, shiny .... **V. arboreum**
  3 Corolla urceolate, constricted at the apex; anthers not awned 5
    5 Corolla and fruit densely pubescent .... **V. hirsutum**
    5 Corolla and fruit glabrous or nearly so 6
      6 Plants 8 dm or less tall 7
        7 Stems and leaves densely pubescent .... **V. myrtilloides**
        7 Stems and leaves glabrous or only slightly pubescent 8
          8 Leaves pale green or glaucous beneath .... **V. pallidum**
          8 Leaves bright green on both surfaces .... **V. angustifolium**
      6 Plants more than 8 dm tall 9
        9 Leaves densely pubescent beneath .... **V. fuscatum**
        9 Leaves glabrous beneath (the midrib sometimes slightly pubescent) .... **V. corymbosum**

V. angustifolium Ait., Lowbush B.; dry woods; occasional; May–Jun; TN, VA.

V. arboreum Marsh., Farkleberry; dry woods; occasional; Apr–Jun; GA, NC, SC, TN. This species is quite tolerant of basic soils.

V. corymbosum L., Highbush B.; damp or dry woodlands; common; Mar–Jun; ALL. *V. constablaei* Gray, *V. simulatum* Small.

V. erythrocarpum Michx., Bearberry; rich woods at high elevations; occasional; May–Jul;.GA, NC, TN, VA.

V. fuscatum Ait.; damp woodlands; occasional; Mar–May; ALL. *V. atrococcum* Heller.

V. hirsutum Buckley; dry woods at mid to upper elevations; infrequent; Apr–May; GA, NC, TN.

V. macrocarpon Ait., Cranberry; bogs and lowlands; occasional; May–Jul; NC, TN, VA.

V. myrtilloides Michx., Velvetleaf B.; bogs and damp woods; infrequent; Apr–Jun; VA.

V. pallidum Ait., Lowbush B.; dry woods at mid to upper elevations; common; Mar–May; ALL. *V. vacillans* Torr.

V. stamineum L., Squaw Huckleberry, Deerberry; dry woodlands; common; Apr–Jun; ALL. The var. *melanocarpum* Mohr with pubescent branches, leaves, and fruits is recognized by some authors.

Vaccinium erythrocarpum, *V. macrocarpon*, *V. stamineum*, *V. arboreum*, and *V. hirsutum* are distinct taxa easily recognized either in the field or from herbarium material. The remaining taxa, however, are extremely polymorphic as a result of extensive hybridization between and within the low- and high-bush types. A conservative approach seems appropriate until additional biological data are available.

EUPHORBIACEAE (Spurge Family)

1 Leaves palmately lobed; stinging hairs present .... **(Cnidoscolus)**
1 Leaves not palmately lobed; stinging hairs present or absent 2
  2 Plants with a milky sap; flowers enclosed in a cuplike cyathium (appearing as a single flower) .... **Euphorbia**
  2 Plants with a watery sap; flowers solitary, spicate, or cymose 3
    3 Seeds 2/locule .... **Phyllanthus**
    3 Seeds 1/locule 4
      4 Fruits indehiscent, 1-locular, seeds solitary .... **Crotonopsis**
      4 Fruits dehiscent, 3-locular, seeds 3 5
        5 Plants with stinging trichomes .... **Tragia**
        5 Plants without stinging trichomes 6
          6 Pistillate flowers surrounded by conspicuous leafy bract; plants pubescent but not stellate .... **Acalypha**
          6 Pistillate flowers not surrounded by a conspicuous leafy bract; plants with stellate pubescence .... **Croton**

Acalypha L. (Three-seeded Mercury)

1 Flowering spikes unisexual, the female terminal .... **A. ostryifolia**
1 Flowering spikes bisexual, strictly axillary 2
  2 Leaves ovate-rhombic, more than 2.5 cm wide .... **A. rhomboidea**
  2 Leaves linear-lanceolate, rarely more than 2.5 cm wide 3
    3 Bracts of female flowers with long, spreading trichomes, bract lobes lanceolate .... **A. virginica**
    3 Bracts of female flowers without long, spreading trichomes, bract lobes ovate to deltoid .... **A. gracilens**

A. gracilens Gray; disturbed sites; occasional; Jun–Oct; NC, SC, TN, VA.

A. ostryifolia Ridd.; disturbed sites; infrequent; Jun–Oct; GA, NC, TN, VA.

A. rhomboidea Raf.; disturbed sites; common; Jun–Oct; ALL.
A. virginica L.; disturbed sites; frequent; Jun–Oct; ALL.

A. ostryifolia is a clearly defined species. The remaining taxa, however, present taxonomic problems not clearly resolved.

Cnidoscolus Pohl (Tread Softly)

C. stimulosus (Michx.) Engelm. & Gray; sandy fields; infrequent; Apr–Aug; NC (Piedmont). A very appropriate specific epithet for this taxon.

Croton L.

1 Leaves toothed .... **C. glandulosus**
1 Leaves entire 2
  2 Leaf blades more than 3 cm long; fruit erect .... **C. capitatus**
  2 Leaf blades less than 3 cm long; fruit pendulous .... **C. monanthogynus**

C. capitatus Michx.; fields and disturbed sites, often on basic soils; infrequent; Aug–Oct; VA.
C. glandulosus L.; disturbed sites; occasional; May–Oct; ALL. Represented in our area by var. **septentrionalis** Muell.-Arg.
C. monanthogynus Michx.; disturbed sites on basic soils; infrequent; Jun–Oct; NC, TN, VA.

Crotonopsis Michx.

C. elliptica Willd.; granite and sandstone outcrops; infrequent; Jun–Oct; NC, SC.

Euphorbia L. (Spurge)

1 Stem leaves oblique at base and entirely opposite 2
  2 Capsule glabrous; stems erect .... **E. nutans**
  2 Capsule pubescent; stems decumbent or prostrate .... **E. maculata**
1 Stem leaves symmetrical at base and usually alternate, rarely opposite 3
  3 Glands of the involucre with reduced or conspicuous white, petallike appendages 4
    4 Appendages reduced; stem leaves greatly reduced or scalelike .. **E. mercurialina**
    4 Appendages conspicuous; stem leaves normally developed .... **E. corollata**
  3 Glands of the involucre without petallike appendages 5
    5 Bracteal leaves coarsely dentate .... **E. dentata**
    5 Bracteal leaves entire 6
      6 Principal stem leaves serrate, especially toward the apex 7
        7 Seeds smooth or obscurely reticulate .... **E. obtusata**
        7 Seeds distinctly alveolate .... **E. spathulata**
      6 Principal stem leaves entire 8
        8 Stem leaves 1–3 mm wide .... **E. cyparissias**
        8 Stem leaves more than 3 mm wide 9
          9 Stem leaves opposite or subopposite; capsule more than 1 cm wide .... **E. lathyris**
          9 Stem leaves alternate; capsule less than 1 cm wide 10
            10 Stem leaves elliptic to oblong, 5–10 cm long .... **E. purpurea**

10 Stem leaves oblanceolate to obovate, up to 1.5 cm long .... **E. commutata**

E. commutata Engelm., Wood S.; deciduous woodlands, often on basic soils; occasional; Mar–May; NC, TN, VA.

E. corollata L., Flowering S.; woodlands and disturbed sites; common; May–Sep; ALL. *E. zinniiflora* Small.

E. cyparissias L.*, Cypress S.; waste areas; occasional; Mar–May; NC, VA.

E. dentata Michx.; disturbed sites; occasional; Jul–Oct; GA, NC, TN, VA.

E. lathyris L.*, Caper S.; waste places; occasional; Jun–Aug; NC, TN, VA.

E. maculata L., Wartweed; disturbed sites; common; May–Oct; ALL. *E. supina* Raf.

E. mercurialina Michx.; woodlands, usually on basic soils; infrequent; May–Jun; TN.

E. nutans Lagasca; disturbed sites; common; May–Oct; ALL. *E. preslii* Guss.; *E. maculata* has also been applied to this taxon.

E. obtusata Pursh; alluvial woods and disturbed sites; infrequent; Apr–Jul; NC.

E. purpurea (Raf.) Fern.; dry or moist woodlands on basic soils; infrequent; May–Aug; NC, VA.

E. spathulata Lam.; rocky woodlands; occasional; May–Jun; VA. *E. dictyosperma* Fisch. & Mey.

Phyllanthus L.

P. caroliniensis Walt.; fields and disturbed sites; infrequent; Jul–Oct; NC, SC, VA.

Tragia L.

1 Leaves subsessile, cuneate at base .... **T. urens**

1 Leaves distinctly petiolate, truncate to subcordate at base .... **T. urticifolia**

T. urens L.; sandy fields and disturbed sites; infrequent; May–Oct; SC.

T. urticifolia Michx.; woodland borders and fields on basic soils; infrequent; May–Oct; SC.

FABACEAE (Bean Family)

Mahler, W. F. 1970. Manual of the legumes of Tennessee. *J. Tennessee Acad. Sci.* 45:65–96.

1 Trees, shrubs, or woody vines 2

2 Leaves simple, cordate; flowers appearing before or with the leaves .... **Cercis**

2 Leaves compound; flowers appearing with or after the leaves 3

3 Leaves bipinnately compound 4

4 Stamens 15 or more; flowers in spherical heads; leaflets asymmetrical; plants without thorns .... **Albizia**

4 Stamens 10 or less; flowers in racemes; leaflets symmetrical; plants thorny ... . **Gleditsia**

3 Leaves trifoliolate or once-pinnate 5

5 Leaves trifoliolate, the uppermost rarely 1-foliolate 6

6 Vines .... **Pueraria**

6 Shrubs 7

7 Petals yellow .... **Cytisus**

7 Petals purplish .... **Lespedeza**

5 Leaves once-pinnate with more than 3 leaflets 8
8 Vines; petals bluish-violet or rarely white . . . . **Wisteria**
8 Trees or shrubs; petals white, pink, or purple 9
9 Petal 1; leaves with or without glands . . . . **Amorpha**
9 Petals 3–5; leaves not glandular 10
10 Leaves even-pinnate . . . . **Gleditsia**
10 Leaves odd-pinnate 11
11 Filaments separate; flowers in panicles; leaflets 7–9 . . . . **Cladrastis**
11 Filaments fused; flowers in racemes; leaflets 11 or more . . . . **Robinia**
1 Herbs 12
12 Leaves simple . . . . **Crotalaria**
12 At least some leaves compound 13
13 Leaves with more than 3 leaflets and either palmately or bipinnately compound 14
14 Leaves palmately compound . . . . **Lupinus**
14 Leaves bipinnately compound 15
15 Stamens 5, whitish; stems without prickles . . . . **Desmanthus**
15 Stamens 9–13, pinkish; stems with prickles . . . . **Schrankia**
13 Leaves either 2- or 3-foliolate or once-pinnate with numerous leaflets 16
16 Leaves 2-foliolate or once-pinnate with more than 3 leaflets 17
17 Leaves 2-foliolate, the terminal leaflet modified into a tendril . . . . **Lathyrus**
17 Leaves once-pinnate, tendril present or absent 18
18 Leaves odd-pinnate 19
19 Inflorescence umbellike 20
20 Flowers yellow; leaflets 5, the lowermost pair resembling stipules . . . . **Lotus**
20 Flowers pinkish; leaflets more than 5 . . . . **Coronilla**
19 Inflorescence racemose 21
21 Twining vines, usually more than 1 m long . . . . **Apios**
21 Plants erect, decumbent, or trailing, but not twining and rarely as long as 1 m 22
22 Corolla white; inflorescence axillary . . . . . . **Astragalus**
22 Corolla yellowish, pink, or crimson; raceme terminal or lateral and opposite the leaves . . . . **Tephrosia**
18 Leaves even-pinnate (excluding tendrils, if present) 23
23 Flowers yellow, actinomorphic or nearly so; tendrils absent . . . . **Cassia**
23 Flowers white, pink, purple, blue, or, rarely, yellowish, distinctly zygomorphic; tendrils present 24
24 Leaflets averaging less than 1 cm wide . . . . **Vicia**
24 Leaflets averaging more than 1 cm wide . . . . **Lathyrus**
16 Leaves palmately or pinnately 3-foliolate 25
25 Leaflets serrulate or crenate-serrulate 26
26 Inflorescence capitate; petals mostly persistent; petals attached to the staminal tube; terminal leaflet usually sessile or subsessile . . . . **Trifolium**

26 Inflorescence racemose or spicate; petals soon deciduous; petals free from the staminal tube; terminal leaflet distinctly stalked 27

27 Flowers bluish-purple .... **Medicago**

27 Flowers white or yellow 28

28 Flowers in elongate racemes; fruit straight ...... **Melilotus**

28 Flowers in dense racemes; fruit reniform .... **Medicago**

25 Leaflets entire 29

29 Leaves palmately 3-foliolate 30

30 Stamens distinct 31

31 Fruit flat; flowers yellow .... **Thermopsis**

31 Fruit inflated; flowers white, blue, or yellow ..... **Baptisia**

30 Stamens monodelphous or diadelphous .... **Trifolium**

29 Leaves pinnately 3-foliolate 32

32 Flowers 2.5 cm or more long 33

33 Calyx lobes shorter than the tube .... **Clitoria**

33 Calyx lobes longer than the tube .... **Centrosema**

32 Flowers less than 2.5 cm long 34

34 Fruits 1-seeded, or 2–several-seeded loments 35

35 Leaves glandular-punctate .... **Psoralea**

35 Leaves not glandular-punctate 36

36 Petals bluish; pod 1-seeded .... **Psoralea**

36 Petals yellow, white, pink, or purplish (at least not bluish); pod 1-seeded or composed of 2–several, 1-seeded segments 37

37 Stipules united to the petiole and forming a sheath that encircles the stem; flowers bright yellow or orange .... **Stylosanthes**

37 Stipules free; flowers variously colored 38

38 Loment with 2–several segments, usually with hooked hairs; stipels usually present .... **Desmodium**

38 Loment 1-seeded, lacking hooked hairs; stipels absent .... **Lespedeza**

34 Fruits 2–many seeded and never segmented 39

39 Stem erect, flowers yellow .... **Rhynchosia**

39 Stem decumbent, trailing, or climbing; flowers white, pink, purple, or blue 40

40 Keel petals spirally coiled; calyx lobes 5 .... **Phaseolus**

40 Keel petals straight or curved upward; calyx (by fusion) appearing 4-lobed 41

41 Leaflets 10 cm or more long .... **Pueraria**

41 Leaflets less than 10 cm long 42

42 Calyx cylindric, not subtended by a pair of bracteoles .... **Amphicarpaea**

42 Calyx campanulate, subtended by a pair of bracteoles 43

43 Keel petals nearly straight; style glabrous .... **Galactia**

43 Keel petals strongly curved upward; style pubescent along the upper surface .... **Strophostyles**

Albizia Durazzini (Mimosa)

A. julibrissin Durazzini*; roadsides, lawns, and waste places; occasional; May–Jul; ALL.

Amorpha L.

Wilbur, R. L. 1975. A revision of the North American genus *Amorpha* (Leguminosae-Psoraleae). *Rhodora* 77:337–409.

1 Plants less than 1 m tall; petiole usually shorter than the width of the lowermost leaflet .... **A. herbacea**
1 Plants more than 1 m tall; petiole longer than the width of the lowermost leaflet ... 2
  2 Calyx lobes nearly absent, or less than 0.8 mm long and rounded ...... **A. glabra**
  2 At least some calyx lobes acute and 0.8 mm or more long .... **A. fruticosa**

A. fruticosa L.; open woodlands and stream banks; occasional; May–Jun; GA, NC, SC, TN. An extremely variable taxon throughout its entire range.
A. glabra Poir.; woodland margins; frequent; May–Jun; GA, NC, SC, TN. Endemic to the southern portion of our area.
A. herbacea Walt.; dry woods; infrequent; May–Jul; GA, NC. More common in the Piedmont and Coastal Plain.

Amphicarpaea Ell. ex Nutt. (Hog Peanut)

A. bracteata (L.) Fern.; thickets and roadsides; common; Jul–Sep; ALL.

Apios Medicus (Groundnut)

A. americana Medicus; bottomland thickets; common; Jun–Aug; ALL.

Astragalus L. (Canada Milk Vetch)

A. canadensis L.; woodlands and stream banks; frequent; Jun–Aug; NC, SC, TN, VA.

Baptisia Vent. (Wild Indigo)

1 Petals blue .... **B. australis**
1 Petals white or yellow 2
  2 Lowermost bracts 1 cm or more long, persistent; stems pubescent ... **B. bracteata**
  2 Lowermost bracts less than 1 cm long, deciduous; stems glabrous or nearly so ... 3
    3 Petals white; leaflets mostly more than 2 cm long .... **B. alba**
    3 Petals yellow; leaflets mostly less than 2 cm long .... **B. tinctoria**

B. alba (L.) R. Br.; open woodlands; occasional; May–Jul; GA, NC, SC, TN. *B. albescens* Small.
B. australis (L.) R. Br.; open woods and river banks; occasional; Apr–Jun; NC, SC, VA.
B. bracteata Muhl. ex Ell.; sandy open woods; infrequent; Mar–Apr; NC, SC.
B. tinctoria (L.) R. Br.; dry open woodlands; common; Apr–Aug; ALL.

Cassia L.

1 Leaflets 2.5 cm or more long; petiole more than 1 cm long; fruit usually curved and more than 7 cm long 2
2 Flowers single or in axillary clusters; leaflets obovate; fruit tetragonal . . . . **C. obtusifolia**
2 Flowers in terminal or axillary racemes; leaflets narrow to broadly elliptic; fruit flat 3
3 Ovary villous; pod segments nearly square, not wider than long . . . **C. hebecarpa**
3 Ovary with appressed pubescence; pod segments rectangular . . . **C. marilandica**
1 Leaflets less than 2.5 cm long; petiole less than 1 cm long; fruit straight and less than 7 cm long 4
4 Flowers 2–3 cm in diameter; stamens 10 . . . . **C. fasciculata**
4 Flowers 1 cm or less in diameter; stamens 5 . . . . **C. nictitans**

C. fasciculata Michx., Partridge Pea; roadsides and waste places; common; May–Sep; ALL. *C. chamaecrista* auth., non L.

C. hebecarpa Fern., Wild Senna; bottomlands and thickets; frequent; Jul–Aug; ALL.

C. marilandica L., Wild Senna; open woods and thickets; occasional; Jul–Aug; ALL.

C. nictitans L.; roadsides and waste places; common; May–Oct; ALL.

C. obtusifolia L., Sicklepod; bottomlands, fields, and roadsides; infrequent; Jul–Oct; NC, SC, VA.

Centrosema (DC.) Bentham (Butterfly Pea)

C. virginianum (L.) Benth.; open woodlands; occasional; Jun–Aug; ALL.

Cercis L. (Redbud)

C. canadensis L.; woodlands, especially on calcareous sites; common; Mar–May; ALL.

Cladrastis Raf. (Yellowwood)

C. kentukea (Dum.-Cours.) Rudd; rich woods; infrequent; Apr–May; GA, NC, SC, TN. *C. lutea* (Michx. f.) K. Koch.

Clitoria L. (Butterfly Pea)

C. mariana L.; open woodlands; frequent; Jun–Aug; ALL.

Coronilla L. (Crown Vetch)

C. varia L.*; fields and road cuts; occasional; May–Jul; ALL. Often used for erosion control.

Crotalaria L. (Rattlebox)

1 Stems and leaves with appressed pubescence . . . . **C. purshii**
1 Stems and leaves with spreading pubescence . . . . **C. sagittalis**

C. purshii DC.; sandy clearings; infrequent; May–Jul; SC, TN. Our populations represent Coastal Plain disjuncts.

C. sagittalis L.; woodlands and waste areas; occasional; Jun–Aug; NC, SC, TN, VA.

Cytisus L. (Scotch Broom)

C. scoparius (L.) Link*; established along roadsides and waste areas; occasional; Apr–May; ALL.

Desmanthus Willd. (Prairie Mimosa)

D. illinoensis (Michx.) MacM. ex Robins. & Fern.; fields and open woodlands; infrequent; Jun–Aug; SC, TN.

Desmodium Desv. (Beggar Lice, Tick Trefoil)

Isely, D. 1955. The Leguminosae of the north-central United States II. Hedysareae. *Iowa State Coll. J. Sci.* 30:33–118.

1 Calyx lobes shorter than the tube; fruit stipe about 3× as long as the calyx; leaves usually subverticillate 2
  2 Flowering stalk scapose, rarely leafy; pedicel 1 cm or more long . . **D. nudiflorum**
  2 Flowering stalk leafy; pedicel 4–6 mm long 3
    3 Petals white; leaves scattered; terminal leaflet as long as or longer than wide; stems ascending or sprawling . . . . **D. pauciflorum**
    3 Petals pinkish; leaves clustered; terminal leaflet as broad as or broader than long; stems erect . . . . **D. glutinosum**
1 Calyx lobes equal to or longer than the tube; fruit sessile or subsessile; leaves scattered 4
  4 Stipules cordate 5
    5 Plants erect . . . . **D. canescens**
    5 Plants prostrate, viny 6
      6 Flowers purplish; terminal leaflet orbicular . . . . **D. rotundifolium**
      6 Flowers white or yellowish; terminal leaflet ovate . . . . **D. ochroleucum**
  4 Stipules linear to ovate 7
    7 Stems prostrate or trailing . . . . **D. lineatum**
    7 Stems erect 8
      8 Fruits 1–4-segmented, the segments rounded ventrally; flowers generally small, 3–6 mm long 9
        9 Leaflets linear, about 4× longer than wide 10
          10 Leaves subsessile, the petioles less than 5 mm long . . . **D. sessilifolium**
          10 Leaf petiole more than 5 mm long . . . . **D. strictum**
        9 Leaflets elliptic to ovate, less than 4× longer than wide 11
          11 Leaflets up to 3 cm long and 2 cm wide 12
            12 Stems and leaves glabrous or nearly so; pedicels 10–15 mm long . . . . **D. marilandicum**
            12 Stems and leaves pilose; pedicels less than 10 mm long . . **D. ciliare**
          11 Leaflets, at least the terminal one, more than 3 cm long and 2 cm wide 13
            13 Leaves glabrous or with hooked trichomes; stipules not reddish . . . . **D. obtusum**
            13 Leaves pilose beneath; stipules reddish . . . . **D. nuttallii**

8 Fruits 4–many-segmented, the segments angled ventrally; flowers usually longer than 6 mm 14

14 Bracts conspicuous, up to 8 mm long; flowers about 10 mm long 15

15 Petioles of median stem leaves 2 cm or more long . . . . . **D. cuspidatum**

15 Petioles of median stem leaves less than 2 cm long . . . . . **D. canadense**

14 Bracts less than 8 mm long; flowers about 5–8 mm long 16

16 Leaves velvety-tomentose beneath . . . . **D. viridiflorum**

16 Leaves glabrous or sparsely pubescent 17

17 Leaves glabrate and glaucous beneath; pedicels 1–2 cm long . . . . **D. laevigatum**

17 Leaves pubescent but not glaucous beneath; pedicels less than 1 cm long 18

18 Pubescence of lower leaf surface of hooked trichomes restricted primarily to principal veins . . . . **D. fernaldii**

18 Lower leaf surface glabrous or sparsely pubescent, the trichomes not evidently hooked or restricted to principal veins 19

19 Terminal leaflet lanceolate, more than 3× as long as wide . . . . **D. paniculatum** var. **paniculatum**

19 Terminal leaflet ovate, less than 3× as long as wide . . . . **D. paniculatum** var. **dillenii**

D. canadense (L.) DC.; open woods and stream banks; infrequent; Jul–Sep; VA.

D. canescens (L.) DC.; fields and woodlands; frequent; Jun–Sep; NC, SC, TN, VA.

D. ciliare (Willd.) DC.; fields and woodlands; frequent; Jun–Sep; NC, SC, TN, VA.

D. cuspidatum (Willd.) Loudon; woodlands and thickets; occasional; Jun–Sep; NC, SC, VA.

D. fernaldii Schubert; fields and woodlands; infrequent; Jun–Sep; SC. More common on the Coastal Plain.

D. glutinosum (Willd.) Wood; rich woods; frequent; Jun–Sep; NC, SC, TN, VA.

D. laevigatum (Nutt.) DC.; fields and open woods; frequent; Jun–Sep; ALL.

D. lineatum DC.; open woodlands; infrequent; Jun–Sep; NC, SC. More common on the Coastal Plain.

D. marilandicum (L.) DC.; open woodlands; frequent; Jun–Sep; NC, SC, TN, VA.

D. nudiflorum (L.) DC.; woodlands; common; Jul–Sep; ALL.

D. nuttallii (Schindler) Schubert; open woodlands; occasional; Jun–Sep; NC, SC, TN, VA.

D. obtusum (Willd.) DC.; fields and open woodlands; occasional; Jun–Sep; NC, SC, VA. *D. rigidum* (Ell.) DC.

D. ochroleucum M. A. Curtis; roadsides and woodlands; infrequent; Jun–Sep; TN.

D. paniculatum (L.) DC. var. **dillenii** (Darl.) Isely; fields and open woods; common; Jun–Sep; ALL. *D. perplexum* Schubert, *D. glabellum* (Michx.) DC.

D. paniculatum (L.) DC. var. **paniculatum**; fields and open woods; common; Jun–Sep; ALL.

D. pauciflorum (Nutt.) DC.; rich woods; infrequent; Jun–Aug; GA, TN.

D. rotundifolium DC.; open woodlands; common; Jun–Aug; ALL.

D. sessilifolium (Torr.) T. & G.; open woodlands; infrequent; Jul–Sep; VA.

D. strictum (Pursh) DC.; sandy woods; infrequent; Jul–Aug; SC.

D. viridiflorum (L.) DC.; fields and woodlands; occasional; Jun–Sep; NC, SC, TN, VA.

Galactia P. Browne (Milk Pea)

Duncan, W. H. 1979. Changes in *Galactia* (Fabaceae) of the southeastern United States. *Sida* 8:170–180.

1 Mature flowers 7–9 mm long .... **G. regularis**
1 Mature flowers 10–18 mm long .... **G. volubilis**

G. regularis (L.) BSP.; open woods and roadsides; frequent; Jun–Aug; ALL. *G. volubilis* (L.) Britt. of authors.
G. volubilis (L.) Britt.; woodlands and thickets; infrequent; Jul–Sep; SC. *G. regularis* (L.) BSP. of authors, *G. macreei* M. A. Curtis.

Gleditsia L. (Honey Locust)

G. triacanthos L.; bottomlands and woodland margins; frequent; Apr–May; ALL. The fruit contains a honey-flavored pulp.

Lathyrus L.

1 Leaves with 2 leaflets 2
  2 Ovary and legume densely hirsute; flowers 10–13 mm long .... **L. hirsutus**
  2 Ovary and legume glabrous; flowers 2–2.5 cm long .... **L. latifolius**
1 Leaves with more than 2 leaflets, commonly 10 or more .... **L. venosus**

L. hirsutus L.*; waste places; infrequent; Apr–Jul; TN, VA.
L. latifolius L.*, Everlasting Pea; persistent in waste places; common; May–Aug; ALL.
L. venosus Willd.; mesic woodlands and stream banks; occasional; Apr–Jun; NC, SC, TN, VA.

Lespedeza Michx. (Bush Clover)

Clewell, A. F. 1966. Native North American species of *Lespedeza* (Leguminosae). *Rhodora* 68:359–405.

1 Shrubs, usually more than 1 m tall .... **L. bicolor**
1 Herbs, usually less than 1 m tall 2
  2 Stipules ovate to narrowly ovate; annuals 3
    3 Stem hairs downwardly appressed .... **L. striata**
    3 Stem hairs upwardly appressed .... **L. stipulacea**
  2 Stipules linear to bristlelike; perennials 4
    4 Plants prostrate or weakly ascending; peduncle longer than the subtending leaf 5
      5 Stem hairs spreading .... **L. procumbens**
      5 Stem hairs appressed 6
        6 Stem procumbent; keel about as long as the wings .... **L. repens**
        6 Stem weakly erect; keel usually longer than the wings .... **L. violacea**
    4 Plants erect or strongly ascending; peduncle usually shorter than the subtending leaf 7
      7 Petals yellowish-white, often with a purple spot on the standard 8
        8 Flowers solitary or in axillary clusters, the inflorescence shorter than the subtending leaves .... **L. cuneata**
        8 Flowers several in spikelike or globose heads, the inflorescence longer than the subtending leaves 9

9 Leaflets less than 2× as long as wide, generally elliptic to obovate . . . . **L. hirta**
9 Leaflets more than 2× as long as wide, linear to narrowly elliptic 10
10 Calyx 8–13 mm long; leaflets narrowly linear, rachis shorter than the petiole . . . . **L. angustifolia**
10 Calyx 4–9 mm long; leaflets narrowly elliptic, rachis longer than the petiole . . . . **L. capitata**
7 Petals purplish 11
11 Keel longer than the wings; peduncle longer than the subtending leaf . . . . **L. violacea**
11 Keel equal to or shorter than the wings; peduncle shorter than the subtending leaf 12
12 Leaflets linear, mostly 4× or more longer than wide . . . . **L. virginica**
12 Leaflets elliptic, oblong, or obovate, mostly less than 3.5× as long as wide 13
13 Leaflets glabrous above; stems moderately appressed-pubescent . . . . **L. intermedia**
13 Leaflets pilose or appressed-pubescent above; stems densely short pilose . . . . **L. stuevei**

L. angustifolia (Pursh) Ell.; moist or dry barrens; infrequent and disjunct from the Coastal Plain; Aug–Oct; NC.
L. bicolor Turcz.*; roadsides and waste places; occasional; Jun–Sep; ALL.
L. capitata Michx.; fields and disturbed sites; occasional; Aug–Oct; NC, SC, TN, VA.
L. cuneata (Dumont) G. Don*, Sericea; fields and waste places; frequent; Jul–Sep; ALL.
L. hirta (L.) Hornemann; fields and waste areas; frequent; Aug–Oct; ALL. *L. nuttallii* Darl. is considered to be a hybrid between this and the following taxon.
L. intermedia (S. Wats.) Britt.; fields and waste areas; frequent; Jul–Sep; ALL.
L. procumbens Michx.; fields and waste areas; occasional; Jun–Sep; NC, SC, TN, VA.
L. repens (L.) Bart.; fields and waste areas; frequent; May–Oct; ALL.
L. stipulacea Maxim.*, Korean Clover; fields and waste areas; frequent; Jul–Sep; NC, SC, TN, VA. *Kummerowia stipulacia* (Maxim.) Makino.
L. striata (Thunb.) H. & A.*, Japanese Clover; fields and waste areas; frequent; Jul–Sep; NC, SC, TN, VA. *Kummerowia striata* (Thunb.) Schindl.
L. stuevei Nutt.; fields and open woods; occasional; Jul–Sep; NC, SC, TN, VA.
L. violacea (L.) Pers.; fields and waste areas; infrequent; Jul–Sep; NC, SC, TN, VA.
L. virginica (L.) Britt.; fields and waste areas; frequent; Aug–Sep; ALL.

Numerous hybrids have been reported in this genus.

Lotus L. (Birdsfoot Trefoil)

L. corniculatus L.*; fields and waste areas; infrequent; Jun–Sep; NC, VA.

Lupinus L. (Lupine)

L. perennis L.; dry sandy woodlands; occasional; Apr–Jul; VA.

Medicago L.

1 Flowers bluish-purple, more than 6 mm long . . . . **M. sativa**
1 Flowers yellow, less than 6 mm long . . . . **M. lupulina**

M. lupulina L.*, Black Medic; lawns, fields, and waste areas; common; Apr–Aug; NC, SC, TN, VA.

M. sativa L.*, Alfalfa; fields and waste areas; frequent; Apr–Jul; ALL.

Melilotus Mill. (Sweet Clover)

1 Petals yellow . . . . **M. officinalis**
1 Petals white . . . . **M. alba**

M. alba Medic.*, White S. C.; roadsides and waste areas; common; Apr–Sep; NC, SC, TN, VA.

M. officinalis (L.) Pallas*, Yellow S. C.; roadsides and waste areas; frequent; Apr–Sep; ALL.

Phaseolus L. (Wild Kidney Bean)

1 Leaves unlobed; plants usually climbing . . . . **P. polystachyus**
1 Leaves 3-lobed; plants trailing . . . . **P. sinuatus**

P. polystachyus (L.) BSP.; thickets and margins of woods; occasional; Jul–Sep; NC, SC, TN, VA. *P. polystachios* (L.) BSP.

P. sinuatus Nutt. ex T. & G.; sandy woodlands; infrequent; Jul–Sep; SC.

Psoralea L.

1 Leaflets ovate or ovate-lanceolate, 2–3 cm or more wide; plants rhizomatous 2
  2 Leaves glandular-punctate, obtuse at the apex; petals 8–10 mm long . . . . **P. macrophylla**
  2 Leaves eglandular below, glandular above, acute to acuminate at the apex; petals 5–7 mm long . . . . **P. onobrychis**
1 Leaflets oblong to lanceolate, less than 2 cm wide; plants with a fusiform taproot . . . . **P. psoralioides**

P. macrophylla Rowlee ex Small; wooded slopes of Tryon Mt.; perhaps extinct, at least not seen since the original collection in 1897; Jun–Jul?; NC.

P. onobrychis Nutt.; rich woods; infrequent; Jun–Jul; NC, TN.

P. psoralioides (Walt.) Cory; fields and open woodlands; occasional; May–Jul; NC, SC, TN. Represented in our area by var. **eglandulosa** (Ell.) Freeman. The typical variety, with densely glandular-punctate leaflets and inflorescences, occurs throughout the Piedmont and Coastal Plain and may be expected sporadically in our area.

Pueraria DC. (Kudzu)

P. lobata (Willd.) Ohwi*; a noxious weed of roadsides and waste places; too common; Jul–Oct; ALL.

Rhynchosia Lour.

R. tomentosa (L.) H. & A.; roadsides and open woods; infrequent; Jun–Aug; GA, NC, SC, TN.

Robinia L. (Locust)

1 Petals white; medium to large trees; stems neither bristly nor viscid . . . . **R. pseudoacacia**
1 Petals usually pinkish, rarely white; stoloniferous shrubs or small trees; stems either bristly or viscid 2

2 Branches and peduncles with sessile, viscid glands; bracts 3 mm or more wide . . . . **R. viscosa**
2 Branches and peduncles variously hispid, the trichomes sometimes glandular at the tip; bracts less than 2 mm wide . . . . **R. hispida**

R. hispida L., Bristly L.; disturbed sites and often cultivated; occasional; Apr–Jun; ALL.
R. pseudoacacia L., Black L.; rich woods; frequent; Apr–Jun; ALL.
R. viscosa Vent., Clammy L.; open woods and road banks; infrequent; May–Jun; NC.

This is one of the most intriguing taxonomic complexes in our area. Perhaps hundreds of mostly sterile variants occur on disturbed sites. Names have been applied to but a few of these. For additional information see D. Isely and F. J. Peabody, *Robinia* (Leguminosae: Papilionoideae), *Castanea* 49 (1984): 187–202.

Schrankia Willd. (Sensitive Brier)

S. microphylla (Dry.) MacBride; roadsides and disturbed sites; occasional; Jun–Sep; GA, NC, SC, TN.

Strophostyles Ell.

1 Floral bracts equal to or longer than the calyx tube; legumes about 8 cm long; leaves often lobed . . . . **S. helvula**
1 Floral bracts shorter than the calyx tube; legumes 3–6 cm long; leaves unlobed . . . . **S. umbellata**

S. helvula (L.) Ell.; roadsides and open areas; occasional; Jun–Sep; NC, SC, TN, VA.
S. umbellata (Muhl. ex Willd.) Britt.; roadsides and open areas; occasional; Jun–Sep; NC, SC, TN, VA.

Stylosanthes Swartz (Pencil Flower)

S. biflora (L.) BSP.; roadsides and waste places; common; Jun–Aug; ALL.

Tephrosia Pers.

1 Stems single, erect; inflorescence terminal; standard yellowish . . . . **T. virginiana**
1 Stems decumbent, usually branched, inflorescences terminal and lateral; standard reddish-purple . . . . **T. spicata**

T. spicata (Walt.) T. & G.; open woods and disturbed sites; occasional; Jun–Aug; GA, NC, SC, TN.
T. virginiana (L.) Pers., Goat's Rue; roadsides and clearings; common; May–Jul; ALL.

Thermopsis R. Br.

1 Stipules ovate, clasping the stem; ovary densely villous-tomentose; flowers more than 20/inflorescence . . . . **T. villosa**
1 Stipules linear, not clasping the stem; ovary with appressed trichomes; flowers usually less than 20/inflorescence 2
2 Calyx tube glabrous or essentially so externally; bracts shorter than the pedicels . . . . **T. fraxinifolia**
2 Calyx tube pubescent externally; bracts longer than the pedicels . . . . **T. mollis**

T. fraxinifolia M. A. Curtis; dry woodlands; occasional; Jun–Jul; GA, NC, SC, TN. *T. mollis* (Michx.) M. A. Curtis var. *fraxinifolia* (M. A. Curtis) Isely.
T. mollis (Michx.) M. A. Curtis; open woodlands; occasional; Apr–Jun; ALL.

T. villosa (Walt.) Fern. & Schubert; open woodlands; occasional; May–Jun; GA, NC, TN.

Trifolium L. (Clover)

1 Petals yellow 2
 2 Leaves palmately trifoliolate . . . . **T. aureum**
 2 Leaves pinnately trifoliolate 3
  3 Heads with up to 10 flowers; standard without conspicuous striations . . . . **T. dubium**
  3 Heads with 15–40 flowers; standard distinctly striate . . . . **T. campestre**
1 Petals not yellow, either white, cream, pink, or crimson 4
 4 Pedicels 2 mm or more long, usually recurved in fruit 5
  5 Stem prostrate, rooting at the nodes; peduncles arising at ground level . . . . **T. repens**
  5 Stem erect; peduncles axillary or terminal 6
   6 Flowers 6 mm or less long . . . . **T. carolinianum**
   6 Flowers 7–12 mm long 7
    7 Calyx 5-nerved, the lobes about as long as the tube; plants essentially glabrous . . . . **T. hybridum**
    7 Calyx 10-nerved, the lobes 2× or more as long as the tube; plants variously pubescent 8
     8 Leaflets 3× or more longer than wide; plants rarely more than 1 dm tall . . . . (**T. virginicum**)
     8 Leaflets less than 3× as long as wide; plants 2–5 dm tall . . . . **T. reflexum**
 4 Flowers subsessile, the pedicels less than 2 mm long 9
  9 Flowers 5–7 mm long, white; leaflets about 3× as long as wide . . . . **T. arvense**
  9 Flowers 12–20 mm long, usually pink or crimson; leaflets less than 3× as long as wide 10
   10 Heads nearly spherical, subtended by a pair of leaves; flowers pink, rarely white . . . . **T. pratense**
   10 Heads elongate, not subtended by a pair of leaves; flowers crimson . . . . **T. incarnatum**

T. arvense L.*, Rabbit Foot C.; fields and waste places; frequent; Apr–Aug; NC, SC, TN, VA.

T. aureum Pollich*, Hop C.; fields and waste places; occasional; May–Aug; GA, NC, SC, VA. *T. agrarium* L.

T. campestre Schreb.*, Low Hop C.; fields and waste places; frequent; Apr–Jul; ALL. *T. procumbens* L.

T. carolinianum Michx.; open sites; infrequent; Apr–Jul; GA.

T. dubium Sibthorp*, Low Hop C.; fields and waste areas; occasional; Apr–Sep; ALL.

T. hybridum L.*, Alsike C.; fields and waste places; frequent; Apr–Sep; ALL.

T. incarnatum L.*, Crimson C.; fields and waste places; infrequent; Apr–Jun; ALL.

T. pratense L.*, Red C.; fields and waste places; common; Apr–Aug; ALL.

T. reflexum L., Buffalo C.; open woods and clearings; infrequent; Apr–Aug; NC, SC.

T. repens L.*, White C.; fields, lawns, and waste areas; common; Apr–Aug; ALL.

T. virginicum Small, Kates Mt. C.; shale barrens; infrequent; May–Jun; VA.

Vicia L. (Vetch)

1 Peduncles 5 mm or less long; flowers 1–3/node 2
  2 Corolla yellowish; flowers about 3 cm long .... **V. grandiflora**
  2 Corolla purplish; flowers less than 2 cm long .... **V. angustifolia**
1 Peduncles more than 5 mm long; flowers numerous 3
  3 Corolla white, sometimes tipped with purple .... **V. caroliniana**
  3 Corolla blue or purplish 4
    4 Stems and peduncles with conspicuous, spreading hairs .... **V. villosa**
    4 Stems and peduncles glabrous or with a few appressed hairs 5
      5 Flowers more than 10/raceme .... **V. dasycarpa**
      5 Flowers less than 10/raceme .... **V. americana**

V. americana Willd., American V.; moist woodlands; infrequent; May–Jun; VA.
V. angustifolia L.*, Narrow-leaved V.; fields and waste areas; frequent; Mar–Jun; NC, SC, TN, VA.
V. caroliniana Walt., Carolina V.; open woodlands; common; Apr–Jun; ALL. *V. hugeri* Small.
V. dasycarpa Tenore*; fields and waste places; frequent; May–Aug; NC, SC, TN, VA.
V. grandiflora Scopoli*, Yellow V.; fields and waste places; infrequent; Apr–Jun; NC, SC, TN.
V. villosa Roth*, Hairy V.; fields and waste places; occasional; May–Aug; NC, SC, TN, VA.

Wisteria Nutt.

W. frutescens (L.) Poir.; low woods and streams; infrequent; Apr–May; SC, TN. Primarily a species of Coastal Plain swamps.

FAGACEAE (Beech Family)

1 Fruits enclosed in a spiny or prickly bur 2
  2 Bark smooth; winter buds more than 1 cm long; involucral spines recurved ... **Fagus**
  2 Bark scaly or furrowed; winter buds less than 1 cm long; involucral spines straight .... **Castanea**
1 Fruits partially enclosed in a scaly cup .... **Quercus**

Castanea Miller (Chestnut)

1 Leaves glabrous beneath, usually more than 15 cm long; involucre with more than 1 nut .... **C. dentata**
1 Leaves pubescent beneath, usually less than 15 cm long; involucre usually with 1 nut .... **C. pumila**

C. dentata (Marsh.) Borkh., American C.; mesic slopes; occasional; Jun–Jul; ALL. Once a dominant tree in our area but now persisting as stump sprouts or small trees as a result of the introduction of the chestnut blight. Fruits occur sporadically and infrequently.
C. pumila (L.) Miller, Allegheny Chinkapin; dry woodlands; frequent; Jun–Jul; ALL.

Fagus L. (American Beech)

F. grandifolia Ehrh.; rich woods; common; Mar–Apr; ALL. *F. grandifolia* Ehrh. var. *caroliniana* (Loud.) Fern. & Rehd.

Quercus L. (Oak)

1 Leaves toothed or lobed, but not bristle-tipped; inner surface of acorn shell glabrous 2
2 Leaves shallowly to deeply lobed 3
3 Acorn cup bowl-shaped, conspicuously fringed and covering more than 1/2 of the nut . . . . **Q. macrocarpa**
3 Acorn cup saucer-shaped, enclosing less than 1/2 of the nut 4
4 Leaves glabrous and usually glaucous beneath, with 7–10 shallow lobes . . . . **Q. alba**
4 Leaves pubescent beneath, with 3–5 lobes, often cross-shaped . . . **Q. stellata**
2 Leaves with coarse teeth or scalloped (resembling chestnut leaves) but not distinctly lobed 5
5 Acorn stalks 2.5–10 cm long, longer than the leaf petiole . . . . **Q. bicolor**
5 Acorns nearly sessile or on stalks shorter than the leaf petiole 6
6 Acorn 2.5–4 cm long 7
7 Cup scales free to the base; blades commonly tomentose beneath; plants of lowlands and mesic sites . . . . **Q. michauxii**
7 Cup scales united to near the tip; blades often pubescent but not tomentose; plants of dry uplands . . . . **Q. montana**
6 Acorn small, 1.5–2.5 cm long 8
8 Medium-sized tree; leaves with 9–15 pairs of lateral veins . . . . **Q. muhlenbergii**
8 Shrubs, often thicket-forming; leaves with 4–8 pairs of lateral veins . . . . **Q. prinoides**
1 Leaves entire, toothed, or lobed, always with bristlelike tips extending beyond the leaf margin; inner surface of acorn shell pubescent 9
9 Leaves entire or somewhat 3–5-lobed 10
10 Leaves widest above the middle 11
11 Mature leaves about as long as wide; blades more than 10 cm long, rusty tomentose beneath . . . . **Q. marilandica**
11 Mature leaves about 2× as long as wide; blades less than 10 cm long, not rusty tomentose beneath . . . . **Q. nigra**
10 Leaves uniformly wide or sometimes wider at or slightly below the middle (willowlike) 12
12 Leaves glabrous beneath, light green above, less than 3 cm wide . . . . **Q. phellos**
12 Leaves pubescent beneath, dark green above, more than 3 cm wide . . . . **Q. imbricaria**
9 Leaves distinctly lobed 13
13 Acorn cups saucer-shaped, enclosing about 1/4 of the nut 14
14 Leaf blades more than 12 cm long; acorn 1.2–2.5 cm long . . . . **Q. rubra**
14 Leaf blades less than 12 cm long; acorn 1–1.4 cm long 15
15 Leaves 5–7-lobed; petioles 2–5 cm long; acorns sessile . . . **Q. palustris**

15 Leaves 3–5-lobed; petioles less than 2 cm long; acorns subsessile .... **Q. georgiana**

13 Acorn cups bowl-shaped, enclosing 1/3 or more of the nut 16

16 Leaves with white to gray hairs beneath 17

17 Leaves usually 5-lobed, less than 10 cm long; shrubs or small trees .... **Q. ilicifolia**

17 Leaves dimorphic, either 3-lobed at the tip or 5–11-lobed, more than 10 cm long; medium to large trees .... **Q. falcata**

16 Leaves green beneath, sometimes with tufts of hairs in the axils of the main veins 18

18 Petioles 7–15 cm long; leaves yellowish-brown beneath ... **Q. velutina**

18 Petioles up to 6.5 cm long; leaves pale green beneath 19

19 Acorn with 1 or more concentric grooves near the apex; leaves bright green above .... **Q. coccinea**

19 Acorn without concentric grooves at the apex; leaves dark green above .... **Q. shumardii**

Q. alba L., White O.; dry to mesic woodlands; common; Apr–May; ALL.

Q. bicolor Willd., Swamp White O.; lowlands and wet meadows; occasional; Apr–May; TN, VA.

Q. coccinea Muenchh., Scarlet O.; uplands; frequent; Apr–May; ALL.

Q. falcata Michx., Southern Red O.; uplands; occasional; Apr–May; ALL.

Q. georgiana M. A. Curtis, Georgia O.; granite outcrops; infrequent; Apr–May; GA.

Q. ilicifolia Wang., Bear O.; dry barrens and uplands; occasional; Apr–May; NC, VA.

Q. imbricaria Michx., Shingle O.; lowlands; occasional; Apr–May; NC, TN, VA.

Q. macrocarpa Michx., Bur O.; bottomlands and uplands, often on calcareous sites; infrequent; Apr–May; VA.

Q. marilandica Muenchh., Blackjack O.; dry barrens and uplands; common; Apr–May; ALL.

Q. michauxii Nutt., Swamp Chestnut O.; bottomlands; infrequent; Apr–May; NC, SC, TN.

Q. montana Willd., Chestnut O.; dry upland slopes; common; Apr–May; ALL. *Q. prinus* L.

Q. muhlenbergii Engelm., Chinkapin O.; dry calcareous uplands; occasional; Apr–May; ALL.

Q. nigra L., Water O.; lowlands; occasional; Apr–May; GA, NC, SC.

Q. palustris Muenchh., Pin O.; lowlands; infrequent; Apr–May; VA.

Q. phellos L., Willow O.; lowlands; infrequent; Apr–May; NC, SC, TN, VA.

Q. prinoides Willd., Dwarf Chinkapin O.; dry slopes and barrens; occasional; Apr–May; ALL.

Q. rubra L., Northern Red O.; rich woodlands; common; Apr–May; ALL. *Q. borealis* Michx. f.

Q. shumardii Buckl., Shumard O.; lowlands and rich woods; infrequent; Apr–May; NC, TN, VA.

Q. stellata Wang., Post O.; dry uplands; frequent; Apr–May; ALL.

Q. velutina Lam., Black O.; dry slopes and uplands; common; Apr–May; ALL.

A taxonomically difficult genus requiring mature leaves and acorns for positive identifications.

FUMARIACEAE (Fumitory Family)

1 Viny biennials; petals persistent . . . . **Adlumia**
1 Perennial herbs; petals deciduous 2
  2 Corolla 1-spurred; zygomorphic . . . . **Corydalis**
  2 Corolla 2-spurred; bisbilateral . . . . **Dicentra**

Adlumia Raf. ex DC. (Climbing Fumitory)

A. fungosa (Ait.) Greene; rich coves and cliffs, often on disturbed sites; infrequent and sporadic; Jun–Sep; NC, TN, VA.

Corydalis Vent.

1 Petals pink with yellow tips . . . . **C. sempervirens**
1 Petals yellow 2
  2 Flowers 6–8 mm long; fruits drooping . . . . **C. flavula**
  2 Flowers 9–13 mm long; fruits erect . . . . **C. micrantha**

C. flavula (Raf.) DC., Yellow Fumitory; rocky alluvial woods and floodplains, often on calcareous sites; frequent; Mar–Apr; NC, TN, VA.

C. micrantha (Engelm.) Gray, Slender Fumewort; disturbed rocky slopes; infrequent; Mar–Apr; NC, SC. Represented in our area by subsp. **australis** (Chapman) Ownbey.

C. sempervirens (L.) Pers., Pale Corydalis; rocky open clearings; occasional; Apr–Jun; ALL.

Dicentra Bernh.

1 Inflorescence a panicle; flowers dark pink, more than 1.8 cm long . . . . **D. eximia**
1 Inflorescence a raceme; flowers white or light pink, less than 1.8 cm long 2
  2 Corolla rounded at base; plants with yellow cornlike tubers at the base; flowers with a sweet odor . . . . **D. canadensis**
  2 Corolla spurred or spreading at base; plants with white tubers at base; flowers odorless . . . . **D. cucullaria**

D. canadensis (Goldie) Walp., Squirrel Corn; rich woods; occasional; Apr–May; GA, NC, TN, VA.

D. cucullaria (L.) Bernh., Dutchman's Breeches; rich woods; frequent; Apr–May; GA, NC, TN, VA.

D. eximia (Ker-Gawl.) Torr., Bleeding Heart; rich, often rocky woods and cliffs; occasional; Apr–Jun; ALL.

GENTIANACEAE (Gentian Family)

1 Leaves whorled; plant 1–2 m tall . . . . **Frasera**
1 Leaves alternate or opposite; plants less than 1 m tall 2
  2 Leaves scalelike, 5 mm or less long . . . . **Bartonia**
  2 Leaves not scalelike, more than 5 mm long 3
    3 Corolla rotate, the lobes longer than the tube . . . . **Sabatia**
    3 Corolla variously tubular, the lobes shorter than the tube 4
      4 Calyx lobes 2; plants less than 1 dm tall; spring flowering . . . . **Obolaria**
      4 Calyx lobes 4 or 5; plants more than 1 dm tall; summer and fall flowering . . . . **Gentiana**

Bartonia Muhl. (Screw Stem)

1 Scale leaves mostly alternate .... **B. paniculata**
1 Scale leaves mostly opposite .... **B. virginica**

B. paniculata (Michx.) Muhl.; bogs and wet woods; infrequent; Aug–Oct; GA, TN, VA.
B. virginica (L.) BSP.; bogs and wet woods; occasional; Aug–Oct; NC, TN, VA.

Frasera Walt. (Columbo)

F. caroliniensis Walt.; open woodlands, usually on basic soils; infrequent; May–Jun; GA, NC, TN. *Swertia caroliniensis* (Walt.) Kuntze.

Gentiana L. (Gentian)

Pringle, J. S. 1967. Taxonomy of *Gentiana*, sect. *Pneumonanthae*, in eastern North America. *Brittonia* 19:1–32.

1 Calyx and corolla 4-lobed; corolla lobes fringed .... **G. crinita**
1 Calyx and corolla 5-lobed; corolla lobes not fringed 2
  2 Annuals; corolla without pleats or folds between the lobes ...... **G. quinquefolia**
  2 Perennials; corolla with pleats or folds between the lobes 3
    3 Calyx lobes keeled .... **G. alba**
    3 Calyx lobes not keeled 4
      4 Corolla greenish-white; seeds wingless .... **G. villosa**
      4 Corolla blue or bluish-white; seeds winged 5
        5 Margins of leaves and calyx lobes entire or essentially so ..... **G. linearis**
        5 Margins of leaves and calyx lobes scabrous to ciliate 6
          6 Calyx lobes linear-subulate, shorter than the calyx tube; stems densely pubescent .... **G. decora**
          6 Calyx lobes lanceolate to orbicular, shorter or longer than the calyx tube; stems glabrous or pubescent 7
            7 Calyx lobes oblanceolate; corolla open or loosely closed .... **G. saponaria**
            7 Calyx lobes lanceolate to orbicular or obovate; corolla tightly closed 8
              8 Corolla lobes reduced to minute projections less than 1 mm long .... **G. andrewsii**
              8 Corolla lobes larger 9
                9 Stems and calyx tube densely puberulent .... **G. austromontana**
                9 Stems and usually the calyx tube glabrous .... **G. clausa**

G. alba Muhl.; mesic sites; infrequent; Aug–Oct; NC. Apparently extirpated in our area.
G. andrewsii Grisebach; moist woods; infrequent; Sep–Nov; VA.
G. austromontana Pringle & Sharp; grassy balds at high elevations; infrequent; Sep–Oct; NC, TN.
G. clausa Raf.; open mesic sites; occasional; Aug–Oct; NC, TN, VA.
G. crinita Froelich, Fringed G.; open mesic sites; infrequent; Sep–Oct; GA, NC, VA. *Gentianopsis crinita* (Froelich) Ma.
G. decora Pollard; mesic open woodlands; occasional; Sep–Nov; ALL.
G. linearis Froelich; open grassy areas; infrequent; Sep–Oct; Mt. LeConte, TN.

G. quinquefolia L.; road banks and mesic open areas; frequent; Aug–Oct; ALL. *Gentianella quinquefolia* (L.) Small.

G. saponaria L.; road banks and mesic woodlands; occasional; Aug–Nov; NC, SC, TN, VA.

G. villosa L.; mesic to dry woodlands; frequent; Sep–Dec; ALL.

Obolaria L. (Pennywort)

O. virginica L.; rich woods; frequent; Mar–May; ALL.

Sabatia Adanson (Rose Pink)

1 Upper branches of principal stem alternate .... **S. campanulata**
1 Upper branches of principal stem opposite 2
  2 Main stem and branches angled and winged .... **S. angularis**
  2 Main stem and branches round and not winged .... **S. brachiata**

S. angularis (L.) Pursh; fields and woodlands; common; Jul–Aug; ALL.

S. brachiata Ell.; sandy woodlands; infrequent; Jun–Jul; NC.

S. campanulata (L.) Torr.; damp fields and woodlands; infrequent; Jun–Aug; NC, TN, VA.

GERANIACEAE (Geranium Family)

1 Leaves pinnately compound .... **Erodium**
1 Leaves palmately lobed or compound .... **Geranium**

Erodium L'Her. (Storksbill)

E. cicutarium (L.) L'Her.*; waste places; occasional; Apr–Jun; NC, SC, TN, VA.

Geranium L.

1 Petals more than 1 cm long .... **G. maculatum**
1 Petals less than 1 cm long 2
  2 Sepals at maturity with awns 1 mm or more long 3
    3 Fruiting pedicels more than 2× as long as the calyx 4
      4 Pedicels with appressed, retrorse hairs .... **G. columbinum**
      4 Pedicels densely glandular pilose .... **G. bicknellii**
    3 Fruiting pedicels up to 2× as long as the calyx or shorter 5
      5 Leaves completely divided into 3–5 separate divisions ..... **G. robertianum**
      5 Leaves palmately lobed or deeply dissected 6
        6 Petals dark pink to purple; hairs of mature carpels less than 1 mm long .... **G. dissectum**
        6 Petals pale pink to nearly white; hairs of mature carpels more than 1 mm long .... **G. carolinianum**
  2 Sepals at maturity mucronate or with awns less than 1 mm long 7
    7 Mature carpel body glabrous, obliquely wrinkled .... **G. molle**
    7 Mature carpel body pubescent, not wrinkled .... **G. pusillum**

G. bicknellii Britt.; open woods and clearings; infrequent; Jul–Sep; VA.

G. carolinianum L., Carolina Cranesbill; disturbed sites; common; Mar–Jun; ALL.

G. columbinum L.*, Dovesfoot Cranesbill; pastures and waste places; frequent; May–Jul; GA, NC, TN, VA.

G. dissectum L.*; lawns and pastures; infrequent; May–Jun; NC, SC, VA.
G. maculatum L., Wild Geranium; rich woods; common; Apr–May; ALL.
G. molle L.*; disturbed sites; frequent; Apr–Jun; NC, TN, VA.
G. pusillum L.*; disturbed sites; occasional; May–Jun; NC, VA.
G. robertianum L.; rocky woodlands; infrequent; Jun–Oct; VA.

HALORAGACEAE (Water Milfoil Family)

1 Leaves alternate; flowers 3-merous . . . . **Proserpinaca**
1 Leaves whorled; flowers 4-merous . . . . **Myriophyllum**

Myriophyllum L.

Aiken, S. G. 1981. A conspectus of *Myriophyllum* (Haloragaceae) in North America. *Brittonia* 33:57–69.

1 Emergent leaves usually 2.5–3.5 cm long; petioles 5–7 mm long . . . . **M. aquaticum**
1 Emergent leaves usually less than 2 cm long; petiole absent 2
  2 Uppermost flowers and all submerged leaves whorled . . . . **M. spicatum**
  2 Uppermost flowers alternate; submerged leaves alternate or whorled . . . . **M. pinnatum**

M. aquaticum (Vellozo) Verdcourt*, Parrot Feather; aquatic habitats; occasional, seldom flowering, and only pistillate flowers are known in North America; NC, SC, TN, VA. *M. brasiliense* Camb.
M. pinnatum (Walt.) BSP.; aquatic habitats; infrequent; Jun–Oct; VA.
M. spicatum L.*, Eurasian Water Milfoil; aquatic habitats; infrequent; Jul–Sep; VA.

Proserpinaca L. (Mermaid Weed)

P. palustris L.; shallow water and muck; infrequent; Jun–Sep; VA.

HAMAMELIDACEAE (Witch Hazel Family)

1 Large trees; leaves palmately veined and lobed; flowers in dense round heads . . . . **Liquidambar**
1 Shrubs or small trees; leaves pinnately veined; flowers in axillary clusters or terminal spikes 2
  2 Flowers white, in terminal spikes; petals absent; stamens more than 4 . . . . **Fothergilla**
  2 Flowers yellow, in axillary clusters; petals and stamens 4 . . . . **Hamamelis**

Fothergilla Murray (Witch Alder)

F. major (Sims) Lodd.; rocky woodlands; infrequent; Apr–May; NC, TN.

Hamamelis L. (Witch Hazel)

H. virginiana L.; dry or mesic woodlands; common; Sep–Dec; ALL.

Liquidambar L. (Sweet Gum)

L. styraciflua L.; deciduous woods and disturbed sites; common; Apr–May; ALL. Absent from mid and high elevations.

HIPPOCASTANACEAE (Buckeye Family)

Aesculus L. (Buckeye)

1 Large trees .... **A. flava**
1 Shrubs, rarely small trees 2
  2 Petals scarlet; stamens exserted beyond the lateral petals .... **A. pavia**
  2 Petals pale yellow, rarely dull pink; stamens included within the lateral petals .... **A. sylvatica**

A. flava Solander, Yellow B.; rich deciduous forests; common; Apr–Jun; ALL. *A. octandra* Marsh.
A. pavia L., Red B.; open woodlands and roadsides; infrequent; Apr–May; SC, TN.
A. sylvatica Bartram; rich woods and stream banks; occasional; Apr–May; GA, NC, SC, TN.

HYDROPHYLLACEAE (Waterleaf Family)

1 Flowers solitary, opposite the leaves .... **Ellisia**
1 Flowers cymose or in loose racemes 2
  2 Leaf blades mostly more than 8 cm wide; flowers in subcompact, branched, cymelike inflorescences, not elongate at anthesis .... **Hydrophyllum**
  2 Leaf blades mostly less than 8 cm wide; flowers in loose, racemelike inflorescences, becoming elongate at anthesis .... **Phacelia**

Ellisia L.

E. nyctelea L.; moist woods; infrequent; Apr–Jun; VA.

Hydrophyllum L. (Waterleaf)

1 Leaves palmately lobed .... **H. canadense**
1 Leaves pinnately lobed or compound 2
  2 Stems densely pubescent; leaf segments 9–13 .... **H. macrophyllum**
  2 Stems glabrous or slightly pubescent; leaf segments 3–7 .... **H. virginianum**

H. canadense L.; rich moist woods; frequent; May–Jun; ALL.
H. macrophyllum Nutt.; rich moist woods, especially over calcareous soils; occasional; May–Jun; GA, NC, TN, VA.
H. virginianum L.; rich moist woods; occasional; Apr–Jun; NC, TN, VA.

Phacelia Juss. (Phacelia)

1 Corolla lobes fringed 2
  2 Flowers white, rarely blue; stems spreading-pubescent .... **P. fimbriata**
  2 Flowers blue, rarely white; stems appressed-pubescent .... **P. purshii**
1 Corolla lobes entire or essentially so 3
  3 Inflorescence with glandular trichomes; corolla 10 mm or more broad .... **P. bipinnatifida**
  3 Inflorescence without glandular trichomes; corolla less than 10 mm broad 4
    4 Calyx lobes ovate, trichomes mostly appressed and less than 1 mm long .... **P. dubia**
    4 Calyx lobes linear, trichomes spreading and more than 1 mm long ... **P. maculata**

P. bipinnatifida Michx.; moist woods and slopes; frequent; Apr–May; NC, SC, TN, VA.
P. dubia (L.) Trel.; woodlands and fields; frequent; Apr–May; NC, SC, TN, VA.
P. fimbriata Michx., Fringed P.; rich woods; infrequent; Apr–May; NC, TN, VA.
P. maculata Wood; fields and alluvial woods; infrequent; Apr–May; SC.
P. purshii Buckley; rich woods and roadsides; occasional; Apr–May; ALL.

JUGLANDACEAE (Walnut Family)

1 Pith chambered; fruit with indehiscent husk . . . . **Juglans**
1 Pith not chambered; fruit splitting variously along 4 lines . . . . **Carya**

Carya Nutt. (Hickory)

1 Bud scales valvate . . . . **C. cordiformis**
1 Bud scales imbricate 2
  2 Terminal bud 1.2–3.7 cm long 3
    3 Bark scaly or shaggy; branchlets and leafstalks smooth or slightly hairy 4
      4 Branchlets light orange; leaflets usually 7–9 . . . . **C. laciniosa**
      4 Branchlets light red-brown; leaflets usually 5 . . . . **C. ovata**
    3 Bark tight; branchlets and leafstalks hairy . . . . **C. tomentosa**
  2 Terminal buds 0.6–1.2 cm long 5
    5 Overwintering buds and leaflets with silver-white, peltate (lepidote) scales below . . . . **C. pallida**
    5 Overwintering buds and leaflets not lepidote 6
      6 Leaflets usually 5; husk of fruit indehiscent or splitting to about the middle . . . . **C. glabra** var. **glabra**
      6 Leaflets usually 7; husk splitting to base at maturity . . **C. glabra** var. **odorata**

C. cordiformis (Wang.) K. Koch, Bitternut H.; bottomlands and dry woods; common; Apr–May; ALL.
C. glabra (Mill.) Sweet var. **glabra**, Pignut H.; dry woods; frequent; Apr–May; ALL.
C. glabra (Mill.) Sweet var. **odorata** (Marsh.) Little, Red H.; dry woods; occasional; Apr–May; NC, TN, VA. *C. ovalis* (Wang.) Sarg. Perhaps not distinct from the preceding taxon.
C. laciniosa (Michx. f.) Loud., Big Shellbark H.; basic soils of bottomlands and floodplains; infrequent; Apr–May; TN.
C. ovata (Mill.) K. Koch, Shagbark H.; lowlands and rich valleys; frequent; Apr–May; GA, NC, TN, VA.
C. pallida (Ashe) Engl. & Graebn., Sand H.; dry woods; occasional; Apr–May; ALL.
C. tomentosa (Poir.) Nutt., Mockernut H.; dry woods; common; Apr–May; ALL.

Juglans L. (Walnut)

1 Fruit ellipsoid; pith dark brown . . . . **J. cinerea**
1 Fruit nearly round; pith light tan . . . . **J. nigra**

J. cinerea L., Butternut; rich woods; frequent; Apr–May; ALL.
J. nigra L., Black W.; rich woods; common; Apr–May; ALL.

LAMIACEAE (Mint Family)

Contributed by J. L. Collins

1 Anther-bearing stamens 4 2
2 Ovary deeply 4-lobed; nutlets basally attached; style basal 3
3 Central lobe of corolla lower lip entire, retuse, sinuate, or, rarely, crenate 4
4 Stems, at least the sterile, trailing and mat-forming 5
5 Corolla less than 2 cm long; flowers axillary .... **Glecoma**
5 Corolla more than 2 cm long; flowers terminal .... **Meehania**
4 Stems erect or ascending, not trailing 6
6 Calyx zygomorphic, the lobes distinctly unequal 7
7 Calyx bearing a dorsal scoop-shaped shield or crest .... **Scutellaria**
7 Calyx lacking a dorsal scoop-shaped shield or crest 8
8 Corolla 0.5–2.5 cm long 9
9 Inflorescence axillary, racemose, or of racemose glomerules, not headlike 10
10 Calyx gibbous; the pedicel appearing to attach to the calyx laterally .... **Perilla**
10 Calyx not gibbous; the pedicel attached to the calyx basally .. 11
11 Calyx teeth spiny, divergent in fruit .... **Leonurus**
11 Calyx teeth soft, obtuse or acute but not spiny, ascending in fruit 12
12 Lobes of the calyx upper lip wider than long; foliage lemon scented .... **Melissa**
12 Lobes of calyx upper lip longer than wide; foliage not lemon scented 13
13 Leaves 0.4–2.5 cm long .... **Satureja**
13 Leaves 2.5–8.5 cm long .... **Nepeta**
9 Inflorescence terminal, dense, and headlike 14
14 Calyx hirsute .... **Satureja**
14 Calyx puberulent .... **Pycnanthemum**
8 Corolla 2.6–4 cm long .... **Synandra**
6 Calyx actinomorphic or nearly so, the lobes equal or essentially so 15
15 Anthers hispid .... **Lamium**
15 Anthers glabrous or puberulent 16
16 Calyx teeth 8–10 .... **Marrubium**
16 Calyx teeth 5 17
17 Inflorescence headlike .... **Pycnanthemum**
17 Inflorescence racemose or axillary 18
18 Corolla 0.7–2.9 cm long; calyx about 3 mm wide 19
19 Flowers single in the axil of each bract .. **Physostegia**
19 Flowers 2–several in the axil of each bract . . **Stachys**
18 Corolla 0.3–0.6 cm long; calyx about 1 mm wide .... **Mentha**
3 Central lobe of corolla lower lip lacerate 20
20 Pedicels conspicuous in the open thyrse; stamens exserted more than 0.5 cm .... **Collinsonia**

20 Pedicels hidden within the dense spike or head; stamens included or exserted less than 0.5 cm 21
21 Stamens included under the upper corolla lip . . . . **Prunella**
21 Stamens exserted . . . . **Agastache**
2 Ovary slightly 4-lobed; nutlets laterally or obliquely attached; style terminal 22
22 Corolla upper lip greatly reduced, 1/10 or less the length of the lower . . . . **Teucrium**
22 Corolla upper lip about the same length as the lower . . . . **Trichostema**
1 Anther-bearing stamens 2 23
23 Corolla lower lip lacerate; stamens exserted more than 0.5 cm . . . . . . **Collinsonia**
23 Corolla lower lip variously lobed but not lacerate; stamens included or exserted 0.5 cm or less 24
24 Calyx zygomorphic 25
25 Leaves 2.5 cm or less long . . . . **Hedeoma**
25 Leaves 3 cm or more long 26
26 Flowers creamy, pale blue to lavender, with purple spots, 10 or more in each headlike cluster . . . . **Blephilia**
26 Flowers blue to violet, 10 or fewer in each open verticil . . . . . . **Salvia**
24 Calyx actinomorphic or nearly so 27
27 Calyx 4 mm or more long . . . . **Monarda**
27 Calyx less than 3.5 mm long 28
28 Flowers sessile . . . . **Lycopus**
28 Flowers pedicellate . . . . **Cunila**

Agastache Clayton ex Gronov.

1 Corolla yellowish; sepals and bracts uniformly green . . . . **A. nepetoides**
1 Corolla pinkish; sepals and bracts with whitish or pinkish margins . . . . **A. scrophulariaefolia**

A. nepetoides (L.) Kuntze; open woods and disturbed sites, often on calcareous soils; occasional; Jul–Sep; NC, VA.

A. scrophulariaefolia (Willd.) Kuntze; open woods and meadows; frequent; Jul–Sep; GA, NC, TN, VA.

Blephilia Raf.

1 Lobes of calyx lower lip linear, approaching or surpassing sinuses of upper lip; outer bracteoles acute; petioles usually 0.7 cm or less long . . . . **B. ciliata**
1 Lobes of calyx lower lip deltoid, not reaching sinuses of upper lip; outer bracteoles long-acuminate; petioles usually 1 cm or more long . . . . **B. hirsuta**

B. ciliata (L.) Bentham; dry to moist calcareous woods and openings; infrequent; Jun–Jul; VA.

B. hirsuta (Pursh) Bentham; moist woods; occasional; Jun–Aug; NC, TN, VA.

Collinsonia L. (Stoneroot)

Shinners, L. H. 1962. Synopsis of *Collinsonia* (Labiatae). *Sida* 1:76–83.

1 Stamens 4; leaves 2–6, usually 4, congested; spring or early summer flowering . . . . **C. verticillata**

1 Stamens 2, rarely 4 (in *C. serotina*); leaves 6 or more, dispersed distally on the stem; late summer or fall flowering 2
2 Blades of larger leaves 4–10.5 cm long, teeth 5–15/side; rootstocks 4–6 cm long, tuberlike .... **C. tuberosa**
2 Blades of larger leaves 8–25 cm long, teeth 11–42/side; rootstocks 4–15 cm long, rhizomelike 3
3 Flowering calyx 4–7 mm long, greater than 1/2 the corolla tube length .... (**C. serotina**)
3 Flowering calyx 2–4 mm long, less than 1/2 the corolla tube length .... **C. canadensis**

C. canadensis L.; rich moist woods; frequent; Aug–Oct; ALL.

C. serotina Walt.; rich sandy woods; infrequent, known from only one locality and more common on the Piedmont and Coastal Plain; Sep–Oct; GA.

C. tuberosa Michx.; moist woods; infrequent and more common on the Piedmont and Coastal Plain; Jul–Sep; NC.

C. verticillata Baldw. ex Ell.; rich or dry woods; occasional; Apr–Jun; GA, NC, SC, TN.

Cunila L. (Stone Mint)

C. origanoides (L.) Britt.; dry rocky woodlands; frequent; Aug–Sep; ALL.

Glecoma L. (Ground Ivy)

G. hederacea L.*; open disturbed sites; common; Mar–Jun; ALL.

Hedeoma Pers. (Pennyroyal)

H. pulegioides (L.) Pers.; fields and open woods; common; Jul–Oct; ALL.

Lamium L.

1 Upper leaves and bracts sessile .... **L. amplexicaule**
1 Upper leaves and bracts petiolate .... **L. purpureum**

L. amplexicaule L.*, Henbit; fields, lawns, waste places; common; Mar–Jun; NC, SC, TN, VA.

L. purpureum L.*, Dead-nettle; fields, lawns, waste places; common; Mar–Jun; NC, SC, TN, VA.

Leonurus L. (Motherwort)

L. cardiaca L.*; fields and roadsides; frequent; May–Aug; NC, SC, TN, VA.

Lycopus L.

Henderson, N. C. 1962. A taxonomic revision of the genus *Lycopus* (Labiatae). *Amer. Midl. Naturalist* 68:95–138.

1 Calyx lobes obtuse or acute at the apex, shorter than or barely equaling the length of the nutlets 2
2 Plants tuberous; crests of nutlets undulate to scarcely tuberculate; stamens exserted .... **L. uniflorus**
2 Plants not tuberous; crests of nutlets deeply and prominently tuberculate; stamens included .... **L. virginicus**

1 Calyx lobes acuminate to subulate-tipped, much exceeding the nutlets 3
3 Stem acutely 4-angled, minutely winged; nutlets with a rounded, smooth crest along the upper edges; lower part of stem lacking above-ground runners . . **L. americanus**
3 Stem with rounded angles, not minutely winged; nutlets with a tuberculate (toothed) crest along their upper edges; lower part of stem usually with long, above-ground runners 4
4 Leaves with slender or winged petioles . . . . **L. rubellus**
4 Leaves sessile or nearly so 5
5 Leaves ovate-lanceolate, usually rounded at the base, 3–6 cm long, 1.5–3 cm wide, scarcely reduced upward . . . . **L. amplectens**
5 Leaves lanceolate, gradually narrowed to a sessile base, the upper noticeably narrower, lower leaves 6–12 cm long, 1.5–2 cm wide, upper leaves 3–7 cm long, 0.3–1 cm wide . . . . **L. rubellus**

L. americanus Muhl. ex Barton; wet woods, fields, and stream banks; occasional; Jul–Aug; NC, TN, VA.

L. amplectens Raf.; stream banks and edges of wet woods; infrequent; Jul–Aug; NC.

L. rubellus Moench; creek banks and wet areas; infrequent; Jul–Aug; NC, TN. *L. angustifolius* Ell.

L. uniflorus Michx.; bogs and wet woods; infrequent; Jul–Aug; NC, SC, VA. *L. sherardii* Steele (*L. virginicus* × *L. uniflorus*).

L. virginicus L.; wet woods, low fields, and creek banks; frequent; Jul–Aug; ALL.

Marrubium L. (Horehound)

M. vulgare L.*; disturbed sites; occasional; May–Oct; NC, VA.

Meehania Britt.

M. cordata (Nutt.) Britt.; rich woods, often on basic soils; infrequent; May–Jun; NC, TN, VA.

Melissa L. (Lemon Balm)

M. officinalis L.*; disturbed sites; occasional; Jul–Oct; NC, TN, VA. Crushed leaves emit a lemony fragrance.

Mentha L.

1 Inflorescence of axillary cymules subtended by normal or slightly reduced foliage leaves 2
2 Calyx lobes lanceolate; subtending leaves 2–3× as long as cymules . . . . **M. cardiaca**
2 Calyx lobes deltoid; subtending leaves 3–6× as long as cymules . . . . **M. arvensis**
1 Inflorescence of terminal spikes or heads, any subtending leaves greatly reduced . . . 3
3 Leaves pubescent on upper surfaces; calyx tube pubescent; stems densely pubescent . . . . **M. longifolia**
3 Leaves glabrous on upper surfaces or puberulent on veins; calyx tube glabrous; stems glabrous or essentially so 4
4 Calyx 1.5–2.2 mm long; bracteoles ciliate; petioles of stem leaves 0–3 mm long . . . . **M. piperita**

4 Calyx 3–4 mm long; bracteoles not ciliate; petioles of stem leaves 4–15 mm long .... **M. spicata**

M. arvensis L.; infrequent; Jun–Sep; NC, VA.
M. cardiaca Gerarde ex Baker*; infrequent; Jul–Nov; NC, SC, TN, VA.
M. longifolia L.*; infrequent; Jul–Sep; VA.
M. piperita L.*, Peppermint; occasional; Jun–Nov; NC, SC, TN, VA.
M. spicata L.*, Spearmint; infrequent; Jun–Sep; NC, SC, TN, VA.

All of our species occur in similar habitats: wet soils of low woods, meadows, stream banks, ditches, thickets, and seeps.

Monarda L. (Horsemint, Bee Balm)

Scora, R. W. 1967. Interspecific relationships in the genus *Monarda* (Labiatae). *Univ. Calif. Publ. Bot.* 20:1–59.

1 Glomerules 2–many, axillary and terminal; corolla upper lip strongly arcuate; stamens included .... **M. punctata**
1 Glomerules solitary, terminal; corolla upper lip straight or slightly arcuate; stamens exserted 2
2 Corolla upper lip with a tuft of hairs at the tip .... **M. fistulosa**
2 Corolla upper lip uniformly pubescent, not tufted at the tip 3
3 Corolla brilliant crimson .... **M. didyma**
3 Corolla creamy or yellowish .... **M. clinopodia**

M. clinopodia L.; moist or dryish woods; common; Jun–Sep; ALL.
M. didyma L., Crimson B. B.; wet slopes and stream banks; occasional; Jul–Sep; ALL.
M. fistulosa L., Wild Bergamot; open woods and meadows; common; Jun–Sep; ALL. See Scora for a discussion of variability within this taxon.
M. punctata L.; sandy or rocky fields, forest edges; infrequent; Jul–Sep; ALL.

M. × *media* Willd., with reddish-purple corollas, is infrequently encountered in our area. It apparently is a variable group of hybrids and backcrosses of *M. didyma* with *M. fistulosa* or *M. clinopodia*.

Nepeta L. (Catnip)

N. cataria L.*; disturbed sites; occasional; Jun–Aug; NC, SC, TN, VA.

Perilla L. (Beefsteak Plant)

P. frutescens (L.) Britt.*; disturbed sites; frequent; Aug–Oct; ALL.

Physostegia Benth. (Obedient Plant)

P. virginiana (L.) Benth.; bogs and stream banks; occasional; Jul–Oct; NC, SC, TN, VA. Represented in our area by subsp. **praemorsa** (Shinners) Cantino. *P. praemorsa* Shinners, *Dracocephalum virginianum* L.

Prunella L. (Self Heal)

P. vulgaris L.*; disturbed sites; common; Apr–Oct; ALL.

Pycnanthemum Michx. (Mountain Mint)

Boomhour, E. G. 1941. A taxonomic study of the genus *Pycnanthemum*. Ph.D. thesis, Duke University.

Grant, E., and C. Epling. 1943. A study of *Pycnanthemum* (Labiatae). *Univ. Calif. Publ. Bot.* 20:195–240.

1 Stem faces glabrous, the angles glabrous or pubescent 2
 2 Leaves 1–12 mm wide 3
  3 Stems glabrous; leaves 1–4 mm wide . . . . **P. tenuifolium**
  3 Stems pubescent on the angles; leaves 1–12 mm wide . . . . **P. virginianum**
 2 Leaves 15–30 mm wide . . . . **P. montanum**
1 Stem faces and angles pubescent 4
 4 Shortest calyx lobes 1.2–4.5 mm long . . . . **P. flexuosum**
 4 Shortest calyx lobes less than 1.1 mm long 5
  5 Calyx nearly regular, the teeth of essentially uniform length 6
   6 Lower surface of leaves canescent, sharply contrasting with the darker upper surfaces . . . . **P. curvipes**
   6 Lower surface of leaves glabrous to pilose, similar in color to the upper surfaces 7
    7 Leaves ovate to lanceolate, 2–3× as long as wide, usually more than 1.5 cm wide . . . . **P. muticum**
    7 Leaves narrowly ovate to lanceolate, 3–6.5× as long as wide, usually less than 1.5 cm wide . . . . **P. verticillatum**
  5 Calyx bilabiate, the upper tooth about 1/2 or less the length of the lower teeth . . . 8
   8 Lower surfaces of leaves glabrate to pilose, similar in color to the upper surfaces . . . . **P. beadlei**
   8 Lower surfaces of leaves canescent, sharply contrasting with the darker upper surfaces 9
    9 Calyx lobes acute to subacuminate, the longest about 1/3 the length of the calyx tube . . . . **P. incanum**
    9 Calyx lobes acuminate to subulate, the longest about 1/2 the length of the calyx tube 10
     10 Stems and lower leaf surfaces canescent, the short trichomes frequently mixed with longer spreading ones . . . . **P. pycnanthemoides** var. **pycnanthemoides**
     10 Stems and lower leaf surfaces bearing coarse, spreading trichomes only . . . . **P. pycnanthemoides** var. **viridifolium**

P. beadlei Fern.; rocky open woods; infrequent; Aug–Sep; GA, NC.

P. curvipes (Greene) Grant & Epling; rocky open woods and roadsides; infrequent; Jun–Aug; GA, TN.

P. flexuosum (Walt.) BSP.; moist woods and open areas; infrequent; Jul–Sep; NC. *P. hyssopifolium* Benth.

P. incanum (L.) Michx.; upland woods and forest edges; infrequent; Aug–Sep; VA.

P. montanum Michx.; rich woods; occasional; Jul–Aug; GA, NC, SC, TN.

P. muticum (Michx.) Pers.; bogs, meadows, wet woods; occasional; Jun–Aug; NC, SC, TN, VA.

P. pycnanthemoides (Leavenw.) Fern. var. **pycnanthemoides**; forests, old fields, roadsides; frequent; Jul–Aug; ALL. *P. loomisii* Nutt., *P. tullia* Benth.

P. pycnanthemoides (Leavenw.) Fern. var. **viridifolium** Fern.; forests and thickets; infrequent; Jul–Aug; NC, SC, VA.

P. tenuifolium Schrader; bogs, meadows, and pastures; frequent; Jun–Jul; ALL.

P. verticillatum (Michx.) Pers.; upland forests and forest edges; occasional; Jul–Sep; NC, SC, VA.

P. virginianum (L.) Durand & Jackson; moist open areas and forest edges; infrequent; Jun–Aug; NC, TN, VA.

Taxonomy of the *P. incanum-pycnanthemoides* complex is currently unclear. Several previously recognized species (see synonymy) appear to be based on inconsequential characters and are not accepted in the present, fairly conservative, treatment. Additional study may indicate that an even more conservative approach is warranted: recognizing only one species, *P. incanum, sensu lato*.

Salvia L. (Sage)

1 Leaves mostly basal, lobed to divided . . . . **S. lyrata**

1 Leaves cauline, toothed . . . . **S. urticifolia**

S. lyrata L., Lyre-leaved S.; disturbed sites; common; Apr–Jun; ALL.

S. urticifolia L., Nettle-leaved S.; dry woodlands, often on basic soils; frequent; Apr–Jun; NC, SC, TN, VA.

Satureja L.

1 Calyx hirsute externally; flowers numerous in dense, terminal, and subterminal glomerules . . . . **S. vulgaris**

1 Calyx puberulent externally; flowers few to several in loose cymules . . **S. calamintha**

S. calamintha (L.) Scheele*; upland pastures, open woods, and roadsides; occasional; Jul–Sep; TN, VA. Represented in our area by var. **nepeta** (L.) Briquet. *S. nepeta* Fritsch, *S. nepeta* (L.) Scheele, *Calamintha nepeta* (L.) Savi.

S. vulgaris (L.) Fritsch; moist open woods and roadsides; occasional; Jul–Oct; NC, TN, VA.

Scutellaria L. (Skullcap)

Collins, J. L. 1976. A revision of the annulate *Scutellaria* (Labiatae). Ph.D. diss., Vanderbilt University.

Pittman, A. B. 1987. Systematic studies in *Scutellaria* sect. *Mixtae* (Labiatae). Ph.D. diss., Vanderbilt University.

1 Corolla with a ridge or ring of trichomes within the tube at the level of the calyx opening 2

  2 Leaves entire at midstem or above . . . . **S. integrifolia**

  2 Leaves crenate or serrate throughout 3

    3 Corolla lower lip with a white central band bisected by a deep blue stripe; leaves pilose, occasionally sparingly so; corollas 1.4–1.9 cm long 4

      4 Stems pilose throughout or from midstem downward with coarse, usually glandular trichomes 1–2 mm long . . . . **S. elliptica** var. **hirsuta**

      4 Stems with short (usually 1 mm or less long), upwardly curled, eglandular or glandular trichomes throughout or with spreading trichomes basally . . . . **S. elliptica** var. **elliptica**

    3 Corolla lower lip with an unbisected white central band; leaves glabrate or essentially so; corollas 1.7–3.2 cm long 5

      5 Stems glabrous or essentially so below the inflorescence . . . . **S. serrata**

      5 Stems variously pubescent below the inflorescence 6

6 Corolla 2.4–3.2 cm long .... **S. pseudoserrata**
6 Corolla 1.7–2.2 cm long .... **S. incana**
1 Corolla without a ring of trichomes within the corolla tube 7
7 Leaves sessile or essentially so 8
8 Rhizomes uniform, not constricted; leaves 2 cm or more long ...... **S. nervosa**
8 Rhizomes constricted into beadlike segments; leaves less than 2 cm long 9
9 Stems essentially glabrous, minutely pubescent on the angles with ascending eglandular trichomes .... **S. leonardii**
9 Stems hirtellous, the trichomes gland-tipped .... **S. parvula**
7 Leaf petioles 5 mm or more long 10
10 Corolla less than 1 cm long .... **S. lateriflora**
10 Corolla more than 1 cm long 11
11 Stems glabrous or sparsely pubescent with appressed ascending trichomes .... **S. saxatilis**
11 Stems conspicuously pubescent with spreading or retrorse trichomes 12
12 Corolla lower lip flabelliform with large lateral auricles; sessile floral bracts as long as or longer than calyces, usually more than 8 mm long .... **S. ovata** subsp. **bracteata**
12 Corolla lower lip without large lateral auricles; sessile floral bracts usually shorter than their calyces (except in late-summer-flowering specimens in which the new bracts appear to expand into leaves) 13
13 Lower lip uniformly white or with a few scattered blue spots; leaves sparsely capitate-glandular; largest leaf blades usually more than 6 cm long; plants usually more than 4 dm tall ..... **S. ovata** subsp. **ovata**
13 Lower lip light to dark blue with 2 prominent white bands often marked with spots; leaves often densely capitate-glandular; largest leaf blades usually less than 6 cm long; plants usually less than 4 dm tall 14
14 Leaf margins prominently dentate .... **S. arguta**
14 Leaf margins crenate or shallowly dentate .... **S. ovata** subsp. **rugosa**

S. arguta Buckley; moist talus slopes at high elevations; infrequent; Jul–Aug; NC, TN. Pittman considers this taxon to be closely related to *S. ovata* subsp. *rugosa*.

S. elliptica Muhl. var. **elliptica**; mesic to well-drained woods; frequent; May–Jun; ALL.

S. elliptica Muhl. var. **hirsuta** (Short) Fern.; mesic to slightly xeric woods; frequent; May–Jun; GA, NC, TN, VA.

S. incana Biehler; mesic or dry, gravelly or sandy forest margins and open woods; occasional; Jul–Aug; GA, NC, SC, TN. Represented in our area by var. **punctata** (Chapm.) Mohr.

S. integrifolia L.; moist open areas; occasional; Apr–Jun; ALL.

S. lateriflora L.; moist to wet open areas; infrequent; Jul–Sep; ALL.

S. leonardii Fern.; dry woods and openings; infrequent; May–Jun; VA.

S. nervosa Pursh; alluvial woods; infrequent; May–Jun; VA.

S. ovata Hill subsp. **bracteata** (Benth.) Epling; ledges, bluffs, oak-hickory woods; infrequent; May–Aug; GA, NC, SC.

S. ovata Hill subsp. **ovata**; rich mesic uplands; infrequent; May–Sep; TN.

S. ovata Hill subsp. **rugosa** (Wood) Epling; dry, rocky, open areas and oak-hickory woodlands; infrequent; Jun–Sep; VA.

S. parvula Michx.; dry, open, usually calcareous areas; infrequent; May–Jun; VA.
S. pseudoserrata Epling; mesic, usually rocky woods; infrequent; May–Jun; GA, TN.
S. saxatilis Riddell; rocky, dry to mesic wooded or open areas; infrequent; May–Aug; GA, NC, TN, VA.
S. serrata Andr.; mesic woods; occasional; May–Jun; NC, TN, VA.

Stachys L. (Hedge Nettle)

Nelson, J. B. 1981. *Stachys* (Labiatae) in southeastern United States. *Sida* 9:104–123.

1 Leaves sessile or essentially so, petioles less than 1/6 the length of the longest blades .. 2
  2 Plants slender, glabrous to hispid; lower stem sides usually not glandular; leaves linear to narrowly elliptic, usually less than 1 cm wide .... **S. hyssopifolia**
  2 Plants robust, pubescent, the upper stem sides mostly glandular; leaves ovate-lanceolate to lanceolate, more than 1 cm wide .... **S. eplingii**
1 Leaves with distinct petioles 3
  3 Upper stem sides glabrous; leaf bases truncate to rounded, the blades elliptic to lanceolate 4
    4 Calyx lobes about as long as the calyx tube; stem angles glabrous to bristly .... **S. tenuifolia** var. **tenuifolia**
    4 Calyx lobes shorter than the calyx tube; stem angles glabrous to roughened .... **S. tenuifolia** var. **latidens**
  3 Upper stem sides slightly pubescent; leaf bases mostly cordate, the blades ovate ... 5
    5 Inflorescence lax, sometimes nodding; bracts abruptly reduced above the first fertile node .... **S. nuttallii**
    5 Inflorescence stiff; bracts gradually reduced upward .... **S. clingmannii**

S. clingmannii Small; rich coves and clearings, often at high elevations; infrequent; Jul–Aug; NC, SC, TN.
S. eplingii J. Nelson; woodlands, bogs, and meadows; infrequent; Jun–Aug; NC, VA.
S. hyssopifolia Michx.; moist clearings; infrequent; Jun–Aug; NC, VA.
S. nuttallii Shuttlew. ex Benth.; moist woodlands and clearings at lower elevations; occasional; Jun–Aug; NC, TN, VA. *S. riddellii* House.
S. tenuifolia Willd. var. **latidens** (Small) J. Nelson; woodlands, clearings, and low meadows; frequent; Jun–Aug; ALL. *S. latidens* Small.
S. tenuifolia Willd. var. **tenuifolia**; damp, open woodlands and clearings; infrequent; Jun–Aug; GA, NC, SC, TN.

Synandra Nutt.

S. hispidula (Michx.) Baill.; rich woods, often on basic soils; infrequent; Apr–May; NC, TN, VA.

Teucrium L.

T. canadense L.; moist woods and meadows; common; Jun–Aug; All.

Trichostema L.

Lewis, H. 1945. A revision of the genus *Trichostema*. *Brittonia* 5:276–303.

1 Calyx actinomorphic .... **T. brachiatum**

1 Calyx zygomorphic 2
  2 Leaves linear, at least 6× longer than wide .... **T. setaceum**
  2 Leaves oval or oblong, not more than 5× longer than wide .... **T. dichotomum**

T. brachiatum L., False Pennyroyal; dry basic soils; infrequent; Aug–Sep; NC, VA. *Isanthus brachiatus* (L.) BSP.

T. dichotomum L., Blue Curls; well-drained woods, old fields, roadsides; frequent; Aug–Oct; ALL.

T. setaceum Houtt., Blue Curls; well-drained woods; infrequent; Aug–Oct; NC, SC, VA.

LAURACEAE (Laurel Family)

1 Leaves entire; inflorescence subumbellate; fruit red, cupule absent .... **Lindera**
1 Leaves entire, mitten-shaped or 3-lobed; inflorescence racemose; fruit dark blue on an enlarged red cupule .... **Sassafras**

Lindera Thunb. (Spicebush)

L. benzoin (L.) Blume; stream banks and mesic woods; common; Mar–Apr; ALL.

Sassafras Trew (Sassafras)

S. albidum (Nutt.) Nees; woods, fields, and disturbed sites; common; Mar–Apr; ALL.

LENTIBULARIACEAE (Bladderwort Family)

Utricularia L. (Bladderwort)

Crow, G. E., and C. B. Hellquist. 1985. *Aquatic vascular plants of New England:* Part 8. *Lentibulariaceae*. New Hampshire Agric. Exp. Sta. Bull. no. 528.

1 Flowering stems with a whorl of 3–10 inflated, floating petioles .... **U. radiata**
1 Flowering stems without inflated, floating petioles 2
  2 Plants terrestrial or amphibious; leaves simple 3
    3 Bract peltate, bractlets absent; flowers pedicellate .... **U. subulata**
    3 Bract not peltate, bractlets 2; flowers subsessile .... **U. cornuta**
  2 Plants aquatic; leaves dissected 4
    4 Leaflike segments flat; corolla spur less than 3 mm long .... **U. minor**
    4 Leaflike segments round; corolla spur 3 mm or more long 5
      5 Leaflike branches divided 6× or more; plants robust, 3–10 dm long .... **U. geminiscapa**
      5 Leaflike branches divided 1–3×; plants small, less than 3 dm long 6
        6 Spur much shorter than lower lip of corolla; leaves with 2 segments .... **U. gibba**
        6 Spur nearly as long as lower lip of corolla; leaves with 3–many segments .... **U. biflora**

U. biflora Lam.; pools, swamps, lakes; infrequent; Jun–Oct; GA, NC, SC.

U. cornuta Michx.; bogs, ditches, and shorelines; infrequent; Jun–Sep; NC, SC.

U. geminiscapa Benjamin; lakes; infrequent; Jun–Sep; VA.

U. gibba L.; pools, swamps, lakes; infrequent; Jun–Oct; NC, VA.

U. minor L.; bogs and shorelines; infrequent; Jun–Sep; NC.

U. radiata Small; pools and lakes; infrequent; Jun–Sep; VA.
U. subulata L.; bogs, ditches, and shorelines; infrequent; Jun–Sep; NC, SC.

These carnivorous species are primarily Coastal Plain in distribution and are rarely encountered in our area. Several complexes are polymorphic and in need of further investigations.

LIMNANTHACEAE (False Mermaid Family)

Floerkea Willd. (False Mermaid)

F. proserpinacoides Willd.; moist woods; infrequent; Apr–Jun; VA.

LINACEAE (Flax Family)

Linum L. (Flax)

Rogers, C. M. 1963. Yellow flowered species of *Linum* in eastern North America. *Brittonia* 15:97–122.

1 Styles partially united . . . . **L. sulcatum**
1 Styles distinct 2
  2 Septa of fruit ciliate, sometimes sparsely so 3
    3 Capsule ovoid, pointed at the apex . . . . **L. intercursum**
    3 Capsule subglobose, indented at the apex . . . . **L. virginianum**
  2 Septa not conspicuously ciliate 4
    4 Margin of inner sepals with stalked glands; mature fruit in dried specimens usually adhering to the plant . . . . **L. medium**
    4 Margin of inner sepals glandless or with small sessile glands; mature fruit in dried specimens usually falling from the plant 5
      5 Branchlets round or nearly so; inflorescence corymbose . . . . . **L. virginianum**
      5 Branchlets prominently striate; inflorescence paniculate . . . . **L. striatum**

L. intercursum Bickn.; moist soil of open woodlands; occasional; Jun–Sep; ALL.
L. medium (Planch.) Britt.; open fields and mesic sites; occasional; Jun–Sep; ALL. Represented in our area by var. **texanum** (Planch.) Fern.
L. striatum Walt.; moist woods, roadside ditches; common; Jun–Sep; ALL.
L. sulcatum Ridd.; dry, often basic soils; infrequent; Jun–Sep; VA.
L. virginianum L.; fields and open woods; common; Jun–Sep; ALL.

LOGANIACEAE (Logania Family)

1 Woody vines . . . . **Gelsemium**
1 Herbs 2
  2 Corolla scarlet outside, yellow within, 3–5 cm long . . . . **Spigelia**
  2 Corolla white, less than 5 mm long . . . . **Polypremum**

Gelsemium Juss. (Yellow Jessamine)

G. sempervirens (L.) St.-Hil.; clearings, thickets, and open woods; infrequent; Mar–May; SC.

Polypremum L.

P. procumbens L.; fields, roadsides, and waste places; infrequent; Jul–Oct; NC, SC, VA.

Spigelia L. (Indian Pink)

S. marilandica L.; rich woods, often on basic soils; occasional; May–Jun; GA, SC, TN.

LYTHRACEAE (Loosestrife Family)

Graham, S. 1975. Taxonomy of the Lythraceae in the southeastern United States. *Sida* 6:80–103.

1 Floral tube campanulate to globose, about as long as wide 2
  2 Plant woody, more than 1 m tall .... **Decodon**
  2 Plant herbaceous, less than 1 m tall 3
    3 Flower 1/leaf axil; capsule septicidal .... **Rotala**
    3 Flowers (1–) 3–many/leaf axil; capsule dehiscing irregularly ...... **Ammannia**
1 Floral tube cylindric, about twice as long as wide 4
  4 Flowers zygomorphic; stems glandular-pubescent .... **Cuphea**
  4 Flowers actinomorphic; stems glabrous or with nonglandular hairs ..... **Lythrum**

Ammannia L.

1 Style equal to or longer than the ovary; calyx lobes triangular .... **A. coccinea**
1 Style much shorter than the ovary; calyx lobes obtuse .... **A. latifolia**

A. coccinea Rottb.; marshes and wet places; infrequent; Jun–Sep; NC, SC, VA.
A. latifolia L.; slow streams; infrequent; Jul–Sep; VA. *A. teres* Raf.

Cuphea P. Br. (Blue Waxweed)

C. viscosissima Jacq.; pastures and disturbed places; frequent; Jun–Sep; GA, NC, TN, VA. *C. petiolata* (L.) Koehne.

Decodon J. F. Gmelin (Swamp Loosestrife)

D. verticillatus (L.) Ell.; swamps and margins of shallow waters; infrequent; Jun–Sep; VA.

Lythrum L. (Loosestrife)

1 Flowers solitary or paired in the axils; stamens usually 6 .... **L. alatum**
1 Flowers numerous in showy terminal spikes; stamens usually 12 .... **L. salicaria**

L. alatum Pursh; damp open sites; infrequent; May–Sep; GA, VA.
L. salicaria L.*; margins of marshes, lakes, and rivers; infrequent; Jun–Sep; NC, TN, VA.

Rotala L. (Tooth Cup)

R. ramosior (L.) Koehne; damp depressions; occasional; Jun–Sep; NC, SC, TN, VA.

MAGNOLIACEAE (Magnolia Family)

Hardin, J. W. 1972. Studies of the southeastern United States flora. III. Magnoliaceae and Illiciaceae. *J. Elisha Mitchell Sci. Soc.* 88:30–32.

1 Leaves entire, with or without basal auricles . . . . **Magnolia**
1 Leaves 4-lobed, apex notched . . . . **Liriodendron**

Liriodendron L. (Tulip Poplar)

L. tulipifera L.; coves and rich woods; common; Apr–May; ALL.

Magnolia L.

1 Leaves evergreen, aromatic . . . . **M. virginiana**
1 Leaves deciduous, nonaromatic 2
  2 Leaves auriculate or deeply cordate at base 3
    3 Leaves whitened beneath; buds and twigs pubescent . . . . **M. macrophylla**
    3 Leaves green beneath; buds and twigs glabrous . . . . **M. fraseri**
  2 Leaves cuneate to slightly cordate at base 4
    4 Leaves clustered at tips of branches, base long cuneate; petals more than 8 cm long; buds glabrous . . . . **M. tripetala**
    4 Leaves scattered, base slightly cordate; petals less than 8 cm long; buds pubescent . . . . **M. acuminata**

M. acuminata L., Cucumber Tree; coves and rich woods; common; Apr–May; ALL.

M. fraseri Walt., Fraser Magnolia; coves and rich woods; frequent; Apr–May; ALL.

M. macrophylla Michx., Bigleaf Magnolia; alluvial woods and valleys; infrequent; Apr–May; TN. This tree produces our largest simple leaf.

M. tripetala L., Umbrella Magnolia; coves and rich woods; frequent; Apr–May; GA, NC, TN, VA.

M. virginiana L., Sweet Bay; swamps and low woods; infrequent; Apr–May; NC, TN, VA. More common on the Coastal Plain.

MALVACEAE (Mallow Family)

1 Flowers subtended by involucral bracts 2
  2 Petals 5–8 cm long; calyx subtended by 5 or more bracts . . . . **Hibiscus**
  2 Petals less than 5 cm long; calyx subtended by 3 bracts 3
    3 Carpels 1-seeded . . . . **Malva**
    3 Carpels 2–4-seeded . . . . **Iliamna**
1 Flowers not subtended by involucral bracts 4
  4 Carpels 1-seeded . . . . **Sida**
  4 Carpels 3–9-seeded . . . . **Abutilon**

Abutilon Miller (Velvet Leaf)

A. theophrasti Medicus*; fields and disturbed sites; occasional; Jun–Oct; ALL.

Hibiscus L. (Mallow)

1 Shrubs or small trees . . . . **H. syriacus**
1 Annual or perennial herbs 2
  2 Annuals; fruiting calyx inflated . . . . **H. trionum**
  2 Perennials; fruiting calyx not inflated 3
    3 Leaves glabrous beneath . . . . **H. laevis**
    3 Leaves soft-pubescent beneath 4

4 Petals usually white and with a red-purple base; peduncles usually with a leaf near the middle .... **H. moscheutos** subsp. **moscheutos**
4 Petals usually pink, with or without a red-purple base; peduncles usually leafless .... **H. moscheutos** subsp. **palustris**

H. laevis All., Halberd-leaved Marsh M.; open river and stream banks; infrequent; Jun–Aug; VA. *H. militaris* Cavanilles.
H. moscheutos L. subsp. **moscheutos**, Rose M.; moist woods and meadows; occasional; Jun–Sep; NC, SC, TN, VA.
H. moscheutos L. subsp. **palustris** (L.) Clausen; marshes; infrequent; Jun–Sep; NC.
H. syriacus L.*, Rose of Sharon; persistent after cultivation; Jun–Sep; NC, SC, TN, VA.
H. trionum L.*; fields and roadsides; occasional; Jun–Oct; NC, TN, VA.

Iliamna Greene

I. remota Greene; dry woods; infrequent; Jun–Aug; VA.

Malva L. (Mallow)

1 Flowers 2.5 cm or more broad; plants erect 2
2 Leaves deeply divided; petals 5–8× as long as the calyx .... **M. moschata**
2 Leaves lobed; petals up to 4× as long as the calyx .... **M. sylvestris**
1 Flowers up to 1.5 cm broad; plants decumbent .... **M. neglecta**

M. moschata L.*, Musk M.; fields and roadsides; infrequent; May–Aug; NC, TN, VA.
M. neglecta Wallroth*, Common M.; disturbed sites; frequent; Apr–Oct; ALL.
M. sylvestris L.*, High M.; disturbed sites; infrequent; May–Jul; SC, TN, VA.

Sida L.

1 Leaves 1–2 dm long, 3–7 lobed; petals white .... **S. hermaphrodita**
1 Leaves up to 6 cm long; petals pale yellow 2
2 Peduncles about as long as the subtending petiole .... **S. spinosa**
2 Peduncles 2× or more longer than the subtending petiole .... **S. rhombifolia**

S. hermaphrodita (L.) Rusby; loose, rocky or sandy sites; very rare and local; Jul–Aug; TN?, VA.
S. rhombifolia L.*; disturbed sites; infrequent; Jun–Oct; SC, TN.
S. spinosa L.*; disturbed sites; frequent; Jun–Oct; ALL.

MELASTOMATACEAE (Melastoma Family)

Rhexia L. (Meadow Beauty)

Kral, R., and P. E. Bostick. 1969. The genus *Rhexia* (Melastomataceae). *Sida* 6:387–440.

1 Stem faces at midstem nearly equal, the faces flattened 2
2 Midstem angles conspicuously winged .... **R. virginica**
2 Midstem angles wingless or essentially so .... **R. mariana** var. **ventricosa**
1 Stem faces unequal, the broader pair convex, the narrower pair concave .... **R. mariana** var. **mariana**

R. mariana L. var. **mariana**; open, usually damp, sandy soils; common; May–Sep; ALL.
R. mariana L. var. **ventricosa** (Fern. & Grisc.) Kral & Bostick; sandy, wet clearings; infrequent and more common on the Piedmont and Coastal Plain; May–Sep; VA.
R. virginica L.; open, usually damp, sandy soils; common; May–Sep; ALL.

MENISPERMACEAE (Moonseed Family)

1 Stamens 6; fruit red; petiole attached to edge of leaf blade .... **Cocculus**
1 Stamens 12–24; fruit black; petiole attached to the underside of leaf blade .... **Menispermum**

Cocculus DC. (Coralbeads)

C. carolinus (L.) DC.; thickets and woodland margins; occasional; Jun–Aug; NC, SC, TN.

Menispermum L. (Moonseed)

M. canadense L.; rich thickets and low woods; common; Jun–Jul; ALL.

MENYANTHACEAE (Buckbean Family)

Menyanthes L. (Buckbean)

M. trifoliata L.; bogs and marshes; infrequent; Apr–Jun; NC, VA.

MOLLUGINACEAE (Indian Chickweed Family)

Mollugo L. (Carpetweed)

M. verticillata L.*; common in waste areas; Jun–Oct; NC, SC, TN, VA.

MORACEAE (Mulberry Family)

1 Leaves entire; branches sometimes spiny .... **Maclura**
1 Leaves toothed or lobed or both; spines absent 2
  2 Twigs and petioles soft pubescent .... **Broussonetia**
  2 Twigs and petioles glabrous or soft pubescent .... **Morus**

Broussonetia L'Her. (Paper Mulberry)

B. papyrifera (L.) Vent.*; spreading from waste places; occasional; Apr–May; GA, NC, VA. Apparently not fruiting in our area.

Maclura Nutt. (Osage Orange)

M. pomifera (Raf.) Schneider*; old fields, fencerows, and waste areas on basic soils; occasional; Apr–May; NC, SC, TN. Native of the southcentral U. S. and naturalized in our area.

Morus L. (Mulberry)

1 Leaves glabrous beneath or pubescent only on the main veins .... **M. alba**
1 Leaves pubescent throughout the lower surface .... **M. rubra**

M. alba L.*, White M.; waste places; occasional; Apr–May; NC, SC, TN, VA.
M. rubra L., Red M.; low rich woods and disturbed sites; common; Apr–May; ALL.

MYRICACEAE (Wax Myrtle Family)

1 Leaves pinnatifid; bracts 6–8, longer than the fruit .... **Comptonia**
1 Leaves entire or variously toothed; bracts 2–4 and shorter than the fruit or absent .... **Myrica**

Comptonia L'Her. (Sweet Fern)

C. peregrina (L.) Coult.; dry woodlands, clearings, and roadsides; frequent; Apr–May; GA, NC, SC, VA.

Myrica L.

1 Fruits covered with a white wax; leaves mostly more than 1.5 cm wide .... **M. heterophylla**
1 Fruits not covered with a white wax; leaves mostly less than 1.5 cm wide .... **M. gale**

M. gale L., Sweet Gale; bogs; infrequent; Apr–May; NC. *Gale palustris* (Lam.) Chev.
M. heterophylla Raf., Bayberry; bogs and wet woods; infrequent; Apr–May; SC.

NELUMBONACEAE (Lotus Family)

Nelumbo Adanson (Lotus, Yellow Nelumbo)

N. lutea (Willd.) Pers.; ponds and slow streams; infrequent; Jun–Aug; VA.

NYCTAGINACEAE (Four-O'Clock Family)

Mirabilis L. (Four-O'Clock)

M. nyctaginea (Michx.) MacM.*; weedy sites, especially along railroad embankments; occasional; Jun–Oct; NC, VA.

NYMPHAEACEAE (Water Lily Family)

1 Flowers white to pink; sepals 4 .... **Nymphaea**
1 Flowers yellowish; sepals 6–9 .... **Nuphar**

Nuphar Sm. (Spatter Dock)

N. luteum (L.) Sibth. & Sm.; ponds, slow streams, and marshes; occasional; Apr–Oct; NC, SC, TN, VA. Represented in our area by subsp. **macrophyllum** (Small) Beal.

Nymphaea L. (Water Lily)

N. odorata Ait.; ponds, pools, and lakes; infrequent; Jun–Sep; NC, SC, TN, VA.

NYSSACEAE (Sour Gum Family)

Nyssa L. (Gum)

1 Leaf blade obovate to elliptic; fruiting peduncle 3–6 cm long; occurring on well-drained sites . . . . **N. sylvatica** var. **sylvatica**
1 Leaf blade oblanceolate; fruiting peduncle 1–3 cm long; occurring in wetlands . . . . **N. sylvatica** var. **biflora**

N. sylvatica Marsh. var. **biflora** (Walt.) Sarg., Swamp Tupelo; wet woods and swamps; infrequent and more typical of Coastal Plain swamps; Apr–May; SC. Trunk bases usually swollen when occurring on frequently inundated sites.

N. sylvatica Marsh. var. **sylvatica**, Black Gum; moist to dry woodlands, old fields; common; Apr–May; ALL.

OLEACEAE (Olive Family)

Hardin, J. W. 1974. Studies of the southeastern United States flora. IV. Oleaceae. *Sida* 5:274–285.

1 Leaves pinnately compound; fruit a samara; petals absent . . . . **Fraxinus**
1 Leaves simple; fruit a drupe or capsule; petals present 2
  2 Flowers yellow . . . . **Forsythia**
  2 Flowers white 3
    3 Corolla lobes linear, united at base only; inflorescence drooping . . **Chionanthus**
    3 Corolla funnelform; inflorescence erect or spreading . . . . **Ligustrum**

Chionanthus L. (Fringe Tree)

C. virginicus L.; dry rocky slopes and sandy stream banks; frequent; Apr–May; ALL.

Forsythia Vahl (Golden Bells)

F. viridissima Lindl.*; often persistent in waste areas; Mar–Apr; ALL.

Fraxinus L. (Ash)

1 Lateral leaflets sessile . . . . **F. nigra**
1 Lateral leaflets stalked 2
  2 Leaflets papillose beneath, margin entire or nearly so; wing of samara terminal or only slightly decurrent along upper 1/3 of the body . . . . **F. americana**
  2 Leaflets not papillose beneath, margin often serrate; wing of samara decurrent to near middle of body or beyond 3
    3 Lowermost leaflets decurrent along most of the leaflet stalk; samara wing usually less than 7 mm wide, the body less than 2 mm wide . . . . **F. pennsylvanica**
    3 Lowermost leaflets wingless at base or nearly so; samara wing more than 7 mm wide, the body more than 2 mm wide . . . . **F. profunda**

F. americana L., White A.; rich woods and dry slopes; common; Apr–May; ALL. Several varieties, including var. *biltmoreana* (Beadle) J. Wright ex Fern., and segregate species belong to this complex and are in need of further study.

F. nigra Marsh., Black A.; wet woods and swamps; infrequent; Apr–May; VA.

F. pennsylvanica Marsh., Green A.; moist soils along streams; occasional; Apr–May; GA, TN, VA. *F. pennsylvanica* Marsh. var. *subintegerrima* (Vahl) Fern.

F. profunda (Bush) Bush, Pumpkin A.; swamps and river bottoms; infrequent and more common on the Coastal Plain; Apr–May; NC. *F. tomentosa* Michx. f.

Ligustrum L. (Privet)

L. sinense Lour.*; often persistent in waste places; Apr–May; ALL. This is our only privet that appears to be somewhat naturalized.

ONAGRACEAE (Evening Primrose Family)

1 Petals 2, deeply divided; stamens 2 .... **Circaea**
1 Petals 4 or absent; stamens 4–12 2
 2 Fruit indehiscent, 1–4-seeded .... **Gaura**
 2 Fruit dehiscent, many-seeded 3
  3 Petals pink or white 4
   4 Leaves entire or serrate; seeds comose .... **Epilobium**
   4 Leaves irregularly lobed (at least basally) to pinnatifid; seeds without a coma .... **Oenothera**
  3 Petals yellow or absent 5
   5 Calyx tube extended beyond the summit of the ovary .... **Oenothera**
   5 Calyx tube not extended beyond the summit of the ovary .... **Ludwigia**

Circaea L. (Enchanter's Nightshade)

1 Mature plants rarely more than 2.5 dm tall; fruits not furrowed; petiole narrowly wing-margined when pressed and dried .... **C. alpina**
1 Mature plants usually more than 2.5 dm tall; fruits furrowed; petiole not wing-margined when pressed and dried .... **C. lutetiana**

C. alpina L.; moist woods at high elevations; occasional; Jun–Sep; GA, NC, TN, VA.

C. lutetiana (L.) A. & M.; rich woodlands; frequent; Jun–Aug; ALL. Represented in our area by subsp. **canadensis** (L.) A. & M.

C. × *intermedia* Ehrh., a sterile hybrid, has been reported from NC and VA.

Epilobium L.

1 Petals 1 cm or more long, entire; stigma 4-lobed .... **E. angustifolium**
1 Petals less than 1 cm long, notched at apex; stigma not lobed 2
 2 Leaves linear, entire, margins revolute .... **E. leptophyllum**
 2 Leaves lanceolate, serrate, margins not revolute 3
  3 Seeds beakless; petals pink; coma tan .... **E. coloratum**
  3 Seeds short beaked; petals whitish; coma whitish .... **E. ciliatum**

E. angustifolium L., Fireweed; clearings and recently burned areas; occasional; Jul–Aug; NC, TN, VA.

E. ciliatum Raf.; moist open sites; occasional; Jun–Sep; NC, TN, VA. Closely related to and perhaps not distinct from the following taxon.
E. coloratum Biehler; moist open areas; frequent; Jun–Sep; ALL.
E. leptophyllum Raf.; bogs; infrequent; Jun–Sep; NC, TN, VA.

Gaura L.

1 Fruits sessile or essentially so . . . . **G. biennis**
1 Fruits with pedicels 2–4 mm long . . . . **G. filipes**

G. biennis L.; roadsides and waste places; frequent; Jun–Oct; NC, SC, TN, VA.
G. filipes Spach.; roadsides and waste places; infrequent; Jun–Oct; TN.

Ludwigia L.

1 Leaves opposite; petals absent . . . . **L. palustris**
1 Leaves alternate; petals present 2
  2 Stamens 8–10 3
    3 Flowers 4-merous; leaves decurrent . . . . **L. decurrens**
    3 Flowers 5(6)-merous; leaves not decurrent 4
      4 Stems erect . . . . **L. leptocarpa**
      4 Stems, especially the lower portion, decumbent and rooting at the nodes . . . 5
        5 Bractlets ovate or deltoid; flowering stems usually slightly ascending; stems and leaves glabrous or nearly so . . . . **L. peploides**
        5 Bractlets lanceolate; flowering stems more or less erect; stems and leaves sparsely to densely pubescent . . . . **L. uruguayensis**
  2 Stamens 4 6
    6 Leaf bases cuneate; capsule glabrous or sparsely pubescent . . . . . **L. alternifolia**
    6 Leaf bases rounded; capsule hirsute . . . . **L. hirtella**

L. alternifolia L., Rattlebox; along streams and lowlands; common; May–Oct; ALL.
L. decurrens Walt.; swamps and wet clearings; occasional; Jun–Oct; NC, SC, TN, VA.
L. hirtella Raf.; wet ditches and bogs; infrequent; Jun–Sep; NC.
L. leptocarpa (Nutt.) Hara.; marshes and ditches; infrequent; May–Sep; SC.
L. palustris (L.) Ell.; shorelines and wet clearings; frequent; May–Oct; ALL.
L. peploides (HBK.) Raven; possibly introduced to wet clearings, shorelines, and sometimes as floating mats; infrequent; May–Sep; TN. *L. peploides* (HBK.) Raven var. *glabrescens* (Kuntze) Shinners.
L. uruguayensis (Camb.) Hara.; ponds and sluggish streams; infrequent; May–Sep; SC.

Oenothera L. (Evening Primrose)

Munz, P. A. 1965. Onagraceae. *N. Amer. Fl.* II. 5:1–278.
Straley, G. B. 1977. Systematics of *Oenothera* sect. *Kneiffia* (Onagraceae). *Ann. Missouri Bot. Gard.* 64:381–424.

1 Flowers pink or white . . . . **O. speciosa**
1 Flowers yellow 2
  2 Capsules elongate, subcylindric, not angled 3
    3 Stem leaves sinuate-pinnatifid . . . . **O. laciniata**
    3 Stem leaves entire to dentate 4
      4 Sepal tips subterminal, separated and divergent with a small auricle at the base at anthesis; young stem tips often recurved 5

5 Petals 3–4 cm long; anthers 8–12 mm long .... (**O. argillicola**)
5 Petals 1–2 cm long; anthers 4–7 mm long .... **O. parviflora**
4 Sepal tips terminal, without a small basal auricle at anthesis; young stem tips erect 6
6 Petals 2–2.5 cm long; sepal tips 3–4 mm long .... **O. austromontana**
6 Petals 0.8–2 cm long; sepal tips 1–2 mm long .... **O. biennis**
2 Capsules 4-angled, obovoid 7
7 Leaves linear, less than 1 mm wide .... **O. linifolia**
7 Leaves more than 1 mm wide 8
8 Petals 5–10 mm long; inflorescence usually nodding .... **O. perennis**
8 Petals 15–30 mm long; tip of inflorescence usually erect 9
9 Capsule widest above the middle; hairs of ovary and capsule mostly non-glandular .... **O. fruticosa** subsp. **fruticosa**
9 Capsule widest at the middle; hairs of ovary and capsule mostly glandular, or the ovary glabrous .... **O. fruticosa** subsp. **glauca**

O. argillicola Mackenzie; shale barrens; infrequent; Jul–Sep; VA.

O. austromontana (Munz) Raven, Dietrich & Stubbe; clearings at high elevations; infrequent; Jun–Sep; NC, TN, VA. *O. biennis* L. subsp. *austromontana* Munz.

O. biennis L.; disturbed sites; common; Jun–Oct; ALL.

O. fruticosa L. subsp. **fruticosa**; meadows and open woodlands; frequent; Apr–Aug; ALL.

O. fruticosa L. subsp. **glauca** (Michx.) Straley; meadows and open woodlands; frequent; May–Aug; ALL. *O. tetragona* Roth.

O. laciniata Hill; disturbed sites; occasional; Mar–Jul; ALL.

O. linifolia Nutt.; sandy lake margins; infrequent; May–Jun; TN.

O. parviflora L.; sandy clearings; infrequent; Jun–Sep; VA.

O. perennis L.; low woodlands; occasional; May–Aug; ALL.

O. speciosa Nutt.*; waste places; infrequent; May–Jun; NC, TN, VA.

OROBANCHACEAE (Broom Rape Family)

1 Stem branched .... **Epifagus**
1 Stem unbranched 2
2 Stem 1.5–3 cm thick, yellow-brown .... **Conopholis**
2 Stem less than 1 cm thick, bluish-white .... **Orobanche**

Conopholis Wallroth (Cancer/Squaw Root)

C. americana (L.) Wallroth; dry woods; common; Apr–May; ALL. A root parasite primarily on oaks.

Epifagus Nutt. (Beech Drops)

E. virginiana (L.) Barton; rich woods under beech; common; Aug–Nov; NC, SC, TN, VA. Parasitic on beech roots.

Orobanche L. (One-flowered Cancer Root)

O. uniflora L.; rich woods; frequent; Apr–May; ALL. Parasitic on various hosts.

OXALIDACEAE (Wood Sorrel Family)

Oxalis L. (Wood Sorrel)

Eiten, G. 1963. Taxonomy and regional variation of *Oxalis* section *Corniculata*. I. Introduction, keys, and synopsis of the species. *Amer. Midl. Naturalist* 69:257–309.

1 Plants acaulescent; flowers not yellow 2
  2 Petals rose-purple; sepals with a yellow gland at the tip . . . . . . **O. violacea**
  2 Petals white, veined with pink; sepals not glandular at the tip . . . . . . **O. acetosella**
1 Plants caulescent; flowers yellow 3
  3 Petals 12 mm or more long; leaves with a narrow reddish-brown margin . . . . **O. grandis**
  3 Petals 10 mm or less long; leaves without a colored margin 4
    4 Hairs of stems, petioles, and/or pedicels septate (sometimes mixed with nonseptate hairs); inflorescence cymose . . . . **O. stricta**
    4 Hairs of stems, petioles, and/or pedicels nonseptate; inflorescence umbellate 5
      5 Capsule pubescent . . . . **O. dillenii** subsp. **dillenii**
      5 Capsule glabrous or nearly so . . . . **O. dillenii** subsp. **filipes**

O. acetosella L.; rich moist woods, usually at high elevations; occasional; May–Jun; NC, TN, VA. *O. montana* Raf.

O. dillenii Jacq. subsp. **dillenii**; open disturbed sites; frequent; Mar–Sep; ALL. *O. florida* Salisb.

O. dillenii Jacq. subsp. **filipes** (Small) Eiten; open disturbed sites; frequent; Mar–Sep; GA, NC, VA. *O. florida* Salisb. var. *filipes* (Small) Ahles. Scarcely distinct from the preceding subspecies.

O. grandis Small; woodlands; frequent; May–Jun; ALL.

O. stricta L.; woodlands and disturbed sites; common; May–Sep; ALL. *O. europea* Jord.

O. violacea L.; woodlands; frequent; Apr–May; ALL.

PAPAVERACEAE (Poppy Family)

1 Petals 8 or more; flowers scapose, solitary; leaves usually 1 . . . . **Sanguinaria**
1 Petals 4; flowers on a leafy stem; leaves several 2
  2 Flowers red to white . . . . **Papaver**
  2 Flowers yellow 3
    3 Capsule bristly; style distinct; petals 2–3 cm long . . . . **Stylophorum**
    3 Capsule glabrous; style absent; petals 1 cm long . . . . **Chelidonium**

Chelidonium L. (Celandine)

C. majus L.*; occasionally escaping from cultivation; May–Aug; NC, VA.

Papaver L. (Poppy)

P. dubium L.*; occasionally escaping from cultivation; May–Aug; NC, TN, VA.

Sanguinaria L. (Bloodroot)

S. canadensis L.; rich woods; common; Mar–Apr; ALL.

Stylophorum Nutt. (Celandine Poppy)

S. diphyllum (Michx.) Nutt.; rich woods, often on calcareous sites; occasional; Mar–Apr; TN, VA.

PASSIFLORACEAE (Passion Flower Family)

Passiflora L.

1 Leaves deeply 3–5-lobed, serrate; petals whitish, corona purple; fruit about 5 cm long .... **P. incarnata**
1 Leaves shallowly 3-lobed, entire; petals yellowish; fruit about 1 cm long ... **P. lutea**

P. incarnata L., Maypops; dry fields, roadsides, and open disturbed sites; common; Jun–Sep; ALL.

P. lutea L., Yellow Passion Flower; thickets and open disturbed sites; frequent; Jun–Sep; ALL.

PHRYMACEAE (Lopseed Family)

Phryma L. (Lopseed)

P. leptostachya L.; rich mesic woodlands; common; Jun–Aug; ALL. A monotypic family.

PHYTOLACCACEAE (Pokeweed Family)

Phytolacca L. (Pokeweed)

P. americana L.; open, usually disturbed sites; common; May–Sep; ALL.

PLANTAGINACEAE (Plantain Family)

Plantago L. (Plantain)

1 Leaves narrowly linear, less than 1 cm wide 2
  2 Bracts longer than the flowers, conspicuous .... **P. aristata**
  2 Bracts shorter or equal to the flowers, not conspicuous 3
    3 Leaves less than 5 mm wide .... **P. heterophylla**
    3 Leaves more than 5 mm wide .... **P. hookeriana**
1 Leaves lanceolate to broadly ovate, at least the larger ones 1 cm or more wide 4
  4 Lateral veins arising from the midrib; flowering stem hollow .... **P. cordata**
  4 Lateral veins arising from the leaf base; flowering stem solid 5
    5 Leaves lanceolate to oblanceolate 6
      6 Scape glabrous to slightly pubescent; corolla lobes spreading to reflexed at anthesis .... **P. lanceolata**
      6 Scape moderately to densely pubescent; corolla lobes erect and enclosing the fruit at anthesis .... **P. virginica**
    5 Leaves broadly elliptic to ovate 7
      7 Capsule 3–6 mm long, ellipsoid .... **P. rugelii**
      7 Capsule 2–4 mm long, ovoid .... **P. major**

P. **aristata** Michx., Buckhorn; waste places and disturbed sites; common; May–Jul; NC, SC, TN, VA.
P. **cordata** Lam.; marshes and streams; infrequent; Mar–Apr; VA.
P. **heterophylla** Nutt.; waste places; infrequent; Mar–May; SC.
P. **hookeriana** Fisch. & Mey.; waste places; infrequent; Apr–Jul; SC. Represented in our area by var. **nuda** (Gray) Poe.
P. **lanceolata** L.*, English P.; lawns and waste places; occasional; May–Oct; NC, SC, TN, VA.
P. **major** L.; possibly introduced to lawns and waste places; occasional; May–Oct; GA, NC, VA.
P. **rugelii** Dcne., Broadleaved P.; lawns and waste places; common; May–Oct; ALL.
P. **virginica** L., lawns and waste places; common; Mar–Jun; NC, SC, TN, VA.

PLATANACEAE (Plane Tree Family)

Platanus L. (Sycamore)

P. **occidentalis** L.; mesic to wet woods, especially along streams and bottomland forests; common; Apr–May; ALL.

PODOSTEMACEAE (Riverweed Family)

Podostemum Michx. (Threadfoot)

P. **ceratophyllum** Michx.; submerged aquatic on rocks of clear streams; frequent; May–Aug; ALL.

POLEMONIACEAE (Phlox Family)

1 Leaves simple .... **Phlox**
1 Leaves compound 2
  2 Corolla blue to nearly white .... **Polemonium**
  2 Corolla scarlet, rarely yellow .... **Ipomopsis**

Ipomopsis Michx. (Standing Cypress)

I. **rubra** (L.) Wherry; sandy sites; infrequent; Jun–Aug; NC, SC.

Phlox L. (Phlox)

1 Stems woody, trailing or decumbent; leaves mostly persistent 2
  2 Corolla lobes notched; stamens partly exserted .... **P. subulata**
  2 Corolla lobes entire or essentially so; stamens included within the corolla tube .... **P. nivalis**
1 Stems herbaceous, erect or decumbent; leaves primarily deciduous 3
  3 Style short, about as long as the stigma and not extending beyond the lowest anther 4
    4 Inflorescence compact; calyx with jointed, nonglandular hairs .... **P. amoena**
    4 Inflorescence spreading; calyx with glandular hairs 5
      5 Leaves lanceolate to narrowly ovate; corolla tube glabrous externally .... **P. divaricata**

5 Leaves linear to narrowly elliptic; corolla tube usually pubescent externally .... **P. pilosa**

3 Style elongate, much longer than the stigma and extending well beyond the lowest anther 6

6 Prostrate stems well developed, rooting at the nodes .... **P. stolonifera**

6 Prostrate stems essentially absent 7

7 Lateral leaf veins distinct 8

8 Inflorescence glandular-pubescent; corolla tube glabrous .... **P. amplifolia**

8 Inflorescence with short, stiff nonglandular hairs; corolla tube pubescent .... **P. paniculata**

7 Lateral leaf veins obscure 9

9 Cymes forming a subcylindric panicle (stems usually red spotted) .... **P. maculata**

9 Cymes corymbiform 10

10 Nodes below the inflorescence 7 or fewer 11

11 Inflorescence densely glandular-pubescent .... **P. buckleyi**

11 Inflorescence with nonglandular hairs .... **P. ovata**

10 Nodes below the inflorescence 8 or more 12

12 Calyx 6–11 mm long .... **P. carolina**

12 Calyx 5–8 mm long .... **P. glaberrima**

P. amoena Sims; dry woodlands and roadsides; occasional; Apr–Jun; GA, NC, SC, TN.

P. amplifolia Britt.; mesic woodlands; occasional; Jul–Aug; GA, NC, TN, VA.

P. buckleyi Wherry; open rocky slopes; infrequent; May–Jun; VA.

P. carolina L.; woodlands and clearings; frequent; May–Jul; ALL.

P. divaricata L., Blue P.; mesic or dry woodlands; frequent; Apr–May; NC, TN, VA.

P. glaberrima L.; mesic woods and stream banks; frequent; May–Jul; ALL.

P. maculata L.; moist woods and stream banks; frequent; Jun–Sep; ALL. *P. maculata* L. subsp. *pyramidalis* (Smith) Wherry.

P. nivalis Lodd; sandy woodlands; infrequent; Mar–May; GA, NC, SC.

P. ovata L.; mesic woods and roadsides; frequent; May–Jun; NC, TN, VA.

P. paniculata L.; mesic woods and stream banks; frequent; Jul–Sep; ALL.

P. pilosa L.; dry open woodlands and fields; infrequent; May–Jun; SC, TN, VA.

P. stolonifera Sims, Creeping P.; rich woods; frequent; Apr–May; ALL.

P. subulata L., Moss Pink; rocky slopes; occasional; Apr–May; NC, SC, VA.

Numerous poorly understood subspecific taxa involving *P. carolina*, *P. glaberrima*, *P. ovata*, *P. amplifolia*, and *P. paniculata* have been described and are in need of detailed field and experimental study.

Polemonium L. (Jacob's Ladder)

P. reptans L.; woodlands, often on basic soils; infrequent; Apr–May; TN, VA.

POLYGALACEAE (Milkwort Family)

Polygala L.

1 Leaves opposite or whorled 2

2 Racemes 8 mm or more broad; sepals pinkish; bracts persistent .... **P. cruciata**

2 Racemes 5 mm or less broad; sepals white or greenish; bracts deciduous . . . . **P. verticillata**
1 Leaves alternate 3
3 Flowers 1.4–2 cm long, 1–4/stem . . . . **P. paucifolia**
3 Flowers less than 1.4 cm long, in relatively dense racemes 4
4 Perennials, usually branched at base 5
5 Stem leaves 1 cm or less wide, entire; flowers pinkish . . . . **P. polygama**
5 Stem leaves more than 1 cm wide, serrulate; flowers white or greenish . . . . . . **P. senega**
4 Annuals, often branched above 6
6 Corolla elongate, much longer than the sepals (wings); plants glaucous . . . . . . **P. incarnata**
6 Corolla not longer than the wings; plant not glaucous 7
7 Wings 3 mm or more wide; aril 1/2 or more the length of the seed . . . . **P. sanguinea**
7 Wings less than 3 mm wide; aril less than 1/2 the length of the seed 8
8 Racemes 1 cm or more broad . . . . **P. curtissii**
8 Racemes 7 mm or less broad . . . . **P. nuttallii**

P. cruciata L.; bogs and wet woods; occasional; Jun–Oct; ALL.
P. curtissii Gray; moist open woods and fields; frequent; Jun–Oct; ALL.
P. incarnata L.; dry woodlands and fields; infrequent; Jun–Jul; GA, NC, VA.
P. nuttallii T. & G.; moist open woodlands; infrequent; Jun–Jul; VA.
P. paucifolia Willd., Gay Wings; dry or moist open woods; occasional; Apr–Jun; ALL.
P. polygama Walt.; dry open woodlands; occasional; May–Jul; ALL.
P. sanguinea L.; fields and open woods; frequent; Jun–Aug; NC, SC, TN, VA.
P. senega L., Seneca Snakeroot; dry fields and open woods; frequent; Apr–Jun; ALL.
P. verticillata L.; fields and open woods; common; Jun–Sep; ALL.

POLYGONACEAE (Buckwheat Family)

1 Leaves whorled; ocreae absent; stamens 9; calyx yellow . . . . **(Eriogonum)**
1 Leaves alternate; ocreae present; stamens usually 5–8; calyx not yellow 2
2 Sepals 6, the outer reflexed in fruit, the inner erect and often with an external tubercle or grain . . . . **Rumex**
2 Sepals 4 or 5, all erect and without tubercles 3
3 Achenes triangular and well exserted beyond the sepals; uppermost flowers corymbiform; leaves triangular-hastate . . . . **Fagopyrum**
3 Achenes flat or triangular, enclosed or slightly exserted beyond the sepals; flowers axillary, racemose, or paniculate; leaves various, rarely triangular-hastate . . . . . . **Polygonum**

Eriogonum Michx. (Yellow Buckwheat)

E. allenii Wats.; shale barrens; infrequent; Jul–Aug; VA.

Fagopyrum Mill. (Buckwheat)

F. esculentum Moench*; fields and disturbed sites; occasional; Jun–Oct; ALL.

Polygonum L. (Smartweed, Knotweed)

1 Styles 2-cleft, persistent and recurved at maturity; flowers not overlapping, racemes 15–30 cm long . . . . **P. virginianum**
1 Styles 2- or 3-cleft, deciduous at maturity; flowers in dense racemes, or if flowers remote, the racemes less than 15 cm long 2
2 Stems retrorsely scabrous 3
3 Achenes triangular; leaves sagittate . . . . **P. sagittatum**
3 Achenes flat; leaves hastate . . . . **P. arifolium**
2 Stems not retrorsely scabrous 4
4 Stems twining (often on other plants) or viny, neither erect nor distinctly prostrate 5
5 Ocreae with a retrorse beard at base . . . . **P. cilinode**
5 Ocreae smooth 6
6 Outer sepals winged; achenes shiny . . . . **P. scandens**
6 Outer sepals not winged; achenes dull . . . . **P. convolvulus**
4 Stems erect or distinctly prostrate, but not viny 7
7 Flowers in small axillary clusters of 1–5 flowers 8
8 Leaves plicate; stems angular . . . . **P. tenue**
8 Leaves not plicate; stems round or nearly so 9
9 Leaves more than 5 mm wide; sepals hoodlike . . . . **P. erectum**
9 Leaves less than 5 mm wide; sepals flat . . . . **P. aviculare**
7 Flowers numerous, in terminal and/or axillary spikes or racemes 10
10 Outer sepals broadly winged; coarse perennials up to 2–3 m tall 11
11 Leaves cordate at the base . . . . **P. sachalinense**
11 Leaves truncate at the base . . . . **P. cuspidatum**
10 Outer sepals not winged; annuals or perennials less than 2 m tall 12
12 Leaf sheaths (ocreae) entire at the summit 13
13 Raceme 1 and terminal, rarely 2 or 3 . . . . **P. amphibium**
13 Racemes several, both terminal and axillary 14
14 Peduncles glabrous or with sessile or subsessile glands; racemes mostly nodding . . . . **P. lapathifolium**
14 Peduncles pubescent, the trichomes strigose, stipitate-glandular, or mixed; racemes mostly erect . . . . **P. pensylvanicum**
12 Ocreae with bristles (more than 0.5 mm long) at the summit 15
15 Stems hirsute, especially near the inflorescence . . . . . . **P. orientale**
15 Stems glabrous or with appressed pubescence 16
16 Calyx glandular punctate 17
17 Achene granular and dull . . . . **P. hydropiper**
17 Achene smooth and shiny . . . . **P. punctatum**
16 Calyx without distinct glands 18
18 Perennials with horizontal rootstocks 19
19 Leaves broadly lanceolate, 1.5 cm or more wide; ocreae trichomes spreading . . . . **P. setaceum**
19 Leaves lanceolate to linear-lanceolate, usually less than 1.5 cm wide; ocreae trichomes appressed . . . . **P. hydropiperoides**

18 Annuals with fibrous rootstocks 20
20 Racemes 3–6 mm thick; bristles as long as or longer than the ocreolae (flower bracts); leaves without dark spots .. .. **P. cespitosum**
20 Racemes 6–12 mm thick; bristles shorter than the ocreolae; leaves usually with a dark spot .... **P. persicaria**

P. amphibium L., Water S.; wet sites, sometimes submerged or floating; occasional; Jul–Sep; NC, TN, VA. *P. coccineum* Muhl. ex Willd.

P. arifolium L., Tearthumb; marshes and wet thickets; occasional; Jul–Oct; TN, VA.

P. aviculare L.*; disturbed sites; frequent; May–Oct; NC, SC, TN, VA.

P. cespitosum Blume*; damp shaded sites; frequent; May–Oct; NC, SC, TN, VA. Represented in our area by var. **longisetum** (DeBruyn) Steward.

P. cilinode Michx.; high-elevation clearings; occasional; Jun–Sep; GA, NC, TN, VA.

P. convolvulus L.*; disturbed sites; frequent; May–Sep; ALL.

P. cuspidatum Siebold & Zucc.*; waste areas; infrequent; May–Sep; ALL.

P. erectum L.; disturbed sites; occasional; Jun–Sep; ALL.

P. hydropiper L.; wet sites; frequent; Jul–Oct; ALL.

P. hydropiperoides Michx.; wet sites; occasional; May–Oct; SC, TN, VA.

P. lapathifolium L.; wet sites; infrequent; Aug–Oct; NC, SC, VA.

P. orientale L.*; lawns and waste places; occasional; Jul–Oct; ALL.

P. pensylvanicum L.; disturbed sites; common; Jul–Oct; ALL.

P. persicaria L.*; waste places; common; Jun–Oct; ALL.

P. punctatum Ell.; damp sites; common; Jun–Oct; ALL.

P. sachalinense F. Schmidt*; waste places; infrequent; Jul–Aug; NC, TN.

P. sagittatum L.; wet thickets; common; May–Oct; ALL.

P. scandens L.; disturbed sites; common; Jul–Oct; ALL. Intergrades with and perhaps not distinct from *P. cristatum* Engelm. & Gray.

P. setaceum Bald. ex Ell.; low woods and swamps; infrequent; Jul–Oct; SC, VA.

P. tenue Michx.; dry disturbed sites; occasional; Jul–Sep; GA, NC, VA.

P. virginianum L., Jumpseed; mesic woodlands; common; Jul–Oct; ALL. *Antenoron virginianum* (L.) Roberty & Vautier, *Tovara virginiana* (L.) Raf. With its closest relatives in eastern Asia, persistent styles, and peculiar mode of seed dispersal, this taxon might be best recognized as a segregate genus.

Rumex L. (Dock)

1 Leaves (especially the lowermost) hastate or sagittate; plants dioecious 2
2 Perennials with slender rhizomes; fruiting calyx (valves) about 1 mm wide, not winged .... **R. acetosella**
2 Winter annuals with taproots; fruiting calyx about 4–5 mm wide, broadly winged .... **R. hastatulus**
1 Leaves not hastate or sagittate; plants monoecious 3
3 Valves conspicuously toothed at maturity 4
4 Pedicels longer than the valves; upper verticils of flowers continuous .... **R. obtusifolius**
4 Pedicels about as long as the valves; upper verticils separate .... **R. pulcher**
3 Valves entire or wavy, not toothed 5
5 Pedicel with a conspicuous disarticulation joint at the base 6

6 Leaves with conspicuous wavy margins .... **R. crispus**
6 Leaves flat, the margin not wavy 7
7 Fruiting pedicels 2–5× as long as the calyx .... **R. verticillatus**
7 Fruiting pedicels about as long as the calyx .... **R. altissimus**
5 Pedicel without a conspicuous disarticulation joint at the base ... **R. orbiculatus**

R. acetosella L.*, Sheep Sorrel; fields and waste areas; common; Mar–Jun; ALL.
R. altissimus Wood; low woods and fields; infrequent; Mar–Jun; NC, SC, VA.
R. crispus L.*, Curly D.; disturbed sites; common; Mar–May; ALL.
R. hastatulus Bald. ex Ell.; fields and openings; infrequent; Mar–May; SC. Common on the Piedmont and Coastal Plain.
R. obtusifolius L.*, Bitter D.; disturbed sites; frequent; May–Jun; ALL. Frequently hybridizing with *R. crispus*.
R. orbiculatus Gray; moist woods and streams; infrequent; Jun–Sep; VA.
R. pulcher L.*; wet disturbed sites; infrequent; May–Jun; SC, VA.
R. verticillatus L., Swamp D.; streams, swamps, and wet ditches; infrequent; Apr–May; VA. More common on the Coastal Plain.

PORTULACACEAE (Purslane Family)

1 Flowers sessile, solitary or glomerulate .... **Portulaca**
1 Flowers in racemes or cymes 2
2 Stem leaves opposite; stamens 5; sepals persistent .... **Claytonia**
2 Stem leaves alternate or basal; stamens more than 5; sepals deciduous ... **Talinum**

Claytonia L. (Spring Beauty)

1 Leaves linear to linear-lanceolate; petiole indistinct .... **C. virginica**
1 Leaves lanceolate, ovate, or oblanceolate; petiole distinct .... **C. caroliniana**

C. caroliniana Michx.; rich deciduous woods; frequent; Mar–May; GA, NC, TN, VA. More common at higher elevations than the following species.
C. virginica L.; rich deciduous woods; common; Mar–Apr; ALL.

Portulaca L. (Common Purslane)

P. oleracea L.*; fields and waste places; frequent; May–Sep; GA, NC, TN, VA.

Talinum Adans. (Fameflower)

T. teretifolium Pursh; wet areas on rock outcrops; frequent; Jun–Sep; GA, NC, SC, VA.

PRIMULACEAE (Primrose Family)

1 Leaves all basal; petals reflexed .... **Dodecatheon**
1 Leaves cauline; petals erect or spreading 2
2 Leaves in a single terminal whorl .... **Trientalis**
2 Leaves alternate, opposite, or in several whorls 3
3 Leaves chiefly alternate 4
4 Flowers in terminal racemes .... **Samolus**
4 Flowers axillary, subsessile .... **Centunculus**

3 Leaves chiefly opposite or whorled 5
5 Flowers yellow; capsule longitudinally dehiscent ..... **Lysimachia**
5 Flowers pink, red, or blue; capsule dehiscing by a horizontal ring ... **Anagallis**

Anagallis L. (Scarlet Pimpernel)

A. arvensis L.*; waste grounds; occasional; Jun–Aug; TN, VA.

Centunculus L. (Chaffweed)

C. minimus L.; moist open areas; infrequent; May–Jun; SC.

Dodecatheon L. (Shooting Star)

D. meadia L.; calcareous woodlands; occasional; Apr–May; NC, TN, VA.

Lysimachia L. (Loosestrife)

Coffey, V. J., and S. B. Jones, Jr. 1980. Biosystematics of *Lysimachia* section *Seleucia* (Primulaceae). *Brittonia* 32:309–322.

1 Plant trailing, rooting at the nodes ..... **L. nummularia**
1 Plant erect or arching, rarely rooting at the nodes 2
2 Flowers in terminal racemes or panicles; bracts much reduced 3
3 Inflorescence paniculate ..... **L. fraseri**
3 Inflorescence racemose ..... **L. terrestris**
2 Flowers both axillary and terminal; bracts leaflike 4
4 Principal stem leaves whorled ..... **L. quadrifolia**
4 Principal stem leaves opposite 5
5 Petioles of median cauline leaves pubescent along entire length 6
6 Median cauline leaves 17–60 mm wide; sepals with reddish-brown veins ..... **L. ciliata**
6 Median cauline leaves 4–23 mm wide; sepals without reddish-brown veins 7
7 Leaf blades pubescent at base, lanceolate to linear; sepals not conspicuously veined ..... **L. lanceolata**
7 Leaf blades glabrous at base, lanceolate to ovate; sepals conspicuously veined ..... **L. hybrida**
5 Petioles of median cauline leaves pubescent only at base 8
8 Rhizome absent, new shoots arising from a crown of rootstocks ... **L. tonsa**
8 Rhizome present 9
9 Plants reclining or trailing and rooting at the nodes ..... **L. radicans**
9 Plants strictly erect 10
10 Leaf blades linear to narrowly lanceolate; sepals not conspicuously veined ..... **L. quadriflora**
10 Leaf blades ovate to lanceolate; sepals conspicuously veined ..... **L. hybrida**

L. ciliata L.; moist woods and stream banks; common; May–Sep; ALL.
L. fraseri Duby; meadows and roadsides; infrequent; May–Jun; GA, NC, SC, TN.
L. hybrida Michx.; wet open areas; infrequent; Jul–Aug; NC, TN, VA.

L. lanceolata Walt.; dry or moist open woods; frequent; Jun–Aug; ALL.
L. nummularia L.*, Moneywort; open seeps and alluvial woods; occasional; May–Jun; NC, SC, TN, VA.
L. quadriflora Sims; wet meadows and stream banks; infrequent; Jul–Sep; GA, VA.
L. quadrifolia L., Whorled L.; dry or moist open areas; common; May–Jul; ALL.
L. radicans Hook.; moist woods and swamps; infrequent; Jun–Aug; VA. A disjunct from the Mississippi embayment.
L. terrestris (L.) BSP., Swamp L.; wet meadows and swamps; occasional; May–Jul; NC, TN, VA.
L. tonsa (Wood) Knuth ex Engler; open woodlands and bluffs; infrequent; May–Aug; GA, NC, TN, VA.

Samolus L. (Brookweed)

S. parviflorus Raf.; moist open sites; infrequent; May–Sep; TN, VA.

Trientalis L. (Star Flower)

T. borealis Raf.; moist woods; occasional; May–Jul; GA, TN, VA.

RANUNCULACEAE (Crowfoot Family)

Keener, C. S. 1975a. Studies in the Ranunculaceae of the southeastern United States. I. *Anemone* L. *Castanea* 40:36–44; 1975b. Studies in the Ranunculaceae of the southeastern United States. III. *Clematis* L. *Sida* 6:33–47; 1976a. Studies in the Ranunculaceae of the southeastern United States. II. *Thalictrum* L. *Rhodora* 78:457–472; 1976b. Studies in the Ranunculaceae of the southeastern United States. IV. Genera with zygomorphic flowers. *Castanea* 41:12–20; 1976c. Studies in the Ranunculaceae of the southeastern United States. V. *Ranunculus* L. *Sida* 6:266–283; 1977. Studies in the Ranunculaceae of the southeastern United States. VI. Miscellaneous genera. *Sida* 7:1–12; 1981. The status of *Thalictrum hepaticum* Greene (Ranunculaceae). *Castanea* 46:43–49.

1 Flowers zygomorphic 2
  2 Upper sepal hooded; petals 2 . . . . . **Aconitum**
  2 Upper sepal spurred; petals 2 or 4 3
    3 Petals 2, united; carpel 1; introduction . . . . . **Consolida**
    3 Petals 4, distinct; carpels 2 or more; native . . . . . **Delphinium**
1 Flowers actinomorphic 4
  4 Petals distinctly spurred . . . . . **Aquilegia**
  4 Petals either not spurred or absent 5
    5 Low shrubs; wood yellow . . . . . **Xanthorhiza**
    5 Herbs or woody vines 6
      6 Sepals or petals yellow 7
        7 Petals present, with a nectar gland at the base; fruit an achene . . **Ranunculus**
        7 Petals absent, sepals showy; fruit a follicle . . . . . **Caltha**
      6 Sepals or petals (if present) not yellow, either white or variously colored . . . 8
        8 Leaves all opposite; mature styles plumose and several times longer than the body of the achene . . . . . **Clematis**

8 At least some leaves alternate, whorled, or basal (involucral leaves occasionally opposite or whorled); mature styles not plumose 9
9 Leaves all basal, 3-lobed, often purple beneath . . . . **Hepatica**
9 Stem leaves present 10
10 Perianth (if present) small and inconspicuous (less than 8 mm long), often early deciduous; stamens often showy 11
11 Leaves simple or palmately lobed 12
12 Flowers solitary; fruit a berry . . . . **Hydrastis**
12 Flowers numerous; fruit a dehiscent, 1-seeded utricle . . . . **Trautvetteria**
11 Leaves 1–3-ternately compound 13
13 Flowers in branched panicles, often unisexual; leaflets entire or 3–5-lobed apically . . . . **Thalictrum**
13 Flowers in simple or 2–4-branched racemes; leaflets coarsely toothed 14
14 Racemes elongate, 1.5 dm or more long; fruit a follicle . . . . **Cimicifuga**
14 Racemes short, less than 5 cm long; fruit a white berry . . . . **Actaea**
10 Perianth showy (about 8 mm or more long), present at anthesis . . 15
15 Leaves ternately compound, the leaflets usually 3-lobed apically . . . . **Thalictrum**
15 Leaves palmately divided, often appearing compound, coarsely toothed . . . . **Anemone**

Aconitum L.

1 Perianth blue; upper sepal about as high as or slightly higher than long . . . . **A. uncinatum**
1 Perianth white; upper sepal about 2× as high as long . . . . **A. reclinatum**

A. reclinatum Gray, Wolfsbane; seepage slopes and boulderfields in rich woods; infrequent; Jun–Sep; NC, VA.

A. uncinatum L., Monkshood; seepages and moist soils in rich woods; frequent; Aug–Sep; ALL. *A. uncinatum* L. subsp. *muticum* (DC.) Hardin.

Actaea L. (Doll's Eyes, Baneberry)

A. pachypoda Ell.; rich woods; frequent; Apr–May; ALL.

Anemone L.

1 Plants usually more than 4 dm tall at anthesis, branched; flowers 2–many 2
2 Involucral bracts sessile; heads globose; achenes glabrous or nearly so . . . . **A. canadensis**
2 Involucral bracts petiolate; heads subcylindric; achenes densely pubescent . . . . **A. virginiana**
1 Plants usually less than 4 dm tall at anthesis, unbranched; 1-flowered 3
3 Sepals 10–20, cream-white, often suffused with pink or lavender; rootstocks short-rhizomatous . . . . **A. caroliniana**
3 Sepals 5(–8), usually white; rootstocks long-rhizomatous 4

4 Terminal leaflet broadest below the middle; sepals 15 mm or more long; stems often more than 2 dm tall .... **A. lancifolia**

4 Terminal leaflet broadest at or above the middle; sepals less than 15 mm long; stems usually less than 2 dm tall 5

5 Lateral leaflets cleft or parted; terminal leaflet broadest above the middle .... **A. quinquefolia**

5 Lateral leaflets variously toothed but not cleft; terminal leaflet broadest at the middle .... **A. minima**

A. canadensis L.; low woods; infrequent; May–Aug; VA.
A. caroliniana Walt.; dry woods; infrequent; Mar–Apr; SC.
A. lancifolia Pursh; rich woods; occasional; Mar–May; NC, SC, VA.
A. minima DC.; rich woods; infrequent; Mar–May; NC, TN, VA.
A. quinquefolia L., Wood Anemone; rich woods; frequent; Mar–May; ALL.
A. virginiana L., Thimbleweed; rich woods; common; May–Jul; NC, SC, TN, VA.

A. lancifolia, *A. minima*, and *A. quinquefolia* represent a highly polymorphic complex perhaps best represented by a single species.

Aquilegia L. (Columbine)

A. canadensis L.; rich rocky woods and cliffs; frequent; Apr–May; ALL.

Caltha L. (Marsh Marigold)

C. palustris L.; marshes and bogs; occasional; Apr–Jun; NC, TN, VA.

Cimicifuga Wernischeck (Black Cohosh)

1 Carpels 3–8, stipitate .... **C. americana**

1 Carpels 1(–2), sessile 2

2 Leaflets 20 or more/leaf; follicles about 7 mm long; early summer flowering .... **C. racemosa**

2 Leaflets 3–9/leaf; follicles about 16 mm long; mid summer to fall flowering .... **C. rubifolia**

C. americana Michx.; coves and rich, rocky slopes, usually at high elevations; occasional; Aug–Sep; ALL.
C. racemosa (L.) Nutt.; rich woods; common; May–Aug; ALL.
C. rubifolia Kearney; rich woods; infrequent; Aug–Oct; TN. More common in the adjacent Ridge and Valley Province.

Clematis L.

1 Flowers numerous, in cymose-paniculate inflorescences; filaments glabrous; sepals whitish 2

2 Leaflets coriaceous, entire or sometimes cleft; flowers bisexual ..... **C. terniflora**

2 Leaflets membranaceous, coarsely serrate; flowers unisexual ...... **C. virginiana**

1 Flowers usually solitary; filaments pubescent; sepals variously colored 3

3 Sepals thin, not connivent .... **C. occidentalis**

3 Sepals thick and leathery, somewhat connivent 4

4 Erect herbs; leaves usually simple 5

5 Leaves glaucous and glabrous beneath, the uppermost commonly pinnate and tendril-bearing .... **C. addisonii**

5 Leaves green and usually pubescent beneath, the uppermost neither compound nor tendril-bearing 6

6 Leaves at anthesis soft-pubescent beneath, 3–9 cm wide; leaves at fruiting prominently reticulate above 7

7 Stems and leaves villous; sepal backs moderately pubescent .... **C. ochroleuca**

7 Stems and leaves densely sericeous; sepal backs densely pubescent .... (**C. coactilis**)

6 Leaves at anthesis slightly pubescent to glabrous beneath, 2–5 cm wide; leaves at fruiting not prominently reticulate above 8

8 Sepal backs villous; mature styles white to pale yellow ... (**C. albicoma**)

8 Sepal backs finely puberulent; mature styles tawny to reddish-brown .... (**C. viticaulis**)

4 Climbing vines or scrambling herbs; leaves pinnately compound or simple below and pinnate above 9

9 Sepals 3–5 cm long and dilated distally into thin, crisped margins up to 6 mm wide; peduncles usually without bracts .... **C. crispa**

9 Sepals up to 2.5 cm long, the margins not dilated distally; peduncles usually bracteate 10

10 Leaves glaucous and glabrous beneath; nodes glabrous ..... **C. addisonii**

10 Leaves green and pubescent beneath; nodes pubescent .... **C. viorna**

C. addisonii Britt.; dry calcareous woods; infrequent; Apr–Jun; VA. Restricted to a 4-county area near Roanoke.

C. albicoma Wherry, Leatherflower; shale barrens; infrequent; May–Jun; VA.

C. coactilis (Fern.) Keener; shale barrens and calcareous woods; infrequent; May–Jun; VA. Perhaps a hybrid derivative of *C. ochroleuca* and *C. albicoma*.

C. crispa L.; low woods and swamps; infrequent; Apr–Aug; NC.

C. occidentalis (Hornemann) DC., Purple Clematis; open woods and calcareous slopes; occasional; May–Jun; NC, VA. *C. verticillaris* DC.

C. ochroleuca Ait., Curlyheads; dry woods and clearings; occasional; Apr–Jun; GA, NC, SC, VA.

C. terniflora DC.*; fencerows and disturbed sites; occasional; Jul–Sep; NC, TN, VA. *C. maximowicziana* Franch. & Savat., *C. dioscoreifolia* Levl. & Vaniot, *C. paniculata* Thunb.

C. viorna L., Leatherflower; open calcareous woodlands; common; May–Sep; ALL.

C. virginiana L., Virgin's Bower; low woods and thickets; common; Jul–Sep; ALL.

C. viticaulis Steele; shale barrens; very local; May–Jun; VA.

Consolida (DC.) S. F. Gray (Larkspur)

C. ambigua (L.) Ball & Heywood*; roadsides and waste areas; occasional; May–Sep; NC, SC, TN, VA. *Delphinium ajacis* L., *D. ambiguum* L.

Delphinium L. (Larkspur)

1 Stems less than 4 dm tall; follicles spreading; spring flowering .... **D. tricorne**

1 Stems more than 4 dm tall; follicles erect; summer flowering .... **D. exaltatum**

D. exaltatum Ait.; open calcareous woods; occasional and becoming quite rare; Jul–Sep; NC, SC, VA.

D. tricorne Michx., Dwarf L.; rich calcareous woods; occasional; Mar–May; NC, TN, VA.

Hepatica Mill. (Liverleaf)

1 Apex of leaf lobes and bracts rounded to obtuse .... **H. americana**
1 Apex of leaf lobes and bracts acute .... **H. acutiloba**

H. acutiloba DC.; rich woods; common; Mar–Apr; ALL. *H. nobilis* Mill. var. *acuta* (Pursh) Steyermark.

H. americana (DC.) Ker-Gawl.; rich woods; common; Mar–Apr; ALL. *H. nobilis* Mill. var. *obtusa* (Pursh) Steyermark.

Although considered as varieties by some authors, these taxa are distinct and show no evidence of intergradation in our area.

Hydrastis Ell. ex L. (Golden Seal)

H. canadensis L.; rich, often calcareous woods; infrequent; Apr–May; GA, NC, TN, VA. Commercially exploited and in need of protection in our area.

Ranunculus L. (Buttercup)

1 Leaves variously lobed or compound 2
2 Basal leaves mostly entire, cauline leaves deeply divided; achenes turgid, without marginal rims 3
3 Petals less than 3.5 mm long, equal to or shorter than the sepals; roots mostly filiform 4
4 Petals more than 1/2 as long as the sepals; achene beaks 0.1–0.3 mm long .. 5
5 Plants glabrous .... **R. abortivus**
5 Plants villous, at least basally .... **R. micranthus**
4 Petals less than 1/2 as long as the sepals; achene beaks 0.6-1 mm long .... **R. allegheniensis**
3 Petals 6–8 mm long, longer than the sepals; some roots thickened .... **R. harveyi**
2 Basal leaves deeply lobed or compound, usually similar to the smaller cauline leaves 6
6 Achenes densely papillose or with short, hooked spines .... **R. parviflorus**
6 Achenes smooth, pubescent, or slightly papillose 7
7 Achenes turgid, marginal rim essentially absent; plants of aquatic habitats .... **R. sceleratus**
7 Achenes flattened, marginal rim evident; plants of terrestrial habitats 8
8 Petals 3.5–6 mm long, about as long as the sepals; achene beaks recurved .... **R. recurvatus**
8 Petals 5–18 mm long, longer than the sepals; achene beaks straight or recurved 9
9 Achene beaks distinctly recurved, usually less than 1.5 mm long 10
10 Sepals reflexed; plants bulbous-thickened at the base .. **R. bulbosus**
10 Sepals spreading; plants not bulbous-thickened at the base 11
11 Stems repent, rooting at the nodes; receptacle hispid .... **R. repens**

11 Stems erect, not rooting at the nodes; receptacle glabrous .... **R. acris**

9 Achene beaks straight or nearly so, 1.5–5 mm long 12

12 Plants erect, not stoloniferous; basal leaves both simple and compound, stipules gradually tapering upward; achene bodies 1.5–3.5 mm long 13

13 At least some roots fusiform-thickened; basal leaves ovate-oblong, often longer than broad .... **R. fascicularis**

13 Roots all fibrous; basal leaves cordate-ovate, often broader than long .... **R. hispidus**

12 Plants lax, usually stoloniferous; basal leaves all compound, stipules rounded or truncate upward; achene bodies 3–5 mm long 14

14 Achenes 15–30, not winged; sepals spreading, almost as long as the petals .... **R. septentrionalis**

14 Achenes 7–15, broadly wing-keeled; sepals usually reflexed, about 1/2 as long as the petals .... **R. carolinianus**

1 Leaves all simple, not lobed or deeply divided 15

15 Petals 1–2 mm long, about as long as the sepals .... **R. pusillus**

15 Petals 3–8 mm long, longer than the sepals 16

16 Basal leaves absent; stems reclining and rooting at the nodes; sepals about 4 mm long; perennials .... **R. ambigens**

16 Basal leaves present; stems erect and not rooting at the nodes; sepals 1.5–2.5 mm long; annuals .... **R. laxicaulis**

R. abortivus L., Crowfoot; woodlands and waste places; common; Mar–Jun; ALL.

R. acris L.*, Tall B.; clearings and disturbed sites; frequent; May–Aug; ALL.

R. allegheniensis Britt.; rich woods and calcareous slopes; occasional; Apr–Jun; NC, TN, VA.

R. ambigens Watson; low woods and shorelines; infrequent; Apr–Jun; TN.

R. bulbosus L.*, Bulbous B.; disturbed sites; frequent; Apr–Jun; NC, SC, TN, VA.

R. carolinianus DC.; low woods and thickets; infrequent; Apr–Aug; GA, NC, TN, VA.

R. fascicularis Muhl. ex Bigel.; dry calcareous woods; infrequent; Mar–Jun; NC, TN, VA.

R. harveyi (Gray) Britt.; river bluffs; infrequent; Apr–May; TN.

R. hispidus Michx., Hispid B.; woodlands; common; Mar–Jun; ALL.

R. laxicaulis (T. & G.) Darby; marshes and wet ditches; infrequent; Apr–Jun; TN.

R. micranthus Nutt.; calcareous woodlands; occasional; Apr–Jun; TN, VA.

R. parviflorus L.*; waste places and lawns; infrequent; Feb–Jul; ALL.

R. pusillus Poir.; low grounds; occasional; Apr–Jun; NC, SC, TN, VA.

R. recurvatus Poir., Hooked B.; rich woods; common; Apr–Jun; ALL.

R. repens L.*, Swamp B.; low open areas; May–Jul; NC, SC, TN, VA. *R. repens* L. var. *pleniflorus* Fern.

R. sceleratus L.; marshes and shorelines; infrequent; Jun–Sep; VA.

R. septentrionalis Poir., Northern Swamp B.; low woods and marshes; occasional; Apr–Aug; NC, TN, VA.

A taxonomically difficult genus often requiring mature achenes, basal leaves, and rootstocks for positive identification. The complex involving *R. hispidus*, *R. fascicularis*, *R. septentrionalis*, and *R. carolinianus* is poorly understood in our area.

Thalictrum L. (Meadow Rue)

1 Inflorescence umbellate .... **T. thalictroides**
1 Inflorescence paniculate 2
  2 Flowers bisexual; fruit scimitar-shaped .... **T. clavatum**
  2 Flowers unisexual; fruit plump, not scimitar-shaped 3
    3 Upper leaves long-petiolate; early spring flowering .... **T. dioicum**
    3 Upper leaves sessile or subsessile; summer flowering 4
      4 Achenes stipitate; filaments colored, filiform 5
        5 Plants rhizomatous; terminal leaflet shorter than wide; anthers 3.2–5 mm long .... **T. steeleanum**
        5 Plants not rhizomatous; terminal leaflet longer than wide; anthers 2–3.5 mm long .... **T. coriaceum**
      4 Achenes sessile or subsessile; filaments usually white, clavate 6
        6 Stipitate or sessile glands present on achenes and lower leaf surface, rarely muriculate or whitish papillose .... **T. revolutum**
        6 Stipitate or sessile glands absent on achenes and lower leaf surface 7
          7 Leaves more or less pubescent beneath ... **T. pubescens** var. **pubescens**
          7 Leaves glabrous beneath 8
            8 Largest leaflets entire or rarely 3-lobed apically, less than 17 mm wide .... **T. macrostylum**
            8 Largest leaflets 3-lobed apically, 20–40 mm wide .... **T. pubescens** var. **hepaticum**

T. clavatum DC., Mountain M. R.; damp woods and seeps; frequent; May–Jul; ALL.
T. coriaceum (Britt.) Small; rich woods; occasional; May–Jul; NC, SC, TN, VA.
T. dioicum L., Early M. R.; rich woods and seeps; frequent; Mar–Apr; ALL.
T. macrostylum Small & Heller; rich woods and meadows; infrequent; May–Jun; GA, NC, SC.
T. pubescens Pursh var. **hepaticum** (Greene) Keener; mesic woods, swamps, and balds; infrequent; May–Jul; GA, NC, TN.
T. pubescens Pursh var. **pubescens**; rich woods and meadows; frequent; May–Jul; ALL. *T. polygamum* Muhl.
T. revolutum DC.; dry woods and thickets; frequent; May–Jul; ALL.
T. steeleanum Boivin; moist thickets; infrequent; May–Jul; VA.
T. thalictroides (L.) Eames & Boivin, Rue Anemone; rich woods; common; Mar–May; ALL. *Anemonella thalictroides* (L.) Spach.

Trautvetteria Fischer and Meyer (False Bugbane)

T. carolinensis (Walt.) Vail; seeps, stream banks, and bogs; frequent; Jun–Jul; ALL.

Xanthorhiza Marshall (Yellowroot)

X. simplicissima Marshall; stream banks and mesic woods; frequent; Apr–May; ALL.

RHAMNACEAE (Buckthorn Family)

1 Leaves 3-nerved at base; flowers white .... **Ceanothus**
1 Leaves pinnately veined; flowers greenish-yellow .... **Rhamnus**

Ceanothus L. (New Jersey Tea)

C. americanus L.; open woods, roadsides, and waste areas; common; May–Jun; ALL.

Rhamnus L. (Buckthorn)

1 Sepals 4 . . . . **R. lanceolata**
1 Sepals 5 2
  2 Flowers bisexual; petals present; buds naked . . . . **R. caroliniana**
  2 Flowers unisexual; petals absent; buds scaly . . . . **R. alnifolia**

R. alnifolia L'Her., Alder-leaved B.; swamps and low woods; infrequent; May–Jul; VA.
R. caroliniana Walt., Carolina B.; rich woods, slopes, and roadsides; frequent; May–Jun; ALL.
R. lanceolata Pursh, Lance-leaved B.; thickets, river banks, and woodland borders; infrequent; Apr–May; VA.

ROSACEAE (Rose Family)

Robertson, K. R. 1974. The genera of Rosaceae in the southeastern United States. *J. Arnold Arbor.* 55:303–332, 344–401, 611–662.

1 Trees or shrubs 2
  2 Leaves simple 3
    3 Ovary or ovaries inferior, surrounded by and adnate to the hypanthium (floral cup) 4
      4 Petals 2× or more longer than broad; flowers racemose; ovary 6–10-locular . . . . **Amelanchier**
      4 Petals less than 2× as long as broad; flowers solitary or corymbose; ovary 2–5-locular 5
        5 Leaves with a row of glands along the midrib on the upper surface; summit of ovary densely woolly-pubescent; thorns absent . . . . **Aronia**
        5 Leaves eglandular along the midrib on the upper surface; summit of ovary glabrous or with short trichomes; thorns present or absent 6
          6 Styles coherent at base . . . . **Malus**
          6 Styles separate to base 7
            7 Flowers about 2.5 cm broad; leaves finely serrulate, shiny above . . . . **Pyrus**
            7 Flowers 1–2 cm broad; leaves coarsely toothed, often lobed, not shiny above . . . . **Crataegus**
    3 Ovary or ovaries superior, the hypanthium (floral cup) flattened or cup-shaped and free from the carpel(s) 8
      8 Pistil 1 . . . . **Prunus**
      8 Pistils 2–many 9
        9 Flowers purple, more than 2 cm broad; fruit an aggregate of drupelets . . . . **Rubus**
        9 Flowers white or pink, less than 2 cm broad; fruit a follicle 10
          10 Inflorescence paniculate or racemose . . . . **Spiraea**
          10 Inflorescence corymbose 11
            11 Stipules present . . . . **Physocarpus**
            11 Stipules absent . . . . **Spiraea**

2 Leaves compound 12
12 Trees; stems not prickly; petals less than 6 mm long . . . . . **Sorbus**
12 Shrubs; stems prickly, either upright, arching, or trailing; petals usually more than 6 mm long 13
13 Ovaries concealed within the hypanthium and appearing inferior; stipules adnate to the petiole 1/3 or more their length . . . . . **Rosa**
13 Ovaries distinctly superior; stipules not adnate to the petiole . . . . . **Rubus**
1 Herbs (if woody at the base, then less than 0.5 m tall) 14
14 Leaves simple, sometimes deeply lobed 15
15 Stamen 1; petals absent . . . . . **Alchemilla**
15 Stamens 10–many; petals present (at least on some flowers) 16
16 Petals purple; fruit juicy . . . . . **Rubus**
16 Petals white or yellow; fruit dry 17
17 Pistils numerous . . . . . **Geum**
17 Pistils 10 or fewer 18
18 Petals yellow . . . . . **Waldsteinia**
18 Petals white (sometimes absent) . . . . . **Dalibarda**
14 Leaves compound or deeply dissected 19
19 Leaves palmately compound with 5 or more leaflets 20
20 Petals yellow . . . . . **Potentilla**
20 Petals white or pink . . . . . **Rubus**
19 Leaves pinnately or ternately compound or dissected 21
21 Leaves pinnate with more than 3 leaflets 22
22 Petals absent; sepals 4, white . . . . . **Sanguisorba**
22 Petals present; sepals 5, green 23
23 Hypanthium armed at the summit with several rows of hooked bristles . . . . . **Agrimonia**
23 Hypanthium without hooked bristles 24
24 Achenes hook-beaked; plants usually less than 1 m tall . . **Geum**
24 Achenes not hook-beaked; plants 1–2 m tall . . . . . **Filipendula**
21 Leaves ternately divided or compound 25
25 Leaves twice- or thrice-ternately compound; flowers unisexual . . . . . **Aruncus**
25 Leaves once-ternately compound; flowers bisexual 26
26 Calyx lobes alternating with 5 similar bracts (epicalyx), thus appearing as 10 sepals 27
27 Epicalyx 3–5-toothed at the apex . . . . . **Duchesnea**
27 Epicalyx entire 28
28 Petals yellow . . . . . **Potentilla**
28 Petals white 29
29 Leaflets with numerous teeth along the margin . . . . . **Fragaria**
29 Leaflets with 3–5 teeth at the apex . . . . . **Potentilla**
26 Calyx lobes 5; epicalyx absent 30
30 Pistils 2–6 31
31 Petals white . . . . . **Porteranthus**
31 Petals yellow . . . . . **Waldsteinia**
30 Pistils more than 6 . . . . . **Geum**

Agrimonia L. (Agrimony)

1 Primary leaflets 11–15 .... **A. parviflora**
1 Primary leaflets 9 or fewer 2
  2 Inflorescence branches densely glandular; mature fruits 6–8 mm long; roots not fusiform-thickened .... **A. gryposepala**
  2 Inflorescence branches eglandular or occasionally with a few glands; mature fruits 1.5–5 mm long; roots fusiform-thickened 3
    3 Stems nearly glabrous or with a few spreading hairs; lower leaf surface glabrous or pubescent along the primary veins .... **A. rostellata**
    3 Stems and lower leaf surface densely pubescent 4
      4 Leaflets 5–7(–9); mature fruits 2.5–5 mm long .... **A. pubescens**
      4 Leaflets 3–5(–7); mature fruits 1.5–3 mm long .... **A. microcarpa**

A. gryposepala Wallroth; moist thickets, woodlands, and meadows; frequent; Jul–Aug; NC, SC, TN, VA.

A. microcarpa Wallroth; woodlands; infrequent; Jul–Sep; NC, VA. Closely related to and scarcely distinct from *A. pubescens*. *A. pubescens* Wallroth var. *microcarpa* (Wallroth) Ahles.

A. parviflora Ait.; moist open marshes and thickets; common; Jul–Sep; ALL.

A. pubescens Wallroth; dry woodlands; frequent; Jul–Sep; ALL.

A. rostellata Wallroth; open woodlands; frequent; Jul–Sep; ALL.

Alchemilla L.

A. microcarpa Boiss. & Reuter*; lawns and disturbed sites; infrequent but probably overlooked; Apr–May; NC, SC, TN, VA.

Amelanchier Medicus (Serviceberry, Shadbush)

1 Ovary glabrous or slightly pubescent at the summit 2
  2 Inflorescence nodding; leaves acuminate at maturity 3
    3 Leaves unfolded and densely tomentose at anthesis; fruits maroon-purple, dry .... **A. arborea**
    3 Leaves about 1/2 expanded and nearly glabrous at anthesis; fruits purple to black, juicy .... **A. laevis**
  2 Inflorescence erect; leaves rounded or slightly mucronate at maturity .... **A. canadensis**
1 Ovary densely pubescent at the summit 4
  4 Plants stoloniferous; leaves finely toothed; inflorescence erect .... **A. stolonifera**
  4 Plants not stoloniferous; leaves coarsely toothed; inflorescence nodding .... **A. sanguinea**

A. arborea (Michx. f.) Fern.; woodlands; common; Mar–May; ALL. *A. arborea* (Michx. f.) Fern. var. *austromontana* (Ashe) Ahles.

A. canadensis (L.) Medicus; woodlands; infrequent; Mar–May; GA, SC, TN, VA.

A. laevis Wiegand; openings and woodland margins; common; Mar–May; ALL. *A. arborea* (Michx. f.) Fern. var. *laevis* (Wiegand) Ahles.

A. sanguinea (Pursh) DC.; woodlands, slopes, and stream banks; infrequent; Apr–May; NC, TN, VA.

A. stolonifera Wiegand; rocky or sandy woods; infrequent; Apr–May; VA. *A. spicata* (Lam.) K. Koch.

Aronia Medicus (Chokeberry)

Hardin, J. W. 1973. The enigmatic chokeberries (*Aronia*, Rosaceae). *Bull. Torrey Bot. Club* 100:178–184.

1 Fruits bright red; twigs and leaves densely pubescent .... **A. arbutifolia**
1 Fruits black; twigs and leaves glabrous .... **A. melanocarpa**

A. arbutifolia (L.) Ell., Red C.; low woods and swamps; frequent; Mar–May; ALL. *Sorbus arbutifolia* (L.) Heynold, *Pyrus arbutifolia* (L.) L. f.

A. melanocarpa (Michx.) Ell., Black C.; lowlands and dry clearings; frequent; Mar–May; ALL. *Sorbus melanocarpa* (Michx.) Schneid., *Pyrus melanocarpa* (Michx.) Willd., *A. prunifolia* (Marsh.) Rehder.

Aruncus Schaeffer (Goat's Beard)

A. dioicus (Walt.) Fern.; rich woods; common; May–Jun; ALL. This species is strikingly similar in general aspect to *Astilbe biternata* (Vent.) Britt. (Saxifragaceae). The following key may be used for separation:

1 Trichomes simple; stamens 20; carpels 3–4, free; follicles 2–4-seeded; seeds less than 2.5 mm long .... **Aruncus**
1 Trichomes glandular; stamens 10; carpels 2, partly united; follicles many-seeded; seeds about 4 mm long .... **Astilbe**

Crataegus L. (Hawthorn)

Contributed by J. B. Phipps

Crataegus is a taxonomically difficult genus in our area, and the following notes and terms are provided as an aid to identifications:

Bark: mature bark of trunk more than 10 years old.
Branchlets: 2–3-year-old twigs unless stated otherwise.
Flower size: diameter of fully open (cupped) flowers; pressed specimens are of similar size because flattening compensates for shrinkage.
LII: leaf incision index; depth of sinus as percentage of width of 1 side of leaf.
Leaves: the short-shoot leaf is most generally useful for identification; "leaf" unqualified means short-shoot leaves; long-shoot leaves are those on shoots of rapid elongation and are often differently shaped.
Thorns: either "simple" thorns of definite growth, or sharp-tipped short-shoots; color and form terms refer only to simple thorns; *C. calpodendron*, often, and older trees or slowly growing branches of other species are more or less thornless.

1 All leaves deeply lobed; veins extending to all major sinuses 2
  2 Leaves medium-sized, tending toward conspicuously trilobed, sinuses more than 30 degrees; fruits usually with 3 nutlets .... **C. phaenopyrum**
  2 Leaves small, mostly ovate to deltoid in outline, sinuses mostly parallel-sided; fruits with 1–5 nutlets 3
    3 Mature bark exfoliating; fruit orbicular and usually less than 5 mm in diameter; nutlets 3–5 .... **C. spathulata**
    3 Mature bark rough, not exfoliating; fruit 1.2–2× as long as broad, 6–12 mm in diameter; nutlets 1–3 4

4 Inflorescence pubescent; simple thorns present; fruit glossy, bright red, usually at least 1.5× as long as broad; native .... **C. marshallii**

4 Inflorescence glabrous; simple thorns scarce or absent but short-shoots often sharp-tipped; fruit not especially glossy, dark deep red to purple-red, spherical to a little longer than broad; introduced .... **C. monogyna**

1 Short-shoot leaves with LIIs of less than 25%; sinuses of long-shoot leaves sometimes greater than 25% 5

5 Veins extending to the sinuses; at least some of the leaves of long shoots with LIIs of greater than 40% 6

6 Large bush with flattened, tablelike (tabulate) branching; leaves small, narrow-obovate or elliptic, lobes rounded; fruit small, spherical, usually less than 5 mm in diameter; mature bark flaking .... **C. spathulata**

6 Bushes of various sizes but without tabulate branching; fruits larger; bark rough 7

7 Generally small trees; leaves usually more than 2 cm wide, elliptic-ovate or narrowly so, conspicuously and evenly serrate; margins not black-glandular .... **C. viridis**

7 Small to large bushes; leaves less than 1.5 cm wide, if leaves serrate, very finely so; margins commonly black glandular (especially when young) 8

8 Leaves very small, often obovate; inflorescences compact, mostly axillary along the branches; branchlets often zigzag at the nodes, very thorny with short thorns, often drooping in large mature specimens; commonly found on sandy soils .... **C. flava**

8 Leaves larger, broad-elliptic to ovate; branches horizontal to ascending, branchlets straightish 9

9 Inflorescence 1(–3)-flowered; calyx lobes deeply glandular-serrate (almost pectinate); fruit yellow .... **C. uniflora**

9 Inflorescence 3–5-flowered; calyx lobe margins glandular-serrate but not deeply incised; fruit red .... **C. aprica**

5 Veins never extending to the sinuses; LIIs of greater than 25% confined to occasional long-shoot leaves 10

10 Leaves elliptic to oblanceolate or obovate, finely serrate or crenate, unlobed except sometimes on vegetative shoots; mature foliage sometimes conspicuously glossy and more or less coriaceous 11

11 Leaves and inflorescences pubescent; mature leaf surface dull, not shiny; veins conspicuously impressed above; short-shoot leaves occasionally jaggedly serrate 12

12 Leaves variable in shape, narrow-elliptic to obovate, often shallowly lobed; calyx lobes entire .... **C. punctata**

12 Leaves elliptic to rhombic, sometimes obovate, lobing barely discernable; calyx lobes with glandular-serrate margins .... **C. collina**

11 Leaves and inflorescences glabrous or pubescent; veins not strikingly impressed above; leaf margins entire to very finely serrate or crenate (except in 1 form); leaf surfaces often glossy at maturity .... **C. crus-galli**

10 Leaves otherwise 13

13 Leaves and inflorescences quite glabrous; fruit highly pruinose; fruiting calyx elevated on a short collar; flowering calyx more or less entire margined .... **C. pruinosa**

13 Not as above 14

14 Inflorescences few (-3)-flowered; flowers very large, 2–3 cm wide; leaves ovate to broadly elliptic, densely pubescent, with black marginal and petiolar glands . . . . **C. triflora**

14 Not as above 15

15 Thorns generally fine and straight, blackish, 4–6 cm long; inflorescences conspicuously glandular-bracteate; leaf margins and petioles glandular, but leaves thin and not tomentose . . . . **C. intricata**

15 Thorns stouter, often recurved; often brownish at least when young, 4–7 cm long; if finer, then other characters different 16

16 Bud scales large-conspicuous and coral-red at leaf expansion; nutlets radially (ventrally) pitted; thorns, when present, medium to very large, stout 17

17 Bush thorny; 2-year-old thorns stout, often long, dark mahogany-brown to blackish; calyx lobes glandular-serrate; fruit spherical, bright red and succulent when ripe; leaves broad-rhombic, ovate to broadly obtrullate, often slightly lobed, bright or dark green at maturity . . . . **C. succulenta**

17 Bush often with few or no thorns; 2-year-old thorns a little less stout, shorter, paler; calyx lobes finely glandular-serrate; fruit small, orbicular to pyriform, orange-red to vermillion when ripe, not very succulent; leaves usually larger, medium green, duller at maturity . . . . **C. calpodendron**

16 Bud scales with little expansion, early deciduous, less conspicuous; nutlets smooth on radial surfaces; thorns generally smaller . . . 18

18 Leaves 4–7 cm long; shrubs and flowers larger than the following; calyx lobes glandular-serrate; stamens 5–8 . . . . **C. pedicellata**

18 Leaves 3–4 cm long; shrubs and flowers smaller than the preceding; calyx lobes entire; stamens 10–20 (sometimes 5–8 in *C. macrosperma*) 19

19 Leaves generally ovate to rhomb-ovate, thin; thorns stout to medium in thickness 20

20 Stamens 10 . . . . **C. macrosperma**

20 Stamens 20 . . . . **C. basilica**

19 Leaves deltoid to broad-ovate with truncate bases, occasionally rhomboid, firmer, to subcoriaceous; thorns thin, blackish . . . . **C. iracunda**

C. aprica Beadle, Sunny Thorn; sunny open places, open woodlands, or rocky slopes below 2,000 ft.; occasional; Apr–May; GA, NC, SC, TN.

C. basilica Beadle, Princely Thorn; pastures, fencerows, roadsides above 2,000 ft.; infrequent; May–Jun; NC, TN, VA.

C. calpodendron (Ehrh.) Medic., Urn-tree/Pear Haw; pastures, fencerows, roadsides above 2,000 ft.; occasional; May–Jun; GA, NC, TN, VA.

C. collina Chapm., Hillside Thorn; open brushy places, fencerows, roadsides; infrequent; Apr–May; GA, NC, TN.

C. **crus-galli** L., Cockspur Thorn; fields, roadsides, fencerows, and open woodlands mostly below 2,000 ft.; frequent; Apr–May; ALL.

C. **flava** auctt. Amm., non Aiton; sunny open places, open woodlands or rocky slopes below 2,000 ft.; Mar–Apr; GA, NC, SC, TN.

C. **intricata** Lange, Intricate Haw/Biltmore H.; middle elevations but extending to the Coastal Plain; occasional; Apr–May; ALL. *C. biltmoreana* Beadle and other segregates.

C. **iracunda** Beadle, Angry Thorn; open brushy places and understory of open woods; Apr–May; GA, NC, SC, TN.

C. **macrosperma** Ashe, Common Thorn; pastures, fencerows, and roadsides between 1,500 and 4,500 ft.; occasional; May–Jun; NC, TN, VA. *C. flabellata* (Bosc.) K. Koch.

C. **marshallii** Egglest., Parsley Haw; oak-pine woodlands and disturbed areas at low elevations; frequent; Mar–Apr; GA, NC, SC, TN.

C. **monogyna** Jacq.*, English H.; an occasional escape from cultivation and more common north of our area; May–Jun; VA.

C. **pedicellata** Sarg., Scarlet H.; pastures, fencerows, and roadsides above 2,000 ft.; occasional; May–Jun; NC, TN, VA.

C. **phaenopyrum** (L. f.) Medic., Washington Thorn; alluvial soils, marsh edges; infrequent; May–Jun; NC.

C. **pruinosa** (Wendl. f.) K. Koch, Frosted Thorn; middle elevations; occasional; Apr–May; ALL. *C. georgiana* Sarg.

C. **punctata** Jacq., White Haw/Dotted Thorn; pastures, fencerows, and roadsides at about 2,000 ft.; occasional; May–Jun; NC, TN, VA.

C. **spathulata** Michx., Littlehip H.; open brushy places, fencerows, open woods, and pastures, lower elevations; locally common; Apr–May; GA, NC, SC, TN.

C. **succulenta** Link, Succulent Haw; pastures, fencerows, and roadsides above 2,000 ft.; occasional; May–Jun; NC, TN, VA. *C. neofluvialis* Ashe, *C. macrantha* Lodd.

C. **triflora** Chapm., Three-flowered Thorn; open woodlands, open brushy places, rocky slopes; very rare; Apr–May; GA, VA.

C. **uniflora** Muenchh., One-flowered Thorn; open places and woodlands; occasional; Apr–May; GA, NC, SC.

C. **viridis** L., Green Haw; ditches and low ground, in the open or below open canopy woodlands at low elevations; infrequent; Apr–May; GA, NC, SC, TN.

Many of the common names used here represent English translations of the specific epithets, and some are not found in other regional manuals.

Dalibarda L. (False Violet)

D. repens L.; bogs and moist rhododendron thickets; infrequent; Jun–Sep; NC, VA.

Duchesnea Smith (Indian Strawberry)

D. indica (Andr.) Focke*; lawns and waste areas; common; Mar–May; ALL.

Filipendula P. Miller (Queen-of-the-Prairie)

F. rubra (Hill) Robinson; bogs and wet meadows; infrequent; Jun–Jul; NC, VA.

Fragaria L. (Strawberry)

1 Achenes embedded in the receptacle; petals 7–10 mm long .... **F. virginiana**
1 Achenes not embedded in the receptacle; petals 5–7 mm long .... **F. vesca**

F. vesca L.; rich woodlands; infrequent; May–Jul; NC, SC, TN, VA. Represented in our area by subsp. **americana** (Porter) Staudt.

F. virginiana Duchesne; field and woodland borders; common; Mar–Jun; ALL.

Geum L. (Avens)

1 Basal leaves with a large, reniform terminal lobe; stem leaves reduced; style straight, not jointed; flowers 3–4 cm broad .... **G. radiatum**

1 Basal leaves with large lateral lobes, or the stem leaves not greatly reduced; style jointed near the middle or above; flowers less than 3 cm broad 2

2 Sepals reflexed at anthesis; sepallike bracts absent; aggregate of achenes elevated on a stipe above the calyx .... **G. vernum**

2 Sepals not reflexed at anthesis; sepallike bracts present; aggregate of achenes sessile 3

3 Petals spatulate; styles longer than the sepals .... **G. geniculatum**

3 Petals elliptic, ovate, oblanceolate to nearly orbicular; styles shorter than the sepals 4

4 Petals white 5

5 Receptacle densely pubescent; petals distinctly exceeding the sepals .... **G. canadense**

5 Receptacle essentially glabrous; petals equal to or slightly longer than the sepals .... **G. laciniatum**

4 Petals creamy, yellow, or orange 6

6 Petals creamy, $1/2$–$2/3$ as long as the sepals .... **G. virginianum**

6 Petals deep yellow to orange, equal to or slightly longer than the sepals .... **G. aleppicum**

G. aleppicum Jacq.; wet meadows; infrequent; Jun–Jul; NC.

G. canadense Jacq.; moist woods and stream banks; common; May–Jul; ALL.

G. geniculatum Michx.; open moist thickets at high elevations; infrequent; Jun–Aug; Roan Mt., NC–TN, Grandfather Mt., NC.

G. laciniatum Murray; wet meadows; infrequent; Jun–Jul; NC, TN, VA.

G. radiatum Michx.; exposed high-elevation rock outcrops; infrequent; Jun–Jul; NC, TN.

G. vernum (Raf.) T. & G.; moist woodlands; infrequent; Apr–May; NC, TN, VA.

G. virginianum L.; moist woodlands; frequent; Jun–Aug; NC, SC, TN, VA.

Malus Miller (Apple)

1 Leaves involute in bud, unlobed; anthers yellow; ovary fused throughout its length to the fleshy floral tube .... **M. pumila**

1 Leaves folded in the bud, often lobed; anthers pinkish; ovary in fruit pointed and free from the floral cup 2

2 Leaves lanceolate to narrowly elliptic, obtuse to short mucronate at the apex .... **M. angustifolia**

2 Leaves broadly lanceolate-ovate; acute to acuminate at the apex .... **M. coronaria**

M. angustifolia (Ait.) Michx., Wild Crab A.; fields and woodland margins; occasional; Apr–May; ALL. *Pyrus angustifolia* Ait.

M. coronaria (L.) Miller, Wild Crab A.; fields and woodland margins; occasional; Apr–May; ALL. *Pyrus coronaria* L.

M. pumila Miller*, Common A.; persistent after cultivation; occasional; Apr–May; NC, SC, TN, VA. *Pyrus malus* L.

Physocarpus (Camb.) Maxim. (Ninebark)

P. opulifolius (L.) Maxim.; bogs, bluffs, and stream banks; frequent; May–Jul; ALL.

Porteranthus Britt. (Indian Physic)

1 Stipules linear, 6–8 mm long; follicles scarcely wrinkled; seeds about 5 mm long .... **P. trifoliatus**
1 Stipules foliaceous, 10–20 mm long; follicles distinctly wrinkled and veined; seeds about 3.5 mm long .... **P. stipulatus**

P. stipulatus (Muhl. ex Willd.) Britt.; rich woods, often on basic soils; infrequent; May–Jun; GA, TN. *Gillenia stipulata* (Muhl. ex Willd.) Baill.
P. trifoliatus (L.) Britt.; rich woods, often on acidic soils; common; Apr–Jun; ALL. *Gillenia trifoliata* (L.) Moench.

Potentilla L. (Cinquefoil, Five Finger)

1 Flowers axillary, solitary; stems rooting at the nodes 2
2 Leaflets of mature leaves usually less than 1/2 as wide as long and toothed below the middle; first flower of the season in axil of the second fully developed cauline leaf .... **P. simplex**
2 Leaflets of mature leaves more than 1/2 as wide as long and toothed above the middle; first flower of the season in axil of the first fully developed cauline leaf .... **P. canadensis**
1 Flowers cymose, few to many; stems not rooting at the nodes 3
3 Flowers white; leaves evergreen; carpels pubescent .... **P. tridentata**
3 Flowers yellowish; leaves deciduous; carpels glabrous 4
4 Leaflets 3 .... **P. norvegica**
4 Leaflets 5–7 5
5 Leaves densely white-tomentose beneath, margins revolute .... **P. argentea**
5 Leaves green beneath, margins not revolute .... **P. recta**

P. argentea L.*; fields and waste areas; infrequent; May–Aug; NC, SC, VA.
P. canadensis L.; lawns, fields, and woodlands; common; Mar–May; ALL.
P. norvegica L.; lawns, fields, and woodlands; common; May–Oct; ALL.
P. recta L.*; open disturbed sites; common; Apr–Jul; NC, SC, TN, VA.
P. simplex Michx.; lawns, fields, and woodlands; common; Mar–May; ALL.
P. tridentata (Solander) Ait.; open rocky meadows, balds, and exposed rock outcrops at high elevations; occasional; Jun–Aug; GA, NC, TN, VA.

Prunus L.

1 Flowers numerous (20 or more), distinctly racemose 2
2 Large trees; leaves shiny above, crenate-serrate with callous tips; calyx lobes acute; floral cup persistent; drupes nearly black; bark aromatic .... **P. serotina**
2 Shrubs or small trees; leaves dull above, sharply serrate; calyx lobes obtuse; floral cup dehiscent; drupes red to purple; bark not aromatic .... **P. virginiana**
1 Flowers solitary or in few-flowered (fewer than 20) racemes, fascicles, umbels, or corymbs 3

3 Flowers and fruits solitary, sessile or nearly so; fruit velvety .... **P. persica**
3 Flowers and fruits more than 1, pedicellate; fruit glabrous or nearly so 4
4 Leaf blades less than 1.5× as long as wide .... **P. mahaleb**
4 Leaf blades more than 1.5× as long as wide 5
5 Flower clusters subtended by leafy bracts (cultivated cherries) 6
6 Leaves glabrous beneath; fruit sour .... **P. cerasus**
6 Leaves pubescent beneath; fruit sweet .... **P. avium**
5 Flower clusters not leafy bracted at base, subtended by bud scales only (native cherries and plums) 7
7 Leaf teeth with a distinct gland or callous thickening 8
8 Leaves 6–12 cm long; sepals glabrous on the inner surface; pedicels 1–1.5 cm long .... **P. pensylvanica**
8 Leaves 3–6 cm long; sepals pubescent on the inner surface, especially at base; pedicels 3–8 mm long .... **P. angustifolia**
7 Leaf teeth glandless 9
9 Low shrubs rarely more than 1 m tall; leaves essentially entire below the middle or with teeth 1–4 mm apart; sepals with distinct marginal glands .... **P. pumila**
9 Shrubs over 1 m tall or small trees; leaves serrate to near the base; sepals usually without marginal glands 10
10 Leaf tips acuminate; sepals glabrous on the outer surface; fruit red .... **P. americana**
10 Leaf tips acute or obtuse; sepals pubescent on the outer surface; fruit dark purple .... **P. alleghaniensis**

P. alleghaniensis Porter, Allegheny Sloe; dry uplands; infrequent; Apr–May; VA.

P. americana Marsh., Wild Plum; thickets, stream banks, and woodlands; frequent; Mar–Apr; ALL.

P. angustifolia Marsh., Chickasaw Plum; thickets, fencerows, and woodlands; occasional; Mar–Apr; ALL.

P. avium L.*, Sweet Cherry; persistent after cultivation; occasional; Apr–May; NC, TN, VA.

P. cerasus L.*, Sour Cherry; persistent after cultivation; infrequent; Apr–May; ALL.

P. mahaleb L.*, Mahaleb Cherry; occasionally escaping to stream and river banks; infrequent; Apr–May; NC, SC, TN, VA.

P. pensylvanica L. f., Fire/Pin Cherry; clearings and burned areas, usually above 3,000 ft.; frequent; Apr–May; GA, NC, TN, VA.

P. persica (L.) Batsch*, Peach; persistent and occasionally escaping; Mar–Apr; NC, SC, TN, VA.

P. pumila L., Sand Cherry; dry sandy woods; infrequent; Apr–May; NC, VA. *P. cuneata* Raf., *P. susquehanae* Willd.

P. serotina Ehrhart, Wild Black Cherry; fields and woodlands; common; Apr–May; ALL.

P. virginiana L., Choke Cherry; woodlands and open balds; occasional; Apr–Jun; NC, SC, TN, VA.

Pyrus L. (Pear)

P. communis L.*; persistent after cultivation; occasional; Apr–May; NC, SC, TN, VA.

Rosa L. (Rose)

1 Styles exserted, forming a distinct column above the summit of the hypanthium, about ½ as long as the stamens 2
2 Leaflets 3(5) . . . . . **R. setigera**
2 Leaflets (5)7–9 3
3 Stipules fimbriate-pectinate; styles glabrous; stems arching or trailing . . . . . **R. multiflora**
3 Stipules dentate; styles pubescent; stems prostrate or trailing . . . **R. wichuraiana**
1 Styles included to scarcely extending beyond the tip of the hypanthium 4
4 Outer calyx lobes pinnate; carpels long-stipitate (introduced taxa) 5
5 Leaflets pubescent and densely glandular beneath 6
6 Styles pubescent . . . . . **R. eglanteria**
6 Styles glabrous . . . . . **R. micrantha**
5 Leaflets glabrous and eglandular beneath . . . . . **R. canina**
4 Outer calyx lobes entire or with only a few linear appendages; carpels inserted at the base of the hypanthium (native taxa) 7
7 Infrastipular prickles thin, round in cross section, and only slightly broadened at the base; internodal prickles and bristles abundant . . . . . **R. carolina**
7 Infrastipular prickles large, stout, usually flattened in cross section, and broadened at the base; internodal prickles and bristles few to absent 8
8 Leaflets finely toothed, mid-leaflet teeth about 0.5 mm high; flowers often solitary; stipular auricles less than 2.5 mm wide . . . . . **R. palustris**
8 Leaflets coarsely toothed, mid-leaflet teeth about 1 mm high; inflorescence usually of 3 or more flowers; stipular auricles more than 2.5 mm wide . . . . . **R. virginiana**

R. canina L.*, Dog R.; waste places; infrequent; May–Jun; NC, VA.
R. carolina L.; open pastures and woodlands; common; May–Jul; ALL.
R. eglanteria L.*, Sweetbrier R.; waste places; occasional; May–Jun; GA, NC, TN, VA.
R. micrantha Smith*, Sweetbrier R.; waste places; infrequent; May–Jun; NC, VA.
R. multiflora Thunb.*; waste places; frequent; May–Jun; NC, SC, TN, VA.
R. palustris Marsh., Swamp R.; swamps and low woods; common; Jul–Sep; ALL.
R. setigera Michx., Prairie R.; open disturbed sites; infrequent; May–Jun; NC, TN, VA.
R. virginiana Mill.; open areas; infrequent; May–Jul; VA.
R. wichuraiana Crepin*; waste places; occasional; May–Jun; NC, SC, VA.

Rubus L. (Raspberry, Dewberry, Blackberry)

1 Leaves simple, palmately lobed; petals purplish; receptacle nearly flat . . . **R. odoratus**
1 Leaves pinnately or palmately compound; petals white or pink; receptacle conical . . 2
2 Petals equal to or shorter than the sepals; fruits separating from the receptacle at maturity (raspberries) 3
3 Pedicels without gland-tipped bristles; fruit black or yellow; stems thorny and usually glaucous . . . . . **R. occidentalis**
3 Pedicels with slender, gland-tipped bristles; fruit red; stems bristly, often with a few thorns and usually not glaucous 4
4 Stems, pedicels, and sepals densely glandular-bristly pubescent; fruits shiny and sticky, not pubescent . . . . . **R. phoenicolasius**

4 Stems, pedicels, and sepals only slightly glandular-pubescent; fruits grayish-pubescent . . . . **R. idaeus**

2 Petals longer than the sepals; fruits retained on the deciduous or persistent receptacle 5

5 Leaves white or grayish-tomentose beneath (blackberries) 6

6 Inflorescence paniculate, flowers numerous . . . . **R. bifrons**

6 Inflorescence cymose, flowers few . . . . **R. cuneifolius**

5 Leaves not white or grayish-tomentose beneath 7

7 Stems trailing (dewberries) 8

8 Petals about 1 cm or less long; stems with bristles; stem prickles, if present, thin and with narrow bases . . . . **R. hispidus**

8 Petals 1.5 cm or more long; bristles present or absent; stem prickles stiff, broad based, and recurved 9

9 Flowers mostly solitary; bristles usually present . . . . **R. trivialis**

9 Flowers mostly 2 or more/branch; bristles absent . . . . **R. flagellaris**

7 Stems erect or arching (blackberries) 10

10 Pedicels with gland-tipped bristles . . . . **R. allegheniensis**

10 Pedicels without gland-tipped bristles 11

11 Leaflets deeply divided or lobed . . . . **R. laciniatus**

11 Leaflets merely toothed, not deeply divided or lobed 12

12 Leaves soft-pubescent beneath . . . . **R. argutus**

12 Leaves glabrous or with a few scattered hairs 13

13 Stems of the current season unarmed or with a few slender prickles . . . . **R. canadensis**

13 Stems of the current season with stiff prickles . . . **R. betulifolius**

R. allegheniensis Porter; thickets and disturbed sites; common; May–Jun; NC, TN, VA.

R. argutus Link; disturbed sites; common; Apr–May; NC, SC, TN, VA. *R. pensilvanicus* Poir.

R. betulifolius Small; disturbed sites; infrequent and more common on the Coastal Plain; Apr–May; NC, SC, TN.

R. bifrons Vest*; waste areas; infrequent; May–Jun; NC, SC, TN, VA.

R. canadensis L.; high-elevation clearings; occasional; Jun–Jul; ALL.

R. cuneifolius Pursh; disturbed sites; infrequent and more common on the Coastal Plain; Apr–Jun; VA.

R. flagellaris Willd.; disturbed sites; common; Apr–May; NC, SC, TN, VA. *R. enslenii* Tratt.

R. hispidus L.; disturbed, often damp clearings; common; Apr–May; ALL.

R. idaeus L.; high-elevation clearings; infrequent; Jun–Aug; NC, TN, VA. Represented in our area by subsp. **sachalinensis** (Levl.) Focke.

R. laciniatus Willd.*; waste areas; infrequent; May–Jun; NC, TN, VA.

R. occidentalis L.; disturbed areas; common; Apr–Jun; ALL.

R. odoratus L.; clearings adjacent to rich woods; frequent; Jun–Aug; ALL.

R. phoenicolasius Maxim.*, Wineberry; disturbed sites; frequent; May–Jun; GA, NC, TN, VA.

R. trivialis Michx.; disturbed sites; infrequent; Mar–Apr; NC, SC, TN.

This treatment is conservative, and I have made no attempts to add any significant taxonomic or nomenclatural confusion to that already available in other floras and taxonomic literature. Long-term field, garden, and experimental data are much needed in this genus.

Sanguisorba L. (American Burnet)

S. canadensis L.; bogs, seeps, and wet meadows; occasional; Jul–Sep; GA, NC, TN, VA.

Sorbus L. (Mountain Ash)

S. americana Marsh.; clearings and woodland margins at high elevations; frequent; Jun–Jul; GA, NC, TN, VA. *Pyrus americana* (Marsh.) DC.

Spiraea L. (Spiraea)

1 Inflorescence an elongate panicle or raceme 2
  2 Leaves densely tomentose beneath; calyx lobes reflexed; flowers usually pinkish . . . . **S. tomentosa**
  2 Leaves glabrous or nearly so; calyx lobes spreading; flowers white or, rarely, pale pink 3
    3 Inflorescence branches pubescent; stems yellowish-brown; leaves finely serrate . . . . **S. alba**
    3 Inflorescence branches glabrous or nearly so; stems reddish; leaves coarsely toothed . . . . **S. latifolia**
1 Inflorescence corymbose 4
  4 Leaves long-acuminate; pedicels and floral cups conspicuously pubescent . . . . **S. japonica**
  4 Leaves acute or obtuse; pedicels and floral cups glabrous or nearly so 5
    5 Leaves obovate, nearly entire; pedicels, floral cups, and lower leaf surfaces glaucous . . . . **S. virginiana**
    5 Leaves elliptic to broadly ovate, coarsely serrate; pedicels, floral cups, and lower leaf surfaces not glaucous . . . . **S. betulifolia**

S. alba Du Roi, Meadowsweet; bogs and moist thickets; occasional; Jun–Sep; NC, TN, VA.

S. betulifolia Pallas; stream banks and rocky woods; occasional; Jun–Aug; NC, VA. Represented in our area by var. **corymbosa** (Raf.) Wenzig.

S. japonica L. f.*; roadsides and thickets; occasional; Jun–Jul; ALL.

S. latifolia (Ait.) Borkh., Meadowsweet; bogs and moist thickets; occasional; Jun–Sep; NC, VA.

S. tomentosa L., Hardhack; swamps and moist woods; frequent; Jul–Sep; ALL.

S. virginiana Britt.; rocky stream banks; infrequent; Jun–Jul; NC, TN, VA.

Waldsteinia Willd. (Barren Strawberry)

1 Leaves ternately compound . . . . **W. fragarioides**
1 Leaves 3–5-lobed . . . . **W. lobata**

W. fragarioides (Michx.) Tratt.; dry to mesic woods; frequent; Mar–May; ALL. *W. donniana* Tratt., *W. parviflora* Small.

W. lobata (Baldw.) T. & G.; moist woods near streams; infrequent; Apr–May; GA, SC.

RUBIACEAE (Madder Family)

1 Shrubs .... **Cephalanthus**
1 Herbs 2
2 Leaves whorled 3
3 Flowers white, yellow, greenish, or purple, in loose cymes or, rarely, dense panicles; corolla tube essentially absent .... **Galium**
3 Flowers pinkish, in involucrate heads; corolla tube evident .... **Sherardia**
2 Leaves opposite 4
4 Leaves evergreen; flowers paired, the ovaries fused; fruit a red berry .. **Mitchella**
4 Leaves not evergreen; flowers not paired, ovary single; fruit a capsule 5
5 Flowers pedicellate, cymose or appearing solitary .... **Houstonia**
5 Flowers sessile, axillary .... **Diodia**

Cephalanthus L. (Button Bush)

C. occidentalis L.; low woods, margins of ponds, and stream banks; frequent; Jun–Jul; ALL.

Diodia L. (Buttonweed)

1 Calyx lobes 2; style divided; flowers white .... **D. virginiana**
1 Calyx lobes 4; style undivided; flowers pink, rarely white .... **D. teres**

D. teres Walt.; fields and waste places; frequent; Jun–Sep; ALL.

D. virginiana L.; wet ditches and shorelines; frequent; Jun–Sep; ALL. Usually in wetter sites than those of the preceding species.

Galium L. (Bedstraw)

1 Fruit fleshy, purplish, 3–3.5 mm long .... **G. hispidulum**
1 Fruit dry, green and becoming dark with age, usually less than 3 mm long 2
2 Fruit distinctly bristly or hairy 3
3 Stem leaves 5–8/node; stems weak and reclining 4
4 Stems retrorsely scabrous; leaves usually 8/node .... **G. aparine**
4 Stems glabrous or weakly retrorsely scabrous; leaves usually 6/node .... **G. triflorum**
3 Stem leaves 4/node; stems erect or ascending 5
5 Flowers sessile; inflorescence few-flowered, arising from the upper 2–4 nodes 6
6 Leaves elliptic-ovate, broadest near the middle .... **G. circaezans**
6 Leaves lanceolate, broadest near the base .... **G. lanceolatum**
5 Flowers distinctly pedicellate; inflorescence several to many-flowered, usually arising from the upper 5–12 nodes 7
7 Flowers in dense panicles, bright white; leaves mostly less than 5 mm wide .... **G. boreale**
7 Flowers in loose panicles, greenish-white or maroon; leaves mostly more than 5 mm wide .... **G. pilosum**
2 Fruit smooth or nearly so 8
8 Stems weakly ascending, decumbent, supported by other plants, or matted ... 9
9 Leaves 4/node 10

10 Stems densely pubescent, rarely more than 2 dm long .... **G. pedemontanum**
10 Stems glabrous or puberulent, 2–8 dm long 11
11 Leaves filiform, 1–2 mm wide .... **G. obtusum** var. **filifolium**
11 Leaves lanceolate to oblanceolate, 2–8 mm wide .... **G. obtusum** var. **obtusum**
9 Leaves 5–8/node 12
12 At least some corollas 3-lobed; upper stem scabrous, but not retrorsely so .... **G. tinctorium**
12 Corollas 4-lobed; upper stem smooth or retrorsely scabrous 13
13 Stem smooth .... **G. concinnum**
13 Stem retrorsely scabrous .... **G. asprellum**
8 Plants erect 14
14 Petals purple or maroon .... **G. latifolium**
14 Petals white, greenish-white, or yellow 15
15 Petals yellow .... **G. verum**
15 Petals white or greenish-white 16
16 Leaves 4/node; flowers in dense panicles, petals bright white .... **G. boreale**
16 Leaves 5–8/node; flowers in loose panicles, petals greenish-white 17
17 Stems retrorsely scabrous, 1–2 dm long .... **G. parisiense**
17 Stems smooth, more than 2 dm long .... **G. mollugo**

G. aparine L.; disturbed sites; common; Apr–May; ALL.

G. asprellum Michx.; moist woodlands; occasional; May–Aug; NC, TN, VA.

G. boreale L.; fields and woodlands; infrequent; May–Aug; VA. Variable in degree of fruit pubescence.

G. circaezans Michx.; rich woods; common; Apr–Jun; ALL.

G. concinnum T. & G.; dry woodlands; infrequent; Jun–Aug; TN, VA.

G. hispidulum Michx.; dry woodlands; infrequent; Jun–Aug; NC.

G. lanceolatum Torr.; rich woodlands; occasional; Jun–Jul; ALL.

G. latifolium Michx.; rich woods; frequent; May–Jul; ALL.

G. mollugo L.*; disturbed sites; occasional; May–Jun; NC, TN, VA.

G. obtusum Bigelow var. **filifolium** (Wieg.) Fern.; low wet areas; infrequent and more common on the Piedmont and Coastal Plain; Apr–Jun; VA.

G. obtusum Bigelow var. **obtusum**; low woods and swamps; occasional; Apr–Jun; TN, VA.

G. parisiense L.*; disturbed sites; occasional; Jun–Aug; VA.

G. pedemontanum (Bellardi) Allioni*; lawns and waste areas; occasional; Apr–Jun; NC, TN, VA.

G. pilosum Ait.; open woodlands; common; May–Aug; ALL.

G. tinctorium L.; wet open sites; frequent; Apr–Jun; ALL.

G. triflorum Michx.; woods and roadsides; frequent; Jul–Aug; ALL.

G. verum L.*; disturbed sites; occasional; Jun–Aug; NC, VA.

Houstonia L. (Bluets)

Terrell, E. E. 1959. A revision of the *Houstonia purpurea* group (Rubiaceae). *Rhodora* 61:157–180, 188–207.

1 Flowers solitary on terminal or axillary pedicels 2
  2 Stems prostrate . . . . **H. serpyllifolia**
  2 Stems erect 3
    3 Corolla throat yellow; perennials . . . . **H. caerulea**
    3 Corolla throat reddish; annuals . . . . **H. pusilla**
1 Flowers several, in terminal cymes 4
  4 Cauline leaves chiefly ovate, widest at or near the base 5
    5 Corolla deep purple; median cauline leaves 1–3 cm long; internodes glabrous . . . . **H. montana**
    5 Corolla white to purplish; median cauline leaves 2.5–6 cm long; internodes variously pubescent . . . . **H. purpurea**
  4 Cauline leaves chiefly linear to elliptic, widest at or above the middle 6
    6 Basal leaves distinctly ciliate and present at anthesis; median cauline leaves elliptic to obovate . . . . **H. canadensis**
    6 Basal leaves either glabrous or absent at anthesis; median cauline leaves linear to lanceolate 7
      7 Leaves mostly 2.5 mm or less wide; internodes 3–9 . . . . **H. tenuifolia**
      7 Leaves mostly more than 2.5 mm wide; internodes 6–13 . . . . . . **H. longifolia**

H. caerulea L.; woodlands and openings; frequent; Apr–May; ALL.

H. canadensis Willd.; woodlands and openings; infrequent; Apr–Jun; VA.

H. longifolia Gaertner; dry open woodlands and disturbed sites; occasional; May–Jul; ALL.

H. montana Small; exposed high-elevation rock outcrops; infrequent; Jun–Aug; NC, TN. Closely related to and sometimes considered to be a variety of or synonymous with *H. purpurea* L.

H. purpurea L.; woodlands; common; Apr–Jul; ALL.

H. pusilla Schoepf; woodlands and fields; infrequent; Mar–Apr; NC, SC.

H. serpyllifolia Michx.; stream banks and rich woods; frequent; Apr–Jun; ALL.

H. tenuifolia Nutt.; rocky slopes and dry woods; frequent; Jun–Jul; ALL.

Mitchella L. (Partridge Berry)

M. repens L.; rich woods; common; May–Jun; ALL.

Sherardia L.

S. arvensis L.*; lawns and waste places; occasional; Apr–Aug; NC, SC, TN, VA.

RUTACEAE (Rue Family)

1 Leaves 3-foliolate . . . . **Ptelea**
1 Leaves odd-pinnate, with 7–13 leaflets . . . . **Zanthoxylum**

Ptelea L. (Wafer Ash)

P. trifoliata L.; rocky woodlands; occasional; Apr–Jun; ALL.

Zanthoxylum L. (Toothache Tree)

Z. americanum Mill.; woodlands and river banks; infrequent; Apr–May; NC, TN, VA. Mostly introduced in our area.

SALICACEAE (Willow Family)

1 Leaves 1.5× or more longer than broad; inflorescence erect or nearly so; bracts entire . . . . **Salix**
1 Leaves about as broad as long; inflorescence drooping; bracts serrate . . . . . . **Populus**

Populus L.

1 Leaves and twigs white-tomentose . . . . **P. alba**
1 Leaves and twigs essentially glabrous 2
2 Petiole round . . . . **P. × gileadensis**
2 Petiole flat, at least at the base 3
3 Leaves triangular . . . . **P. deltoides**
3 Leaves rhombic 4
4 Leaves with more than 12 teeth/side . . . . **P. tremuloides**
4 Leaves with fewer than 12 teeth/side . . . . **P. grandidentata**

P. alba L.*, Silver Poplar; persistent after cultivation; occasional; Mar–Apr; NC, SC, TN, VA.

P. deltoides Bartr. ex Marsh., Cottonwood; lowlands; occasional; Mar–Apr; GA, NC, TN, VA.

P. × gileadensis Roleau*, Balm-of-Gilead; persistent after cultivation; infrequent; Apr–May; GA, NC, TN, VA.

P. grandidentata Michx., Bigtooth Aspen; dry uplands and shrub balds; infrequent; Apr–May; NC, SC, VA.

P. tremuloides Michx., Quaking Aspen; woodlands; infrequent; Apr–May; VA.

Salix L. (Willow)

Argus, G. W. 1986. The genus *Salix* (Salicaceae) in the southeastern United States. *Syst. Bot. Mono.* 9:1–170.

1 Leaves green or pale beneath 2
2 Bud apex sharp-pointed; bud scale margin free and overlapping . . . . **S. nigra**
2 Bud apex blunt; bud scale margin fused 3
3 Leaves lanceolate or elliptic-lanceolate, the margin serrate . . . . . **S. eriocephala**
3 Leaves linear, the margin distinctly glandular-denticulate . . . . **S. exigua**
1 Leaves glaucous beneath 4
4 Bud apex sharp-pointed; bud scale margin free and overlapping . . . **S. caroliniana**
4 Bud apex blunt; bud scale margin fused 5
5 Leaf margin serrate or serrulate 6
6 Native shrubs 7
7 Leaves densely sericeous beneath; stipules reduced to small glands or absent . . . . **S. sericea**
7 Leaves glabrous or essentially so beneath; stipules evident . . . . **S. eriocephala**
6 Introduced trees 8
8 Leaves narrowly lanceolate, the margins serrulate, blades sericeous beneath . . . . **S. alba**
8 Leaves very narrowly lanceolate, the margins spinulose-serrulate, blades nearly glabrous beneath . . . . **S. babylonica**

5 Leaf margin entire or crenate 9
9 Leaves stipulate, blades 5–9 cm long, petioles 3–7 mm long . . . . **S. humilis** var. **humilis**
9 Leaves without stipules, blades 2.5–5 cm long, petioles 0.5–3 mm long . . . . **S. humilis** var. **microphylla**

S. alba L.*, White W.; escaped to moist sites; occasional; Mar–Apr; ALL.
S. babylonica L.*, Weeping W.; persistent in moist soils; occasional; Mar–Apr; ALL.
S. caroliniana Michx., Swamp W.; streams and low areas; infrequent; Mar–Apr; ALL.
S. eriocephala Michx., Heart-leaved W.; stream banks and low woods; infrequent; Apr–May; GA, VA.
S. exigua Nutt., Sandbar W.; sandbars and floodplains; infrequent; Apr–May; VA. *S. interior* Rowlee. More common north and west of our area.
S. humilis Marsh. var. **humilis**, Upland W.; open upland woods and balds; frequent; Mar–May; ALL.
S. humilis Marsh. var. **microphylla** (Andersson) Fern.; open upland woods and balds; occasional; Mar–May; GA, NC, TN, VA. *S. tristis* Ait.
S. nigra Marsh., Black W.; stream banks and low areas; common; Mar–Apr; ALL.
S. sericea Marsh., Silky W.; low areas; frequent; Mar–Apr; ALL.

Salix lucida Muhl. is thought to be a naturalized introduction in Roanoke Co., VA.

SANTALACEAE (Sandalwood Family)

1 Herbs . . . . **Comandra**
1 Shrubs 2
2 Leaves alternate . . . . **Pyrularia**
2 Leaves opposite 3
3 Peduncle of staminate umbel elongate; pistillate flower without bracts; fruit globose . . . . **Nestronia**
3 Peduncle of staminate umbel nearly sessile; pistillate flower with 4 sepallike bracts; fruit ellipsoid . . . . **Buckleya**

Buckleya Torr. (Pirate Bush)

B. distichophylla (Nutt.) Torr.; dry open woods; infrequent; Apr–May; NC, TN, VA. The 2-ranked leaves and green twigs often give the appearance of compound leaves. This parasite is often associated with, but not host-specific to, hemlock.

Comandra Nutt. (Bastard Toadflax)

C. umbellata (L.) Nutt.; dry open woods and roadsides; frequent; May–Jun; ALL.

Nestronia Raf.

N. umbellula Raf.; upland woods; infrequent; May–Jun; SC. More typical of the Piedmont.

Pyrularia Michx. (Buffalo Nut)

P. pubera Michx.; rich woods and woodland margins; common; Apr–May; ALL. The fruits are poisonous.

SARRACENIACEAE (Pitcher Plant Family)

Sarracenia L. (Pitcher Plant)

McDaniel, S. 1971. The genus *Sarracenia* (Sarraceniaceae). *Bull. Tall Timbers Res. Sta.* 9:1–36.

1 Corolla maroon to pink 2
  2 Leaves decumbent to ascending . . . . **S. purpurea**
  2 Leaves erect . . . . **S. jonesii**
1 Corolla yellow 3
  3 Nonpitchered leaves recurved, usually more numerous than the pitchered leaves . . . . **S. oreophila**
  3 Nonpitchered leaves erect, fewer than the pitchered leaves . . . . **S. flava**

S. flava L., Trumpets; bogs and lowlands; infrequent; Apr–May; NC, SC.
S. jonesii Wherry, Sweet P. P.; stream banks; infrequent; May–Jun; NC, SC. *S. rubra* Walt. subsp. *jonesii* (Wherry) Wherry.
S. oreophila (Kearney) Wherry, Green P. P.; open, wet stream banks; infrequent; Apr–May; GA, NC.
S. purpurea L.; boggy areas; infrequent; Apr–May; GA, NC, SC.

SAURURACEAE (Lizard's Tail Family)

Saururus L. (Lizard's Tail)

S. cernuus L.; shorelines of swamps and shallow waters; infrequent; May–Jul; NC, TN, VA.

SAXIFRAGACEAE (Saxifrage Family)

1 Plants woody 2
  2 Vines . . . . **Decumaria**
  2 Shrubs 3
    3 Leaves alternate or clustered on short branches 4
      4 Leaves palmately lobed; ovary inferior, fruit a berry . . . . **Ribes**
      4 Leaves not lobed; ovary superior, fruit a capsule . . . . **Itea**
    3 Leaves opposite 5
      5 Flowers borne singly or in 2s or 3s; petals 4, more than 10 mm long . . . . **Philadelphus**
      5 Flowers borne in dense terminal cymes; petals 5, those of the fertile flowers 2 mm or less long . . . . **Hydrangea**
1 Plants herbaceous 6
  6 Staminodia present; petals more than 10 mm long . . . . **Parnassia**
  6 Staminodia absent; petals less than 10 mm long or absent 7
    7 Leaves ternately decompound; flowers unisexual . . . . **Astilbe**
    7 Leaves simple; flowers bisexual 8
      8 Carpels 5–7 . . . . **Penthorum**
      8 Carpels 2 or 3 9
        9 Petals absent; flowers axillary or in short cymes . . . . **Chrysosplenium**
        9 Petals present; flowers in racemes or panicles 10

10 Stem leaves opposite or subopposite; petals pinnatifid or fringed ...... **Mitella**
10 Stem leaves, if present, alternate; petals not pinnatifid or fringed 11
 11 Ovary 1-celled 12
  12 Stamens 10; inflorescence racemose; carpels unequal at maturity .... **Tiarella**
  12 Stamens 5; inflorescence paniculate; carpels equal at maturity .... **Heuchera**
 11 Ovary 2-celled 13
  13 Stamens 5; leaves palmately lobed .... **Boykinia**
  13 Stamens 10; leaves not palmately lobed .... **Saxifraga**

Astilbe Hamilton (False Goatsbeard)

1 Leaf margin serrate; fruit lanceolate, 4–5 mm long .... **A. biternata**
1 Leaf margin crenate; fruit broadly ovoid, 3 mm long .... **A. crenatiloba**

A. biternata (Vent.) Britt.; rich woods; frequent; May–Jun; ALL.

A. crenatiloba (Britton) Small; endemic to wooded slopes of Roan Mt., NC, TN, and not seen since the original collection in 1888. Probably extinct.

See under *Aruncus* (Rosaceae) for a morphological comparison of these taxa to *Aruncus dioicus*.

Boykinia Nutt.

B. aconitifolia Nutt.; moist woods and stream banks; occasional; Jun–Jul; ALL.

Chrysosplenium L. (Golden Saxifrage)

C. americanum Schweinitz; seeps and small streams; occasional; Mar–Jun; NC, SC, TN, VA. A diminutive plant, easily overlooked.

Decumaria L. (Climbing Hydrangea)

D. barbara L.; open woodlands and thickets; infrequent; May–Jun; GA, NC, SC, TN. More common on the Coastal Plain.

Heuchera L. (Alumroot)

Wells, E. F. 1984. A revision of *Heuchera* (Saxifragaceae) in eastern North America. *Syst. Bot. Mono.* 3:45–121.

1 Calyx villous; plants flowering in midsummer to fall 2
 2 Leaf lobes acute; seeds spiny; pedicels less than 3 mm long .... **H. villosa**
 2 Leaf lobes rounded; seeds smooth; pedicels more than 3 mm long .. **H. parviflora**
1 Calyx glandular-pubescent; plants flowering in spring and early summer 3
 3 Stamens exserted; free portion of the hypanthium less than 2 mm long .... **H. americana**
 3 Stamens included; free portion of the hypanthium more than 2 mm long 4
  4 Styles conspicuously shorter than the calyx; flowers horizontal at anthesis .... **H. longiflora**
  4 Styles equal to or slightly longer than the calyx; flowers ascending at anthesis .... **H. pubescens**

H. americana L.; rich open woods and roadsides; occasional; Apr–Jun; ALL. *H. americana* L. var. *hispida* (Pursh) Wells.
H. longiflora Rydb.; open or shaded calcareous sites; infrequent; May–Jun; NC, TN.
H. parviflora Bartl.; shaded sandstone or limestone ledges; infrequent; Jul–Oct; NC, SC, TN.
H. pubescens Pursh; open sites on basic soils; occasional; May–Jun; NC, VA. Hybridizes with *H. americana*.
H. villosa Michx.; woodlands and rock outcrops; frequent; Jul–Oct; ALL.

Hydrangea L.

Pilatowski, R. E. 1982. A taxonomic study of the *Hydrangea arborescens* complex. *Castanea* 47:84–98.

1 Leaves green beneath, glabrous or puberulent; sterile flowers usually absent, but if present, usually less than 1 cm wide .... **H. arborescens**
1 Leaves gray to nearly white beneath; sterile flowers, if present, usually more than 1 cm wide 2
2 Leaves gray beneath, the trichomes not dense enough to mask the epidermis .... **H. cinerea**
2 Leaves grayish-white beneath, the trichomes masking the epidermis ... **H. radiata**

H. arborescens L.; open slopes, roadsides, and woodland margins; common; May–Jul; ALL.
H. cinerea Small; open slopes, roadsides, and woodland margins; occasional; May–Jul; NC, SC, TN. *H. arborescens* L. subsp. *discolor* (Seringe) McClintock.
H. radiata Walt.; open slopes, roadsides, and woodland margins; frequent; May–Jul; GA, NC, SC, TN. *H. arborescens* L. subsp. *radiata* (Walt.) McClintock.

Itea L. (Virginia Willow)

I. virginica L.; low woods and stream banks; occasional; May–Jun; ALL.

Mitella L. (Bishop's Cap, Miterwort)

M. diphylla L.; rich woods; occasional; Apr–Jun; ALL.

Parnassia L. (Grass-of-Parnassus)

1 Leaves reniform; petals clawed .... **P. asarifolia**
1 Leaves ovate to oblong; petals not clawed .... **P. grandifolia**

P. asarifolia Vent.; bogs and seeps; occasional; Aug–Oct; ALL.
P. grandifolia DC.; seepage areas; infrequent; Sep–Oct; NC, SC, TN, VA.

Penthorum L. (Ditch Stonecrop)

P. sedoides L.; low woods and swamps; occasional; Jun–Oct; ALL. Sometimes placed in the Crassulaceae. For a discussion of anatomical evidence suggesting the placement of this taxon in the monogeneric Penthoraceae, see M. L. Haskins and W. J. Hayden, Anatomy and affinities of *Penthorum, Amer. J. Bot.* 74 (1987): 164–177.

Philadelphus L. (Mock Orange)

1 Calyx hirsute; styles united; seeds tailless .... **P. hirsutus**
1 Calyx glabrous or essentially so; styles separate above; seeds caudate ... **P. inodorus**

P. hirsutus Nutt.; bluffs and stream banks; occasional; Apr–May; ALL.

P. inodorus L.; bluffs and stream banks; occasional; Apr–May; ALL.

A polymorphic complex in need of critical study.

Ribes L.

1 Pedicel jointed at the summit; flowers several in elongate racemes 2
2 Stems with bristles and weak thorns .... **R. lacustre**
2 Stems unarmed 3
3 Ovary and fruit with glandular bristles .... **R. glandulosum**
3 Ovary and fruit smooth .... **R. americanum**
1 Pedicel not jointed at the summit; peduncles few-flowered 4
4 Stamens longer than the sepals at anthesis; ovary and fruit glabrous .... **R. rotundifolium**
4 Stamens about as long as the sepals at anthesis; ovary and fruit with bristles or spines .... **R. cynosbati**

R. americanum Miller, Wild Black Currant; rich woods; infrequent; Apr–Jun; TN, VA.

R. cynosbati L., Gooseberry; rich woods and balds; occasional; May–Jun; GA, NC, TN, VA.

R. glandulosum Grauer, Skunk Currant; rich woods at high elevations, especially boulderfields; occasional; May–Jun; NC, TN, VA.

R. lacustre (Pers.) Poir., Bristly Black Currant; infrequent; May–Jun; VA.

R. rotundifolium Michx., Gooseberry; rich woods at high elevations; occasional; Apr–May; NC, SC, TN, VA.

Saxifraga L.

Lord, L. P. 1961. The genus *Saxifraga* in the southern Appalachians. Ph.D. diss., University of Tennessee, Knoxville.

1 Corolla zygomorphic .... **S. michauxii**
1 Corolla actinomorphic 2
2 Sepals reflexed at late anthesis 3
3 Leaves entire to obscurely serrulate .... **S. pensylvanica**
3 Leaves coarsely dentate 4
4 Leaf blades 3–8× as long as wide .... **S. micranthidifolia**
4 Leaf blades about as long as wide .... **S. caroliniana**
2 Sepals spreading or ascending at late anthesis 5
5 Carpels adnate to the floral cup forming a distinct hypanthium; stamen filaments 1–1.5 mm long .... **S. virginiensis**
5 Carpels not adnate to the floral cup; stamen filaments 2.5–3.5 mm long .... **S. careyana**

S. careyana Gray; moist rocks and cliffs; occasional; May–Jun; NC, SC, TN, VA.

S. caroliniana Gray; moist shaded rocks and slopes; infrequent; May–Jun; NC, TN, VA. Closely related to *S. careyana* but usually separated by reflexed sepals and clavate filaments; this complex is in need of further study.

S. michauxii Britt.; moist, usually open rocks and slopes; frequent; Jun–Aug; ALL.

S. micranthidifolia (Haw.) Steud., Brook Lettuce; seeps and streams; occasional; May–Jun; ALL.

S. pensylvanica L., Swamp Saxifrage; wet meadows and bogs; infrequent; Apr–Jun; NC, VA.

S. virginiensis Michx., Early Saxifrage; dry rocky woodlands, often abundant on basic soils; Mar–Apr; ALL.

Tiarella L. (Foamflower)

T. cordifolia L.; rich woods; common; Apr–Jun; ALL. *T. wherryi* Lakela.

SCHISANDRACEAE (Schisandra Family)

Schisandra Michx. (Wild Sarsaparilla)

S. coccinea Michx.; rich woods; very local; May–Jun; GA. *S. glabra* (Bickn.) Rehd.

SCROPHULARIACEAE (Figwort Family)

1 Stem leaves alternate 2
 2 Flowers nearly actinomorphic; corolla lobes longer than the tube ..... **Verbascum**
 2 Flowers zygomorphic; corolla lobes equal to or shorter than the tube 3
  3 Corolla spurred; calyx lobes 5 4
   4 Inflorescence racemose .... **Linaria**
   4 Inflorescence axillary .... **Chaenorrhinum**
  3 Corolla not spurred; calyx lobes 4, or united and bearing 2 lateral appendages 5
   5 Bracteal leaves trifurcate, usually scarlet .... **Castilleja**
   5 Bracteal leaves pinnately divided, not scarlet .... **Pedicularis**
1 Stem leaves opposite or whorled, bracteal leaves sometimes alternate 6
 6 Trees .... **Paulownia**
 6 Herbs 7
  7 Leaves whorled .... **Veronicastrum**
  7 Leaves opposite 8
   8 Calyx lobes 4, or united and bearing 2 lateral appendages 9
    9 Stamens 2; corolla tube much shorter than the lobes .... **Veronica**
    9 Stamens 4; corolla tube equal to or longer than the lobes 10
     10 Flowers axillary; leaves mostly entire; seeds 4/capsule .... **Melampyrum**
     10 Flowers in terminal spikes; leaves pinnately divided; seeds numerous .... **Pedicularis**
   8 Calyx lobes 5 11
    11 Fertile stamens 2 12
     12 Sterile stamens 2, the anthers absent; bracts absent below the calyx .... **Lindernia**
     12 Sterile stamens absent or minute; bracts present at base of the calyx .... **Gratiola**
    11 Fertile stamens 4 13
     13 Flowers yellow 14
      14 Corolla strongly zygomorphic; plants of wet habitats .... **Mimulus**
      14 Corolla weakly zygomorphic; plants of dry habitats .... **Aureolaria**

13 Flowers not yellow, either white, pink, purple, blue, greenish, or brown 15
15 Sterile stamen present 16
16 Flowers sessile in dense terminal spikes; sterile stamen glabrous . . . . **Chelone**
16 Flowers pedicellate, in panicles or racemes; sterile stamen pubescent . . . . **Penstemon**
15 Sterile stamen absent, or reduced to a callous knob in *Collinsia* 17
17 Corolla strongly zygomorphic 18
18 Corolla tube more than 12 mm long . . . . **Mimulus**
18 Corolla tube less than 12 mm long 19
19 Corolla spurred . . . . **Chaenorrhinum**
19 Corolla not spurred 20
20 Upper leaves sessile; flowers blue and white . . . . **Collinsia**
20 Upper leaves petiolate; flowers greenish-purple . . . . **Scrophularia**
17 Corolla weakly zygomorphic 21
21 Flowers pink; inflorescence racemose; leaves linear . . . . **Agalinis**
21 Flowers blue or purple, rarely white; flowers axillary or in terminal spikes; leaves elliptic to ovate 22
22 Flowers axillary, solitary, subtended by 2 bractlets . . . . **Bacopa**
22 Flowers in terminal spikes, bractlets absent . . . . **Buchnera**

Agalinis Raf.

1 Pedicels less than 2× as long as the calyx 2
2 Stems copiously scabrous; primary leaves with leafy, axillary fascicles . . . . **A. fasciculata**
2 Stems glabrous or scabrescent along the angles; primary leaves usually without axillary fascicles . . . . **A. purpurea**
1 Pedicels 2× or more longer than the calyx 3
3 Calyx tube reticulate veined; leaves rarely more than 1 cm long; plants remaining yellow or green upon drying 4
4 Corolla throat lacking yellow lines . . . . **A. obtusifolia**
4 Corolla throat with 2 yellow lines . . . . **A. decemloba**
3 Calyx tube not reticulate veined; leaves generally 1.5–4 cm long; plants (dark green to red when fresh) usually darker upon drying 5
5 Corolla tube lanate within, upper lip of corolla erect or recurved; leaves filiform . . . . **A. setacea**
5 Corolla tube glabrous within, upper lip of corolla arched over the stamens; leaves up to 4 mm wide . . . . **A. tenuifolia**

A. decemloba (Greene) Pennell; dry open woodlands; infrequent; Sep–Oct; NC, SC, TN, VA. Scarcely distinct from *A. obtusifolia*.

A. fasciculata (Ell.) Raf.; fields and woodlands; infrequent; Aug–Oct; VA. More common on the Coastal Plain.
A. obtusifolia Raf.; fields and woodlands; infrequent; Aug–Oct; SC, VA. More common on the Coastal Plain.
A. purpurea (L.) Pennell; fields and woodlands; frequent; Aug–Oct; ALL.
A. setacea (Walt.) Raf.; sandy fields and woodlands; infrequent; Sep–Oct; NC, SC, TN, VA. More common on the Piedmont and Coastal Plain. *Gerardia gatesii* (Benth.) Pennell.
A. tenuifolia (Vahl) Raf.; fields and woodlands; frequent; Aug–Oct; ALL.

A genus with numerous taxonomic and nomenclatural problems.

Aureolaria Raf.

1 Plants with stipitate-glandular hairs; leaves pectinate .... **A. pedicularia**
1 Plants glabrous or with simple, nonglandular hairs; leaves entire, toothed, or pinnately lobed 2
  2 Stems, leaves, and capsules pubescent .... **A. virginica**
  2 Stems, leaves, and capsules glabrous 3
    3 Leaves entire, rarely lobed; stems not glaucous; pedicels usually less than 5 mm long .... **A. laevigata**
    3 Leaves, especially the lowermost, pinnately divided; stems usually glaucous; pedicels mostly more than 5 mm long .... **A. flava**

A. flava (L.) Farwell; woodlands and open areas; frequent; Aug–Oct; ALL.
A. laevigata (Raf.) Raf.; woodlands and open areas; frequent; Aug–Oct; ALL.
A. pedicularia (L.) Raf.; woodlands and open areas; frequent; Aug–Oct; ALL. Variable in density and distribution of glandular trichomes and includes *A. pectinata* (Nutt.) Pennell.
A. virginica (L.) Pennell; woodlands and open areas; frequent; Jul–Oct; ALL.

Bacopa Aubl.

B. caroliniana (Walt.) Robinson; wet sites; infrequent; May–Sep; SC.

Buchnera L.

B. americana L.; woodlands and roadsides; occasional; Jul–Sep; NC, TN, VA.

Castilleja Mutis ex L. f. (Indian Paint Brush)

C. coccinea (L.) Sprengel; woodlands and open areas; frequent; Apr–May; ALL.

Chaenorrhinum (DC.) Reichenb.

C. minus (L.) Lange*; disturbed sites, especially along railways; infrequent; Jun–Oct; TN, VA.

Chelone L. (Turtlehead)

1 Leaves sessile; flowers distinctly 4-ranked; staminodium purple ...... **C. cuthbertii**
1 Leaves petiolate or rarely subsessile; flowers not 4-ranked; staminodium white or green 2
  2 Petioles 1.5 cm or more long; leaves widest below the middle .... **C. lyonii**
  2 Petioles less than 1.5 cm long; leaves widest at or near the middle 3

3 Corolla white or tinged with purple near the summit; staminodium green .... **C. glabra**
3 Corolla purple throughout; staminodium white .... **C. obliqua**

C. cuthbertii Small; bogs and wet woods; infrequent; Jul–Sep; NC, VA.
C. glabra L.; thickets and low woods; frequent; Aug–Oct; ALL.
C. lyonii Pursh; rich woods, often at high elevations; occasional; Aug–Oct; NC, SC, TN.
C. obliqua L.; low woods and stream banks; infrequent; Aug–Oct; NC, SC.

Nearly a dozen infraspecific taxa have been described in the highly variable *C. glabra-obliqua* complex. More study is needed in this group.

Collinsia Nutt. (Blue-eyed Mary)

C. verna Nutt.; rich woods, often along streams; infrequent; Apr–May; VA.

Gratiola L.

1 Flowers and fruits sessile or subsessile .... **G. pilosa**
1 Flowers and fruits pedicellate 2
2 Pedicels 1 cm or more long, glandular-pubescent 3
3 Sepals 1.5× or less the length of the ovoid capsule; annuals .... **G. neglecta**
3 Sepals 2× or more the length of the globose capsule; perennials ... **G. viscidula**
2 Pedicels less than 1 cm long, glabrous .... **G. virginiana**

G. neglecta Torr.; open wetlands; occasional; Apr–Jun; NC, TN, VA.
G. pilosa Michx.; low woods and ditches; infrequent; Jun–Sep; NC, SC. More common on the Coastal Plain.
G. virginiana L.; streams and low places; infrequent; Mar–May; GA, NC, SC, TN.
G. viscidula Pennell; open wetlands; infrequent; Jun–Oct; NC, SC, TN.

Linaria P. Miller

1 Corolla blue, rarely whitish .... **L. canadensis**
1 Corolla yellow, tinged with orange .... **L. vulgaris**

L. canadensis (L.) Dum., Toadflax; roadsides and disturbed sites; infrequent; Mar–May; NC, SC, TN, VA. More common on the Piedmont and Coastal Plain.
L. vulgaris P. Mill.*, Butter and Eggs; disturbed sites; frequent; Jun–Sep; NC, SC, TN, VA.

Lindernia All.

1 Leaves glandular-punctate; seeds about as long as wide 2
2 Leaves arranged primarily in a basal rosette, stem leaves few and essentially bracteal .... **(L. monticola)**
2 Leaves scattered throughout the stem and not conspicuously reduced above . . . . **L. saxicola**
1 Leaves not glandular-punctate; seeds 2–3× as long as wide .... **L. dubia**

L. dubia (L.) Pennell; wetlands; frequent; May–Oct; ALL. Including *L. anagallidea* (Michx.) Pennell; variability in pedicel length and leaf shape makes separation of these two taxa dubious at best.

L. monticola Muhl. ex Nutt.; wet depressions on granite outcrops; infrequent; Apr–Jun; SC.
L. saxicola M. A. Curtis; mountain streams; very local; Jul–Sep; GA, NC.

Melampyrum L. (Cow Wheat)

M. lineare Desr.; open, often dry woodlands; frequent; May–Jul; ALL.

Mimulus L. (Monkey Flower)

1 Corolla yellow .... **M. moschatus**
1 Corolla blue, lavender or, rarely, white 2
  2 Leaves sessile; pedicels 2–4 cm long; stems wingless .... **M. ringens**
  2 Leaves petiolate; pedicels 1–2 cm long; stems with a narrow wing .... **M. alatus**

M. alatus Ait.; wet woods and marshes; occasional; Jun–Oct; GA, TN, VA. Apparently more common on the Piedmont and Coastal Plain.
M. moschatus Douglas; streams and wet areas; infrequent; Jun–Sep; VA.
M. ringens L.; wet woods and marshes; common; Jun–Oct; ALL.

Paulownia Seibold & Zucc.

Armstrong, J. E. 1985. The delimitation of Bignoniaceae and Scrophulariaceae based on floral anatomy, and the placement of problem genera. *Amer. J. Bot.* 72:755–766.

P. tomentosa (Thunb.) Steudel*, Princess/Empress Tree; persistent and escaping to disturbed sites; occasional; Apr–Jun; ALL.

Pedicularis L.

1 Stems 1–3 dm tall; leaves alternate; stem immediately beneath the inflorescence densely pubescent; spring flowering .... **P. canadensis**
1 Stems 4–8 dm tall; leaves opposite or subopposite; stem immediately beneath the inflorescence glabrous or with a few trichomes; fall flowering .... **P. lanceolata**

P. canadensis L., Lousewort, Wood Betony; dry or moist open woodlands; common; Apr–May; ALL.
P. lanceolata Michx., Swamp Lousewort; bogs and wet woods; infrequent; Aug–Oct; NC, SC, VA. Preferring wetter sites than those of the preceding species.

Penstemon Mitchell (Beard Tongue)

1 Lower lip of the corolla longer than the upper, curved upward and closing the corolla throat .... **P. hirsutus**
1 Lower lip of the corolla spreading or directed forward, the corolla throat open 2
  2 Internodes of mid-cauline leaves glabrous; leaves essentially glabrous beneath .. 3
    3 Corolla white externally, or occasionally tinged with purple; anther sacs often with a tuft of hairs .... **P. digitalis**
    3 Corolla purplish externally; anthers glabrous 4
      4 Sepals 5–10 mm long, linear; corolla 25–35 mm long .... **P. calycosus**
      4 Sepals 3–5 mm long, ovate; corolla 15–25 mm long .... **P. laevigatus**
  2 Internodes of mid-cauline leaves minutely to densely pubescent; leaves usually pubescent beneath, at least along the primary veins 5

5 Leaves glabrous above, leaf bases widely cordate-clasping; lowermost bracts leaf-like, slightly smaller than the cauline leaves .... **P. smallii**
5 Leaves pubescent above, leaf bases narrow to widely cuneate-clasping; lowermost bracts much reduced 6
6 Corolla white or nearly so externally .... **P. pallidus**
6 Corolla light violet to reddish-purple 7
7 Corolla moderately inflated; mid-cauline leaves 2–3.5 cm wide .... **P. canescens**
7 Corolla only slightly inflated; mid-cauline leaves less than 2 cm wide .... **P. australis**

P. australis Small; sandy soils and pinelands; infrequent; May–Jul; GA, SC. A variable species more common on the Piedmont and Coastal Plain.

P. calycosus Small; woodlands; infrequent; May–Jul; SC, VA. Closely related to *P. laevigatus* (L.) Ait.

P. canescens Britt.; roadsides and woodlands; common; May–Jul; ALL. *P. brittonorum* Pennell.

P. digitalis Nutt.; open woods; occasional; May–Jul; NC, SC, TN, VA.

P. hirsutus (L.) Willd.; dry woods and disturbed areas; infrequent; May–Jul; VA.

P. laevigatus (L.) Ait.; damp fields and open woods; common; May–Jun; NC, SC, TN, VA.

P. pallidus Small*; dry woods and openings; occasional; Apr–May; NC, SC, VA. Naturalized in our area.

P. smallii Heller; slopes and river bluffs, often on calcareous soils; frequent; May–Jun; GA, NC, SC, TN.

Scrophularia L. (Figwort)

1 Sterile stamen purple or brown; leaves finely serrate, the teeth less than 3 mm long; flowering from midsummer to fall .... **S. marilandica**
1 Sterile stamen greenish-yellow; leaves often coarsely serrate, the teeth more than 3 mm long; flowering in early summer .... **S. lanceolata**

S. lanceolata Pursh; open woodlands and disturbed sites; infrequent; May–Jun; VA. Scarcely different from the following species.

S. marilandica L.; open woodlands and disturbed sites; frequent; Jul–Oct; ALL.

Verbascum L. (Mullein)

1 Inflorescence racemose, glabrous or glandular-pubescent; fruiting pedicel 8–14 mm long .... **V. blattaria**
1 Inflorescence spikelike, densely stellate; fruiting pedicel less than 8 mm long 2
2 Leaf bases decurrent; flowers up to 2 cm wide; spike mostly uninterrupted .... **V. thapsus**
2 Leaf bases not decurrent; flowers more than 2.5 cm wide; spike interrupted .... **V. phlomoides**

V. blattaria L.*, Moth M.; disturbed sites; frequent; May–Jun; NC, SC, TN, VA. Flowers may be either yellow or white.

V. phlomoides L.*; disturbed sites; infrequent; May–Jun; NC, SC, VA.

V. thapsus L.*, Woolly M.; disturbed sites; common; Jun–Sep; ALL.

Veronica L. (Speedwell)

1 Flowers axillary and solitary or in terminal racemes 2
2 Fruiting pedicels 2–4× as long as the sepals 3
3 Fruits nearly round, seeds black at maturity; corolla 3–5 mm wide; leaves 3–5-lobed . . . . **V. hederifolia**
3 Fruits flattened; seeds tan at maturity; corolla 8–11 mm wide; leaves merely toothed . . . . **V. persica**
2 Fruiting pedicels equal to or shorter than the sepals 4
4 Style elongate, extending well beyond the summit of the ovary . . . . **V. serpyllifolia**
4 Style short, equal to or shorter than the summit of the ovary 5
5 Corolla white; leaves glabrous . . . . **V. peregrina**
5 Corolla blue; leaves pubescent . . . . **V. arvensis**
1 Flowers in axillary racemes 6
6 Plants pubescent . . . . **V. officinalis**
6 Plants glabrous 7
7 Leaves linear to lanceolate, 4–20× as long as wide . . . . **V. scutellata**
7 Leaves elliptic to ovate, 2–4× as long as wide 8
8 Leaves, at least the upper, short-petiolate . . . . **V. americana**
8 Leaves sessile or somewhat clasping . . . . **V. anagallis-aquatica**

V. americana (Raf.) Benth., Brooklime; low woods and stream banks; occasional; May–Jun; NC, TN, VA.
V. anagallis-aquatica L.; bogs, marshes, and stream banks; occasional; May–Jun; NC, TN, VA.
V. arvensis L.*; lawns and disturbed sites; common; Mar–Jun; NC, SC, TN, VA.
V. hederifolia L.*; lawns and disturbed sites; infrequent; Mar–May; NC, SC, VA.
V. officinalis L.*; lawns and disturbed sites; frequent; May–Sep; ALL.
V. peregrina L.*; moist disturbed sites; frequent; Apr–May; NC, SC, TN, VA. Possibly native in our area.
V. persica Poir.*; lawns and disturbed sites; frequent; Feb–Jun; ALL.
V. scutellata L.; bogs and low woods; infrequent; Jun–Sep; TN, VA.
V. serpyllifolia L.*; lawns and disturbed sites; occasional; Apr–Jun; ALL.

Veronicastrum Fabr. (Culver's Root)

V. virginicum (L.) Farwell; woodlands and stream banks; occasional; Jul–Aug; NC, SC, VA.

SIMAROUBACEAE (Quassia Family)

Ailanthus Desf. (Tree-of-Heaven)

A. altissima (Miller) Swingle*; spreading from cultivation; frequent; May–Jun; ALL.

SOLANACEAE (Nightshade Family)

Contributed by E. E. Schilling

1 Corolla more than 3.5 cm long; fruit a spiny capsule . . . . **Datura**
1 Corolla less than 3.5 cm long; fruit a dry or fleshy berry 2

2 Plant woody, trailing or viny shrubs; fruit a dry berry . . . . **Lycium**
2 Plant herbaceous; fruit a fleshy berry 3
3 Fruit not enclosed by calyx; anthers convergent, forming a tube around the style, opening by terminal pores . . . . **Solanum**
3 Fruit enclosed by inflated calyx; anthers separate, longitudinally dehiscent 4
4 Corolla pale blue, more than 2.5 cm long; ovary 3–5-locular . . . . . . **Nicandra**
4 Corolla greenish to yellowish-white or yellow, less than 2.5 cm long; ovary 2-locular . . . . **Physalis**

Datura L. (Jimson Weed)

D. stramonium L.*; fields, roadsides, and waste places; occasional; Jun–Sep; ALL.

Lycium L. (Matrimony Vine)

L. halimifolium Mill.*; old home sites and along railways; infrequent; May–Oct; ALL.

Nicandra Adanson (Apple-of-Peru)

N. physalodes (L.) Gaertner*; fields and waste places; infrequent; Jun–Sep; ALL.

Physalis L. (Ground Cherry)

1 Calyx at anthesis typically 6 mm or more long; anthers 1.8–4.5 mm long, yellow or blue; perennials with rhizomes 2
2 Leaves, pedicels, and calyces glabrous or with only sparse and minute pubescence when young . . . . **P. longifolia**
2 Leaves, pedicels, and calyces villous, pilose, or hirsute with spreading pubescence 3
3 Leaves cuneate or gradually tapering to base; pubescence scattered or sparse on stem; berry reddish . . . . **P. virginiana**
3 Leaves cordate or broadly rounded at base; pubescence very dense on stem, often glandular or viscid; berry yellow . . . . **P. heterophylla**
1 Calyx at anthesis typically 6 mm or less long; anthers 1–2.3 mm long, blue; annuals without rhizomes 4
4 Plant essentially glabrous; corolla throat yellow or only tinged with dark purple . . . . **P. angulata**
4 Plant obviously villous or pilose; corolla throat with 5 large, dark purple spots . . . 5
5 Leaf blades green, drying green; leaf margins entire to dentate . . . **P. pubescens**
5 Leaf blades gray-green, drying orange or with orange patches; leaf margins dentate . . . . **P. pruinosa**

P. angulata L.; fields and waste places; infrequent; Jul–Sep; SC, VA.
P. heterophylla Nees; fields and waste places; infrequent; Jul–Sep; ALL.
P. longifolia Nutt.; fields and waste places; infrequent; Jul–Sep; GA, TN. Represented in our area by var. **subglabrata** (Mackenzie & Bush) Cronquist.
P. pruinosa L.; fields and waste places; infrequent; Jul–Sep; NC, SC, TN.
P. pubescens L.; fields and waste places; infrequent; Jul–Sep; GA, NC, VA.
P. virginiana Mill.; fields and waste places; occasional; Jul–Sep; ALL.

Solanum L.

1 Stems and leaves prickly or spiny; pubescence stellate 2
2 Corolla yellow; berry covered by spinose calyx . . . . **S. rostratum**

2 Corolla white to pale blue; berry not covered by spinose calyx .... **S. carolinense**
1 Stems and leaves not prickly or spiny; pubescence simple or lacking 3
3 Erect herb; corolla white, 6–8 mm broad; berry dark purple, rarely green .... **S. ptycanthum**
3 Vine; corolla purple, 18–22 mm broad; berry crimson .... **S. dulcamara**

S. carolinense L., Horse Nettle; gardens, fields, roadsides, and waste places; frequent; May–Sep; ALL.

S. dulcamara L.*, Bittersweet Nightshade; waste places; infrequent; May–Sep; NC, TN, VA.

S. ptycanthum Dunal, Eastern Black Nightshade; gardens, fields, and waste places; frequent; May–Oct; ALL. *S. nigrum* L., *S. americanum* Miller.

S. rostratum Dunal*, Buffalo Bur; gardens and waste places; infrequent; May–Sep; ALL.

STAPHYLEACEAE (Bladdernut Family)

Staphylea L. (Bladdernut)

S. trifolia L.; rich woods, bluffs, and stream banks, often on calcareous sites; frequent; Apr–Jun; ALL.

STYRACACEAE (Storax Family)

1 Corolla lobes 4; fruit winged .... **Halesia**
1 Corolla lobes 5; fruit nearly round .... **Styrax**

Halesia Ell. ex L. (Silverbell)

H. carolina L.; rich woods and stream banks; frequent; Mar–May; ALL.

Styrax L.

1 Leaves stellate beneath; racemes with 5 or more flowers .... **S. grandifolia**
1 Leaves glabrous beneath or with stellate hairs only on the principal veins; racemes with 4 or fewer flowers .... **S. americana**

S. americana Lam.; low woods and streams; infrequent; Apr–May; NC, SC.

S. grandifolia Ait.; ravines and mesic slopes; occasional; Apr–May; ALL.

SYMPLOCACEAE (Sweetleaf Family)

Symplocos Jacq. (Horsesugar)

S. tinctoria (L.) L'Her.; moist to dry woodlands, bluffs, and ravines; occasional; Apr–May; GA, NC, SC, TN.

THEACEAE (Tea Family)

Stewartia L. (Camellia)

1 Styles united; seeds shiny .... **S. malacodendron**
1 Styles separate; seeds dull .... **S. ovata**

S. malacodendron L., Silky C.; rich mesic woods; infrequent; May–Jun; NC. More common on the Coastal Plain.

S. ovata (Cav.) Weatherby, Mountain C.; rich woods and stream margins; occasional; Jun–Jul; ALL.

THYMELAEACEAE (Mezereum Family)

Dirca L. (Leatherwood)

D. palustris L.; rich, often calcareous woods; infrequent; Mar–Apr; NC, TN, VA.

TILIACEAE (Basswood Family)

Tilia L. (Basswood)

1 Leaves glabrous or essentially so with simple hairs .... **T. americana**
1 Leaves pubescent with stellate hairs, often white beneath .... **T. heterophylla**

T. americana L.; rich woods; occasional; May–Jun; ALL.

T. heterophylla Vent.; rich woods; frequent; May–Jun; ALL.

A polymorphic complex perhaps best recognized as a single species.

ULMACEAE (Elm Family)

1 Leaves with 3 main veins arising from base of leaf; fruit a drupe .... **Celtis**
1 Leaves with 1 main vein; fruit a samara .... **Ulmus**

Celtis L.

1 Leaves of fruiting branches darker above than beneath, teeth numerous except at base; style persistent on young fruits .... **C. occidentalis**
1 Leaves of fruiting branches green on both surfaces, teeth few or absent; style deciduous or nearly so on young fruits 2
2 Leaves lanceolate to lance-ovate; medium to large trees of moist sites .... **C. laevigata**
2 Leaves ovate, at least 1/2 to 3/4 as broad as long; shrubs or small trees of dry exposed sites .... **C. tenuifolia**

C. laevigata Willd., Sugarberry; wet or alluvial soils; infrequent; Apr–May; ALL.

C. occidentalis L., Hackberry; alluvial woods; occasional; Apr–May; GA, NC, TN, VA.

C. tenuifolia Nutt., Georgia Hackberry; dry exposed sites; infrequent; Apr–May; ALL. *C. georgiana* Small.

Ulmus L. (Elm)

1 Flowers in fascicles; samara margin without cilia; upper leaf surface rough, with stiff erect hairs .... **U. rubra**
1 Flowers in short racemes; samara margin conspicuously ciliate; upper leaf surface smooth or nearly so 2
2 At least some branches with corky ridges or wings; samara pubescent on sides; leaves 4–8 cm long .... **U. alata**

2 None of the branches developing corky ridges or wings; samara glabrous on sides; leaves 8–15 cm long . . . . **U. americana**

U. alata Michx., Winged E.; dry woodlands and near streams; occasional; Feb–Mar; ALL.

U. americana L., American E.; mesic woods and along streams and rivers; common; Feb–Mar; ALL.

U. rubra Muhl., Slippery E.; dry woodlands, disturbed sites, often on basic soils; common; Feb–Mar; ALL. The inner bark is mucilaginous.

URTICACEAE (Nettle Family)

1 Stems and leaves with stinging hairs 2
2 Leaves opposite . . . . **Urtica**
2 Leaves alternate . . . . **Laportea**
1 Stems and leaves without stinging hairs 3
3 Leaves alternate . . . . **Parietaria**
3 Leaves opposite 4
4 Flowers on loosely branched axillary panicles; stems succulent . . . . **Pilea**
4 Flowers on unbranched axillary spikes; stems not succulent . . . . **Boehmeria**

Boehmeria Jacq. (False Nettle)

B. cylindrica (L.) Sw.; moist soil of low woodlands and streams; common; Jun–Aug; ALL.

Laportea Gaudin (Wood Nettle)

L. canadensis (L.) Weddell; damp rich woods; common; Jun–Aug; ALL.

Parietaria L. (Pennsylvania Pellitory)

P. pensylvanica Muhl. ex Willd.; roadsides and waste places, often on basic soils; occasional; Apr–Sep; NC, TN, VA.

Pilea Lindl. (Clearweed)

1 Mature achene green, usually with purple-black mottling . . . . **P. pumila**
1 Mature achene dark purple-black, margin pale . . . . **P. fontana**

P. fontana (Lunell) Rydb.; wet soil; infrequent; Jul–Sep; VA.

P. pumila (L.) Gray; wet soil on shaded sites; common; Jul–Sep; NC, SC, TN, VA.

Urtica L. (Stinging Nettle)

U. dioica L.*; waste places; occasional; May–Jun; NC, SC, TN, VA.

VALERIANACEAE (Valerian Family)

1 Stem leaves pinnatifid; fruit 1-celled . . . . **Valeriana**
1 Stem leaves simple; fruit 3-celled . . . . **Valerianella**

Valeriana L.

V. pauciflora Michx.; rich moist woods, often on basic soils; infrequent; May–Jun; VA.

Valerianella Mill. (Corn Salad)

1 Corolla pale blue . . . . **V. locusta**
1 Corolla white 2
  2 Corolla tube 1.5–3 mm long, corolla limb 2.5–4 mm wide . . . . **V. umbilicata**
  2 Corolla tube less than 1.5 mm long, corolla limb less than 2 mm wide . . . **V. radiata**

V. locusta (L.) Betcke*; waste places; occasional; Apr–May; NC, SC, TN, VA.
V. radiata (L.) Dufr.; open disturbed sites; occasional; Apr–May; NC, SC, TN, VA.
V. umbilicata (Sull.) Wood; open disturbed sites; infrequent; Apr–May; NC, VA.

For a discussion of variability in fruit morphology, see D. M. E. Ware, Genetic fruit polymorphism in North American *Valerianella* (Valerianaceae) and its taxonomic implications, *Systematic Botany* 8 (1983): 33–44.

VERBENACEAE (Vervain Family)

1 Shrubs; fruits fleshy . . . . **Callicarpa**
1 Herbs; fruits dry 2
  2 Stems creeping and rooting at the nodes; flowers in dense, nearly round heads . . . . **Phyla**
  2 Stems erect; flowers various, but not in dense heads . . . . **Verbena**

Callicarpa L. (French Mulberry)

C. americana L.; woodland margins and fencerows; infrequent; Jun–Jul; GA, NC, SC, TN.

Phyla Lour. (Fog Fruit)

P. lanceolata (Michx.) Greene; sandy moist sites; infrequent; Jun–Sep; VA. *Lippia lanceolata* Michx.

Verbena L.

1 Leaves pinnatifid . . . . **V. riparia**
1 Leaves not deeply lobed or dissected 2
  2 Stems less than 0.6 m tall; leaves less than 1.5 cm wide . . . . **V. simplex**
  2 Stems more than 0.6 m tall; leaves more than 1.5 cm wide 3
    3 Flowers blue-violet; fruits closely overlapping . . . . **V. hastata**
    3 Flowers white; fruits remotely spaced 4
      4 Leaves densely scabrous above . . . . **V. scabra**
      4 Leaves glabrous to slightly scabrous above . . . . **V. urticifolia**

V. hastata L.; moist fields and swamps; occasional; Jun–Oct; NC, TN, VA.
V. riparia Raf.; river banks; infrequent; Jun–Jul; NC.
V. scabra Vahl; open disturbed sites; infrequent; Jun–Aug; VA.
V. simplex Lehmann; disturbed sites, especially along roadsides; frequent; May–Jun; TN, VA.
V. urticifolia L.; open woodlands and waste places; common; Jun–Oct; ALL.

VIOLACEAE (Violet Family)

1 Petals green, none spurred; sepals not auriculate . . . . **Hybanthus**
1 Petals white, yellow, or blue, the lowermost spurred; sepals auriculate . . . . **Viola**

Hybanthus Jacq. (Green Violet)

H. concolor (Forster) Sprengel; woodlands, usually on basic soils; occasional; Apr–May; ALL.

Viola L. (Violet)

Russell, N. H. 1965. Violets (*Viola*) of central and eastern United States: An introductory survey. *Sida* 2:1–113.

1 Plants with leafy aerial stems (caulescent) 2
2 Annuals with delicate roots 3
3 Petals longer than the sepals . . . . **V. rafinesquii**
3 Petals shorter, to nearly as long as the sepals . . . . **V. arvensis**
2 Perennials with stocky or stolonlike rhizomes 4
4 Petals white, cream, or yellow on the inner surface, often with purple veins or purple tinged 5
5 Petals white or creamy on the inner surface, occasionally purplish on the outer surface 6
6 Stipules conspicuously ciliate, lacerate, or pectinate; petals white on both surfaces . . . . **V. striata**
6 Stipules entire or scarcely ciliate; petals sometimes purplish on the outer surface . . . . **V. canadensis**
5 Petals yellow 7
7 Leaves deeply divided into 3 leaflets . . . . **V. tripartita** var. **tripartita**
7 Leaves not divided 8
8 Leaves triangular or hastate, often mottled with purple; stipules 5 mm or less long . . . . **V. hastata**
8 Leaves ovate, not mottled with purple; stipules 5–15 mm long 9
9 Stem leaves truncate or widely cuneate at the base . . . . **V. tripartita** var. **glaberrima**
9 Stem leaves cordate at the base . . . . **V. pubescens**
4 Petals blue or purple 10
10 Spur 1 cm or more long . . . . **V. rostrata**
10 Spur less than 1 cm long 11
11 Peduncles scabrous; leaves usually purplish beneath; plants often rooting at the nodes . . . . **V. walteri**
11 Peduncles glabrous; leaves green on both surfaces; plants not rooting at the nodes . . . . **V. conspersa**
1 Plants without leafy aerial stems (acaulescent) 12
12 Petals white or yellow 13
13 Petals yellow, flowers appearing before or with the young leaves . . . . **V. rotundifolia**
13 Petals white, usually striped with purple, flowers appearing after some leaves have developed 14
14 Leaf blade 1.5× or more as long as wide, cuneate at the base 15
15 Leaf blade ovate, 1.5–2× as long as broad . . . . **V. primulifolia**
15 Leaf blade lanceolate to linear, 3.5–15× as long as broad . . . . **V. lanceolata**
14 Leaf blade about as broad as long, cordate at the base 16

16 Flower less than 1 cm long; leaves glabrous above .... **V. macloskeyi**
16 Flower 1 cm or more long; leaves glabrous or pubescent above .... **V. blanda**
12 Petals blue or purple 17
17 At least some leaves deeply lobed or dissected 18
18 Rootstock short, vertical; stamens slightly exserted .... **V. pedata**
18 Rootstock horizontal; stamens short, not exserted 19
19 Lower leaf surface and/or petioles moderately to densely pubescent 20
20 Leaves with 5–9 lobes .... **V. palmata**
20 Leaves with 3–5 lobes .... **V. triloba**
19 Lower leaf surface and petioles glabrous or nearly so ... **V. septemloba**
17 Leaves unlobed, or with a few teeth or lobes at the base only 21
21 Mature leaves deeply toothed or lobed at the base only 22
22 Leaves densely pubescent, the petioles shorter than the blades .... **V. fimbriatula**
22 Leaves glabrous to moderately pubescent, the petioles equal to or longer than the blades .... **V. sagittata**
21 Mature leaves not lobed or divided 23
23 Leaves glabrous or essentially so 24
24 Beard of lateral petals conspicuously knobbed; flowers often overtopping the leaves (plants of wet habitats) .... **V. cucullata**
24 Beard of lateral petals not conspicuously knobbed; leaves usually overtopping the flowers 25
25 Spurred petal bearded at the base; leaves often subsagittate at the base and tapering to an acute apex .... **V. affinis**
25 Spurred petal beardless at the base; leaves widely ovate and cordate at the base .... **V. sororia**
23 Leaves distinctly pubescent 26
26 Leaves pubescent above and glabrous below, usually suffused with purple; sepals not ciliate .... **V. hirsutula**
26 Leaves pubescent on both surfaces, green below; sepals often ciliate 27
27 Spurred petal villous at base; petiole at anthesis usually 4× or more longer than the leaf blade (usually at high elevations) .... **V. septentrionalis**
27 Spurred petal glabrous or nearly so at the base; petioles at anthesis usually less than 4× as long as the leaf blade (common, widespread) .... **V. sororia**

V. affinis Le Conte; low woodlands; occasional; Apr–May; NC, SC, TN, VA. *V. floridana* Brainerd.

V. arvensis Murray*; fields and waste places; occasional; Mar–Jun; NC, SC, TN, VA.

V. blanda Willd., Sweet White V.; moist rich woods; frequent; Apr–Jun; ALL. *V. incognita* Brainerd. The flowers are said to have a faint sweet odor.

V. canadensis L., Canada V.; woodlands; frequent; Apr–Jul; ALL.

V. conspersa Reichenb., Dog V.; woodlands, especially along streams; occasional; Mar–May; ALL.

V. cucullata Ait., Bog V.; bogs and along streams; frequent; Apr–Jun; ALL.
V. fimbriatula Sm.; open woods and disturbed sites; occasional; Apr–May; NC, TN, VA.
V. hastata Michx., Halberd-leaf V.; rich deciduous woods; frequent; Mar–May; ALL.
V. hirsutula Brainerd; dry, deciduous or pine woods; frequent; Mar–May; ALL.
V. lanceolata L.; open, sandy wet areas; infrequent; Mar–May; TN, VA.
V. macloskeyi Lloyd; open or shaded wet areas; occasional; Apr–May; ALL. Represented in our area by subsp. **pallens** (Banks) M. S. Baker.
V. palmata L.; dry open woods; occasional; Mar–May; ALL.
V. pedata L., Birdfoot V.; open rocky woods; frequent; Mar–May; ALL. Quite variable in flower color and leaf shape.
V. primulifolia L.; open wet areas; occasional; Mar–May; ALL.
V. pubescens Ait.; moist woods; frequent; Mar–May; ALL. *V. eriocarpa* Schwein., *V. pensylvanica* Michx., *V. pubescens* Ait. var. *leiocarpa* (Fern. & Wieg.) Seymour.
V. rafinesquii Greene, Wild Pansy; lawns and disturbed sites; common; Mar–May; ALL. *V. kitaibeliana* Roemer & Schultes.
V. rostrata Pursh, Long Spurred V.; rich woods; occasional; Apr–May; ALL.
V. rotundifolia Michx., Roundleaf V.; rich woods; occasional; Mar–Apr; ALL. A distinct species not known to hybridize with other violets.
V. sagittata Ait.; woodlands; frequent; Mar–May; ALL. *V. emarginata* (Nutt.) Le Conte.
V. septemloba Le Conte; pinewoods; infrequent; Mar–May; GA, SC, TN, VA. *V. esculenta* Ell.
V. septentrionalis Greene; usually restricted to high-elevation woods; infrequent; May–Jun; NC, TN, VA.
V. sororia Willd.; woods, lawns, and disturbed sites; common; Feb–May; ALL. *V. palmata* L. var. *sororia* (Willd.) Pollard, *V. papilionacea* Pursh pro parte. *V. priceana* Pollard (Confederate Violet) with grayish flowers is a common cultivated variant. This is perhaps the most common and variable violet; probably hybridizing with all other stemless blues.
V. striata Ait.; open or partially shaded alluvial woods; frequent; Mar–Jun; ALL.
V. triloba Schwein.; woodlands; frequent; Mar–May; ALL. *V. palmata* L. var. *triloba* (Schwein.) Ging. ex DC.
V. tripartita Ell. var. **glaberrima** (DC.) Harper; dry woods; infrequent; Mar–May; NC, SC.
V. tripartita Ell. var. **tripartita**; dry woods; infrequent; Mar–May; GA, NC, SC, TN.
V. walteri House; rich woods; infrequent; Mar–May; ALL.

Hybridization and subsequent introgression have produced a vast array of intermediate entities. The stemless blues continue to be the most perplexing complex and are in need of modern biological studies.

VISCACEAE (Mistletoe Family)

Phoradendron Nutt. (Mistletoe)

P. serotinum (Raf.) M. C. Johnston; parasitic on various deciduous trees; common; Oct–Nov; ALL. *P. flavescens* (Pursh) Nutt.

VITACEAE (Grape Family)

1 Leaves palmately compound . . . . **Parthenocissus**
1 Leaves simple or bipinnately, rarely tripinnately, compound 2
  2 Leaves compound . . . . **Ampelopsis**
  2 Leaves simple 3
    3 Inflorescence cymose; pith white, continuous through nodes . . . . . . **Ampelopsis**
    3 Inflorescence paniculate; pith brown, interrupted at node (except *V. rotundifolia*) . . . . **Vitis**

Ampelopsis Michx. (Pepper Vine)

1 Leaves simple . . . . **A. cordata**
1 Leaves compound . . . . **A. arborea**

A. arborea (L.) Koehne; low woodlands and disturbed sites; infrequent; Jun–Oct; NC, TN.

A. cordata Michx.; low woods and river banks; infrequent; Jun–Jul; NC, TN, VA.

Parthenocissus Planch. (Virginia Creeper)

P. quinquefolia (L.) Planch.; woodlands and disturbed sites; common; May–Jun; ALL.

Vitis L. (Grape)

1 Tendrils simple; pith continuous through nodes; bark tight . . . . **V. rotundifolia**
1 Tendrils usually forking; pith diaphragmed at the nodes; bark on older stems shredding 2
  2 Lower leaf surface densely felty-pubescent and concealing the lower leaf surface . . . . **V. labrusca**
  2 Lower leaf surface glabrous or with spreading or cobwebby hairs 3
    3 Stems of the current season, petioles, and lower leaf surfaces with short, spreading hairs . . . . **V. cinerea**
    3 Stems of the current season, petioles, and lower leaf surfaces glabrous or with cobwebby hairs 4
      4 Leaves glaucous or whitish beneath, glabrous or with cobwebby hairs, usually lobed and with rounded sinuses . . . . **V. aestivalis**
      4 Leaves green beneath, glabrous except at vein axils, unlobed or with angled sinuses 5
        5 Leaves broader than long; peduncle not tendril-bearing . . . . . . **V. rupestris**
        5 Leaves usually longer than broad; peduncle usually tendril-bearing 6
          6 Leaves unlobed or shallowly 3-lobed . . . . **V. vulpina**
          6 Leaves lobed 7
            7 Branches of current season green to brown; leaves ciliate on the margins; fruit purple to black, glaucous . . . . **V. riparia**
            7 Branches of current season reddish; leaves scarcely or not ciliate; fruit black, not glaucous . . . . **V. palmata**

V. aestivalis Michx., Summer G.; low woods, streams, and bluffs; common; May–Jun; ALL.

V. cinerea Engelm., Winter G.; low woods and thickets; infrequent; May–Jun; NC, SC, TN.

V. labrusca L., Fox G.; low rich woods; frequent; May–Jun; ALL.
V. palmata Vahl, Cat G.; wet sites; infrequent; May–Jun; TN.
V. riparia Michx., Riverside G.; stream and river banks; frequent; May–Jun; GA, TN, VA.
V. rotundifolia Michx., Muscadine; woodlands and disturbed sites; occasional; May–Jun; GA, NC, SC, TN.
V. rupestris Scheele, Sugar G.; rocky sites; infrequent; May–Jun; VA.
V. vulpina L., Frost G.; low woodlands; common; May–Jun; ALL.

Excluded Taxa

The following list represents introduced taxa that have not and are not likely to become a part of our flora. They are generally known from five or fewer counties and, for the most part, are either persistent after or rarely escaping from cultivation. The inclusion of these taxa in the taxonomic treatment would complicate the keys and add little to our understanding of the native elements of the Blue Ridge flora.

APIACEAE

Aegopodium podagraria L.—SC, TN, VA
Ammoselinum butleri (Engelm.) Coult. & Rose—TN
Anthriscus scandicina (Weber) Mansfeld—NC
Anthriscus sylvestris (L.) Hoffm.—NC, TN
Falcaria vulgaris Bernhardi—VA
Hydrocotyle sibthorpioides Lam.—TN, VA
Hydrocotyle verticillata Thunb.—SC

ASTERACEAE

Acanthospermum australe (Loefl.) Kuntze—SC
Bellis perennis L.—NC
Centaurea calcitrapa L.—VA
Centaurea × *pratensis* Thuiller—VA
Centaurea solstitialis L.—VA
Chrysanthemum parthenium Pers.—SC
Crepis setosa Haller f.—TN
Facelis retusa (Lam.) Schultz-Bip.—SC
Filago germanica (L.) Hudson—VA
Gnaphalium uliginosum L.—VA
Heterotheca villosa (Pursh) Shinners—VA
Hieracium floribundum Wimm. & Grab.—VA
Santolina chamaecyparissus L.—SC
Soliva pterosperma (Juss.) Less.—SC
Tagetes minuta L.—SC

BORAGINACEAE

Myosotis discolor Pers.—SC
Myosotis micrantha Lehmann—NC
Symphytum officinale L.—TN

BRASSICACEAE

Armoracia rusticana (Lam.) Gaert., Mey. & Scherb.—NC
Berteroa incana (L.) DC.—VA
Brassica erucastrum L.—NC
Brassica hirta Moench—NC
Brassica kaber (DC.) Wheller—NC, VA
Brassica nigra (L.) Koch—VA
Calepina irregularis (Asso) Thellung—NC
Camelina sativa (L.) Crantz—VA
Cardamine flexuosa With.—NC
Conringia orientalis (L.) Dum.—NC
Coronopus didymus (L.) Smith—VA
Erysimum repandum L.—NC, VA
Lepidium densiflorum Schrad.—TN
Raphanus sativus L.—NC

BUXACEAE

Pachysandra terminalis Sieb. & Zucc.—SC

CANNABACEAE

Cannabis sativa L.—VA

CAPPARACEAE

Cleome hassleriana Chod.—NC, SC, TN

CAPRIFOLIACEAE

Lonicera fragrantissima Lindl. & Pax.—SC
Lonicera xylosteum L.—VA

CARYOPHYLLACEAE

Dianthus barbatus L.—NC, TN
Dianthus plumarius L.—SC
Lychnis coronaria (L.) Desr.—NC, SC, TN, VA
Spergula arvensis L.—NC, SC, VA

CHENOPODIACEAE

Roubieva multifida (L.) Moq.—SC
Salsola kali L. var. *tenuifolia* Meyer—VA

COMMELINACEAE

Murdannia nudiflora (L.) Brenan—SC

CONVOLVULACEAE

Calystegia pubescens Lindley—TN
Dichondra carolinensis Michx.—SC
Ipomoea quamoclit L.—TN

CRASSULACEAE

Sedum spectabile Boreau—SC

CUCURBITACEAE

Citrullus vulgaris Schrad.—SC
Cucurbita pepo L.—NC

CYPERACEAE

Carex spicata Huds.—VA
Cyperus iria L.—SC, TN
Cyperus rotundus L.—SC, TN
Fimbristylis tomentosa Vahl—TN

DIPSACACEAE

Dipsacus laciniatus L.—VA

ELAEAGNACEAE

Elaeagnus pungens Thunb.—SC, VA

EUPHORBIACEAE

Euphorbia chamaesyce L.—VA
Euphorbia falcata L.—VA
Euphorbia helioscopia L.—SC, VA
Euphorbia humistrata Gray—VA
Euphorbia marginata Pursh—TN, VA
Phyllanthus amarus Schum. & Thonn.—SC
Ricinus communis L.—SC

FABACEAE

Crotalaria spectabilis Roth—GA
Glycine max (L.) Merrill—NC
Lotus americanus (Nutt.) Bischoff—VA
Medicago arabica (L.) Hudson—NC
Medicago orbicularis (L.) Bartalini—TN
Melilotus indica (L.) All.—SC
Pisum sativum L.—SC
Trifolium hirtum All.—SC
Vicia cracca L.—NC, SC
Vicia hirsuta (L.) S. F. Gray—SC
Vicia sativa L.—SC
Vicia tetrasperma (L.) Moench—SC, VA
Wisteria floribunda (Willd.) DC.—NC, TN
Wisteria sinensis (Sims) Sweet—SC, VA

FAGACEAE

Quercus virginiana Mill.—SC

FUMARIACEAE

Fumaria officinalis L.—VA

IRIDACEAE

Gladiolus × grandovensis Van Houtte—NC, VA
Iris germanica L.—NC
Iris sanguinea Donn—NC

JUGLANDACEAE

Carya illinoensis (Wang.) K. Koch—SC, TN

JUNCACEAE

Juncus inflexus L.—VA

LAMIACEAE

Ajuga reptans L.—TN, VA
Galeopsis tetrahit L.—NC
Lamium hybridum Villars—NC
Lamium maculatum L.—VA
Mentha citrata Ehrhart—VA
Mentha gentilis L.—NC
Mentha rotundifolia (L.) Huds.—NC, VA
Mosla dianthera (Buch.) Maxon—TN
Stachys floridana Shuttlew. ex Benth.—SC, TN
Thymus serpyllum L.—NC

LARDIZABALACEAE

Akebia quinata (Houtt.) Dcne.—NC, SC

LILIACEAE

Ipheion uniflorum (Lindl.) Raf.—SC
Lycoris radiata (L'Her.) Herbert—TN
Muscari botryoides (L.) Mill.—VA
Narcissus jonquilla L.—GA
Narcissus poeticus L.—VA
Narcissus pseudo-narcissus L.—NC

LINACEAE

Linum usitatissimum L.—NC, SC

LOGANIACEAE

Buddleja davidii Franchot—NC, SC
Buddleja lindleyana Fortune ex Lindl.—SC

LYTHRACEAE

Lagerstroemia indica L.—GA

MAGNOLIACEAE

Magnolia grandiflora L.—SC

MALVACEAE

Althea rosea (L.) Cav.—NC, SC
Malva rotundifolia L.—VA
Modiola caroliniana (L.) G. Don—SC

MARTYNIACEAE

Proboscidea louisianica (P. Mill.) Thellung—SC

MELIACEAE

Melia azedarach L.—GA, NC, SC

MORACEAE

Ficus carica L.—SC

MYRICACEAE

Myrica cerifera L.—SC

OLEACEAE

Ligustrum amurense Carr.—SC
Ligustrum japonicum Thunb.—SC
Ligustrum ovalifolium Hassk.—SC

ORCHIDACEAE

Epipactis helleborine (L.) Crantz—VA

OROBANCHACEAE

Orobanche minor Smith—VA

OXALIDACEAE

Oxalis corniculata L.—SC, VA
Oxalis rubra Saint-Hilaire—SC, VA

PAPAVERACEAE

Eschscholtzia californica Cham.—SC
Macleaya cordata (Willd.) R. Br.—VA
Papaver hybridum L.—NC
Papaver rhoeas L.—NC
Papaver somniferum L.—NC

PASSIFLORACEAE

Passiflora morifolia Masters—SC

PINACEAE

Larix decidua P. Miller—NC
Picea abies (L.) Karst.—NC

PLANTAGINACEAE

Plantago psyllium L.—VA

POACEAE

Aegilops cylindrica Host—VA
Agrostis gigantea Roth—GA
Aira caryophyllea L.—NC
Aira elegans Gaudin—SC
Alopecurus carolinianus Walt.—NC, TN
Alopecurus geniculatus L.—TN, VA
Anthoxanthum aristatum Boissier—VA
Arundo donax L.—SC, TN, VA
Bromus arvensis L.—VA
Bromus mollis L.—NC, VA
Coix lacryma-jobi L.—TN
Cynosurus cristatus L.—NC
Cynosurus echinatus L.—SC
Dactyloctenium aegyptum (L.) Beauv.—SC
Echinochloa colona (L.) Link—NC, SC
Eragrostis curvula (Schrad.) Nees—GA, TN
Hordeum vulgare L.—NC, TN
Pennisetum glaucum (L.) R. Br.—GA
Phalaris canariensis L.—VA
Phragmites communis Trin.—NC, VA
Poa bulbosa L.—VA
Secale cereale L.—GA
Triticum aestivum L.—TN
Zea mays L.—GA

POLYGONACEAE

Rumex conglomeratus Murray—TN, VA
Rumex patientia L.—NC, TN

PONTEDERIACEAE

Eichornia crassipes (Mart.) Solms—SC

PORTULACACEAE

Talinum paniculatum (Jacq.) Gaertner—NC

PRIMULACEAE

Lysimachia punctata L.—NC

RANUNCULACEAE

Aquilegia vulgaris L.—NC
Ranunculus sardous Crantz —SC
Thalictrum dasycarpum Fis. & Lall.—VA

ROSACEAE

Potentilla anserina L.—TN
Potentilla intermedia L.—NC, VA
Potentilla rivalis Nutt.—VA
Prunus caroliniana (Mill.) Ait.—SC
Rosa bracteata J. C. Wendl.—SC
Rosa damascena Mill.—NC
Rosa gallica L.—NC
Rubus discolor Weihe & Nees—VA
Rubus illecebrosus Focke—NC
Sanguisorba minor Scopoli—NC
Spiraea prunifolia Sieb. & Zucc.—VA
Spiraea salicifolia L.—GA
Spiraea × *vanhouttei* (Broit) Zabel—TN

RUBIACEAE

Richardia brasiliensis Gomez—SC
Richardia scabra L.—NC

RUTACEAE

Poncirus trifoliata (L.) Raf.—SC

SALICACEAE

Populus × *canescens* (Ait.) Sm.—TN
Populus nigra L.—TN
Salix caprea L.—NC
Salix cinerea L. subsp. *cinerea*—NC, SC, VA
Salix cinerea L. subsp. *oleifolia* (Smith) Macreight—NC
Salix pentandra L.—VA
Salix purpurea L.—NC, VA

SAPINDACEAE

Cardiospermum halicacabum L.—SC

SAXIFRAGACEAE

Deutzia scabra Thunb.—VA
Hydrangea paniculata Sieb.—TN
Ribes sativum Syme—NC, TN, VA

SCROPHULARIACEAE

Kickxia elatine (L.) Dumort.—VA
Leucospora multifida (Michx.) Nutt.—VA
Mazus pumilus (Burm. f.) Steenis—SC, TN, VA
Veronica agrestis L.—VA
Veronica chamaedrys L.—NC

SOLANACEAE

Lycopersicon esculentum P. Mill.—SC
Petunia violacea Lindl.—SC
Solanum tuberosum L.—SC

STERCULIACEAE

Firmiana platanifolia (L. f.) Marsili—NC, SC

THYMELAEACEAE

Edgeworthia chrysantha Lindl.—GA

VERBENACEAE

Callicarpa dichotoma (Lour.) K. Koch—NC, SC
Verbena brasiliensis Vell.—SC
Verbena rigida Spreng.—SC
Verbena tenuisecta Briq.—SC
Vitex agnus-castus L.—SC

VIOLACEAE

Viola tricolor L.—NC, SC

VITACEAE

Ampelopsis brevipedunculata (Maxim.) Trautv.—VA

Synopsis

FAMILY	GENERA Native	GENERA Intro.	SPECIES AND LESSER TAXA Native	SPECIES AND LESSER TAXA Intro.
PTERIDOPHYTES				
1. Aspleniaceae	1		10	
2. Blechnaceae	1		2	
3. Dennstaedtiaceae	2		2	
4. Dryopteridaceae	2		8	
5. Equisetaceae	1		4	
6. Hymenophyllaceae	2		3	
7. Isoetaceae	1		3	
8. Lycopodiaceae	1		13	
9. Marsileaceae	1		1	
10. Ophioglossaceae	2		11	
11. Osmundaceae	1		3	
12. Polypodiaceae	1		2	
13. Schizaeaceae	1		1	
14. Selaginellaceae	1		3	
15. Sinopteridaceae	3		9	
16. Thelypteridaceae	1		5	
17. Woodsiaceae	7		14	
GYMNOSPERMS				
18. Cupressaceae	2		3	
19. Pinaceae	4		11	
20. Taxaceae	1		1	
MONOCOTS				
21. Agavaceae	2		2	
22. Alismataceae	2		9	
23. Araceae	5		6	
24. Commelinaceae	2	1	7	2
25. Cyperaceae	13		195	2

(*continued*)

FAMILY	GENERA		SPECIES AND LESSER TAXA	
	Native	Intro.	Native	Intro.
26. Dioscoreaceae	1		1	1
27. Eriocaulaceae	1		3	
28. Haemodoraceae	1		1	
29. Hydrocharitaceae	2	1	3	1
30. Iridaceae	2	1	11	2
31. Juncaceae	2		26	
32. Lemnaceae	2	1	5	1
33. Liliaceae	29	4	66	7
34. Najadaceae	1		3	1
35. Orchidaceae	19		49	
36. Poaceae	45	15	169	44
37. Pontederiaceae	2		3	
38. Potamogetonaceae	1		12	2
39. Smilacaceae	1		12	
40. Sparganiaceae	1		4	
41. Typhaceae	1		2	
42. Xyridaceae	1		4	
43. Zannichelliaceae	1		1	
DICOTS				
44. Acanthaceae	2		5	
45. Aceraceae	1		8	1
46. Amaranthaceae	2		3	3
47. Anacardiaceae	2		7	
48. Annonaceae	1		2	
49. Apiaceae	20	5	36	6
50. Apocynaceae	2	1	5	2
51. Aquifoliaceae	2		5	
52. Araliaceae	2	1	6	1
53. Aristolochiaceae	3		10	
54. Asclepiadaceae	3		16	
55. Asteraceae	47	24	247	53
56. Balsaminaceae	1		2	
57. Berberidaceae	5		6	1
58. Betulaceae	5		13	1
59. Bignoniaceae	2	1	2	2
60. Boraginaceae	6	1	11	4
61. Brassicaceae	7	15	23	24
62. Buxaceae	1		1	
63. Cabombaceae	1		1	
64. Cactaceae	1		1	

(*continued*)

FAMILY	GENERA Native	GENERA Intro.	SPECIES AND LESSER TAXA Native	SPECIES AND LESSER TAXA Intro.
65. Callitrichaceae	1		3	
66. Calycanthaceae	1		2	
67. Campanulaceae	3		13	1
68. Cannabaceae		1		2
69. Capparaceae		1		1
70. Caprifoliaceae	7		22	4
71. Caryophyllaceae	6	7	24	19
72. Celastraceae	3		5	1
73. Ceratophyllaceae	1		1	
74. Chenopodiaceae	3	1	5	4
75. Cistaceae	3		10	
76. Clethraceae	1		1	
77. Clusiaceae	2		19	1
78. Convolvulaceae	4	1	13	4
79. Cornaceae	1		8	
80. Crassulaceae	1		7	2
81. Cucurbitaceae	3		3	
82. Diapensiaceae	2		2	
83. Dipsacaceae		1		1
84. Droseraceae	1		3	
85. Ebenaceae	1		1	
86. Elaeagnaceae		1		1
87. Elatinaceae	1		1	
88. Ericaceae	18		50	
89. Euphorbiaceae	7		21	2
90. Fabaceae	30	7	80	27
91. Fagaceae	3		23	
92. Fumariaceae	3		7	
93. Gentianaceae	5		17	
94. Geraniaceae	1	1	4	5
95. Haloragaceae	2		2	2
96. Hamamelidaceae	3		3	
97. Hippocastanaceae	1		3	
98. Hydrophyllaceae	3		9	
99. Juglandaceae	2		9	
100. Lamiaceae	18	8	62	14
101. Lauraceae	2		2	
102. Lentibulariaceae	1		7	
103. Limnanthaceae	1		1	
104. Linaceae	1		5	
105. Loganiaceae	3		3	

(*continued*)

FAMILY	GENERA		SPECIES AND LESSER TAXA	
	Native	Intro.	Native	Intro.
106. Lythraceae	5		6	1
107. Magnoliaceae	2		6	
108. Malvaceae	3	2	5	8
109. Melastomataceae	1		3	
110. Menispermaceae	2		2	
111. Menyanthaceae	1		1	
112. Molluginaceae		1		1
113. Moraceae	1	2	1	3
114. Myricaceae	2		3	
115. Nelumbonaceae	1		1	
116. Nyctaginaceae		1		1
117. Nymphaeaceae	2		2	
118. Nyssaceae	1		2	
119. Oleaceae	3	1	6	1
120. Onagraceae	5		24	1
121. Orobanchaceae	3		3	
122. Oxalidaceae	1		6	
123. Papaveraceae	2	2	2	2
124. Passifloraceae	1		2	
125. Phrymaceae	1		1	
126. Phytolaccaceae	1		1	
127. Plantaginaceae	1		7	1
128. Platanaceae	1		1	
129. Podostemaceae	1		1	
130. Polemoniaceae	3		15	
131. Polygalaceae	1		9	
132. Polygonaceae	4		20	11
133. Portulacaceae	3		3	1
134. Primulaceae	5	1	13	2
135. Ranunculaceae	14	1	52	6
136. Rhamnaceae	2		4	
137. Rosaceae	20	3	83	19
138. Rubiaceae	5	1	26	4
139. Rutaceae	2		2	
140. Salicaceae	2		10	4
141. Santalaceae	4		4	
142. Sarraceniaceae	1		4	
143. Saururaceae	1		1	
144. Saxifragaceae	14		31	
145. Schisandraceae	1		1	
146. Scrophulariaceae	17	3	45	13

(continued)

FAMILY	GENERA Native	GENERA Intro.	SPECIES AND LESSER TAXA Native	SPECIES AND LESSER TAXA Intro.
147. Simaroubaceae		1		1
148. Solanaceae	2	3	8	5
149. Staphyleaceae	1		1	
150. Styracaceae	2		3	
151. Symplocaceae	1		1	
152. Theaceae	1		2	
153. Thymelaeaceae	1		1	
154. Tiliaceae	1		2	
155. Ulmaceae	2		6	
156. Urticaceae	4	1	5	1
157. Valerianaceae	2		3	1
158. Verbenaceae	3		7	
159. Violaceae	2		27	1
160. Viscaceae	1		1	
161. Vitaceae	3		11	

SUMMARY

	FAMILIES	GENERA	SPECIES AND LESSER TAXA
Pteridophytes	17	29	94
Gymnosperms	3	7	15
Monocots	23	160	657
Dicots	118	531	1625
TOTALS	**161**	**727**	**2391**
Native	154	604	2051
Introduced	7	123	340
TOTALS	**161**	**727**	**2391**
Excluded Taxa	(5)	(76)	(212)
TOTALS	(**166**)	(**803**)	(**2603**)

References

Batson, W. T. 1977. *Genera of eastern plants,* 3d ed. New York: Wiley.

Cronquist, A. 1980. *Vascular flora of the southeastern United States.* Vol. 1, *Asteraceae.* Chapel Hill: Univ. of North Carolina Press.

Duncan, W. H. 1967. Woody vines of the southeastern states. *Sida* 3:1–76.

Duncan, W. H., and J. T. Kartesz. 1981. *Vascular flora of Georgia: An annotated checklist.* Athens: Univ. of Georgia Press.

Elias, T. S. 1980. *The complete trees of North America.* New York: Van Nostrand Reinhold Co.

Fernald, M. L. 1950. *Gray's manual of botany.* 8th ed. New York: American Book Co.

Gleason, H. A. 1952. *New Britton and Brown illustrated flora of the northeastern United States and adjacent Canada,* 3 vols. New York: New York Botanical Garden.

Gleason, H. A., and A. Cronquist. 1963. *Manual of the vascular plants of the northeastern United States and adjacent Canada.* Princeton, N.J.: Van Nostrand.

Godfrey, R. K., and J. W. Wooten. 1979. *Aquatic and wetland plants of southeastern United States. Monocotyledons.* Athens: Univ. of Georgia Press.

———. 1981. *Aquatic and wetland plants of southeastern United States. Dicotyledons.* Athens: Univ. of Georgia Press.

Harvill, A. M., Jr., T. R. Bradley, and C. E. Stevens. 1981. *Atlas of the Virginia flora.* Part 2, *Dicotyledons.* Farmville: Virginia Botanical Associates.

Harvill, A. M., Jr., T. R. Bradley, C. E. Stevens, T. F. Wieboldt, D. M. E. Ware, and D. W. Ogle. 1986. *Atlas of the Virginia flora,* 2d ed. Farmville: Virginia Botanical Associates.

Harvill, A. M., Jr., C. E. Stevens, and D. M. E. Ware. 1977. *Atlas of the Virginia flora.* Part 1, *Pteridophytes through monocotyledons.* Farmville: Virginia Botanical Associates.

Kartesz, J. T., and R. Kartesz. 1980. *A synonymized checklist of the vascular flora of the United States, Canada, and Greenland.* Chapel Hill: Univ. of North Carolina Press.

Luer, C. A. 1975. *The native orchids of the United States and Canada excluding Florida.* Bronx: New York Botanical Garden.

Mohlenbrock, R. H. 1975. *Guide to the vascular flora of Illinois.* Carbondale: Southern Illinois Univ. Press.

Radford, A. E., H. E. Ahles, and C. R. Bell. 1968. *Manual of the vascular flora of the Carolinas.* Chapel Hill: Univ. of North Carolina Press.

Small, J. K. 1933. *Manual of the southeastern flora.* Published by the author.

Strausbaugh, P. D., and E. L. Core. 1952–1964. *Flora of West Virginia.* Introduction and parts 1–4. Morgantown: West Virginia Univ.

Wherry, E. T., J. M. Fogg, Jr., and H. A. Wahl. 1979. *Atlas of the flora of Pennsylvania*. Philadelphia: The Morris Arboretum of the University of Pennsylvania.

White, P. S. 1982. *The flora of Great Smoky Mountains National Park: An annotated checklist of the vascular plants and a review of previous floristic work*. U.S. Dept. of the Interior, National Park Service, Southeast Region Res./Resour. Manage. Rept. SER-55.

Index to Common Names

Index to Scientific Names

Italics refer to synonyms, misapplied names, and other incidental names. Excluded introductions are alphabetized within their respective families in a separate section (see Excluded Taxa, p. 313).

Index to Families

CPSIA information can be obtained at www.ICGtesting.com
Printed in the USA
BVOW070115011211

277227BV00003B/2/A